Ecology of Inland Waters and Estuaries

Ecology of Inland Waters and Estuaries

SECOND EDITION

GEORGE K. REID
ECKERD COLLEGE

RICHARD D. WOOD
UNIVERSITY OF RHODE ISLAND

D. VAN NOSTRAND COMPANY
New York   Cincinnati   Toronto   London   Melbourne

H749

To the memory of
Professor W. C. Allee
and to
Eugenie, Louise, and Philip Reid

Cover photograph by Peter Salwen

D. Van Nostrand Company Regional Offices:
New York Cincinnati Millbrae

D. Van Nostrand Company International Offices:
London Toronto Melbourne

Published by D. Van Nostrand Company
450 West 33rd Street, New York, N. Y. 10001

10 9 8 7 6 5 4 3 2

Preface

The Second Edition of *Ecology of Inland Waters and Estuaries*, like the first, introduces the student to the major principles of aquatic ecology. The scientific and teaching approach originally used in this text has, we believe, stood the test of time. For this reason we have retained the overall structure of the first edition. The four Parts of the book describe, respectively, the history of aquatic ecology and the nature of water; the development and major parameters of basins and channels; the physicochemical variables of natural waters; and the biotic principles (and the plant and animal communities) that make up the living substance of real ecosystems. It is from this reality that ecological science has abstracted its key concepts—trophic structure, biogeochemical cycles, limiting factors, succession, productivity, energetics, and so forth. The final chapter of the text synthesizes the material of the earlier chapters and restates our central theme: to describe the aquatic world, not just as an arena for scientific speculation, but as an actual environment where organisms and communities contend with and adapt to the demands of life in water.

Readers familiar with the original edition of this text will recognize substantial changes. Many of the examples and research findings cited in the text are new, and many of the illustrations are new or updated. The material on estuarine ecology is now complemented by a new chapter, Chapter 6, which describes in some detail the major parameters of marine environments. Material on marine biota and ecosystems has also been added to all the chapters in Part Four, giving the student a more thorough overview of aquatic ecology as an integrated discipline.

We have also stressed the value of ecosystem ecology, a relatively new tool that has facilitated both the design and the execution of quantitative ecological studies. Because the growth of ecosystems analysis promises to make ecology a strongly predictive science in the near future, we include, in Chapter 13, some of the mathematical expressions and models that are commonly used to describe ecological relationships.

In the past fifteen years, ecology, particularly in the area of aquatic studies, has become a field of increasingly fruitful exploration and of growing public concern, and our revised text reflects to a high degree this dynamic character of ecological science. Our increased emphasis on the application of scientific findings to environmental problems underscores an important development: aquatic ecology has now become a matter of serious concern to the educated citizen as well as to the research specialist.

The authors are grateful to reviewers, instructors, and students (many of them our own) whose comments and criticisms have been invaluable in updating and strengthening this new edition. We also express our appreciation to the many authors who have permitted us to reproduce or adapt previously published figures.

George K. Reid
St. Petersburg, Florida

Richard D. Wood
Kingston, Rhode Island

Contents

PART FOUR:
ORGANISMS AND COMMUNITIES IN
AQUATIC ENVIRONMENTS, 259

PART ONE
Introduction

The Background of Aquatic Ecology

an's interest in the waters around him is certainly as old as man himself. At the moment we consciously began to use the natural world for our own purposes, we also began to collect and organize facts about the ponds, lakes, and streams that dot and cut the face of the earth. Some distant ancestor may have discovered that edible oysters could be found clinging to the rocks of a salt marsh, but not under the fresh water of a nearby lake. On the other hand, such freshwater creatures as bass, frogs, and beaver could be found and caught at the margins of lakes and streams, but never along the edge of the sea. But the sea yielded a bounty of its own. At certain times of the year sea turtles would come ashore to bury their eggs in the sand, and some streams would be alive with migrating eels or salmon; this too would be noted, remembered, and used.

To survive in the wild, early man had to become, in his way, an expert naturalist, a full-time "ecologist," and this required an intimate knowledge of the waters that are discussed in this book. The sources and nature of fresh water, its motion and changing condition as it flows to the sea, and the life it supports along the way, are now the subject of **limnology,** the science of fresh waters, and **hydrobiology,** the science of aquatic life. The chemistry, dynamics, and biology of the seas are the subject of **oceanography.** These sciences as they are known today are young disciplines, only about a century old; but they had their roots, as did so much of modern science, with the philosophers of ancient Greece.

In the first section of Book One of his *Historia Animalium,* Aristotle recognized that nature had made various kinds of inland waters different, not only from the sea, but from one another. Aristotle was focusing on the natural history of animals. To understand them, however, he also had to consider their environments. Thus he separated different kinds of species into such categories as "lake dwellers," "river dwellers," and "marsh dwellers." Theophrastus, a student of Aristotle's, carried out a similar study of plants, distinguishing, as Aristotle might have done, "plants of deep fresh water," "plants of shallow lake shores," "plants of wet banks of streams," "plants of the marine shallow zone," and others.

Since those early times the study of aquatic habitats has grown in scope and sophistication. With the accumulation of observation and theory, workers in the field have tended to specialize in relatively narrow areas. Today the entire field may be referred to as **aquatic ecology.** Few researchers, however, can still claim expertise in the field as a whole; most have restricted their studies and researches to such circumscribed subjects as **lake and pond ecology, stream ecology, estuarine ecology,** and **marine ecology.**

THE BACKGROUND OF LIMNOLOGY

Although the study of freshwater systems is varied and old, the field was not established as an area of systematic, scientific exploration until 1869. In that year F. A. Forel, a Swiss physician and naturalist and a professor at the University of Lausanne, published a paper describing the bottom fauna of Lac Léman, now Lake Geneva (Figure 1.1). The publication of Forel's paper is now considered to mark the beginning of the scientific study of lakes. To this study the name **limnology** (from the Greek *limnē,* meaning "marsh") was applied. The term itself was first used, however, in a later report by Forel, entitled *Le Léman, monographie limnologique.* The first two volumes of this monograph described the geology, chemistry, and physics of the Swiss lake, and were published in 1892 and 1895. A third volume, describing the biological aspects of the lake, appeared in 1904. With this monumental treatise, and an impressive list of other publications, Forel earned for himself lasting regard as the founder of limnology. For many years his *Handbook of Limnology* (1901) remained the authoritative work in the field.

In Forel's work we can already see the scientist combining some of the many disciplines that form the core of the study of inland waters: geology, chemistry, physics, and biology. Today these have to be integrated into the broader approach of ecology, which joins these various interests into the study of aquatic environments as integrated, functioning systems. Forel's major contribution was largely restricted to consideration of the water itself, the physical and chemical environment of the organisms inhabiting the lake. His treatment of the plants and animals of Lake Geneva, which consisted only of a partial list of the biota, would not be acceptable by modern standards. But other scientists soon started adding to the work that Forel had begun.

In the United States E. A. Birge, a zoologist and pioneer limnologist at the University of Wisconsin (Figure 1.2), added a stronger biological emphasis to Forel's basically physicochemical limnology. Birge had been studying the assemblage of miscroscopic, floating plants and animals, the **plankton,** of Lake Mendota, Wisconsin. These investigations soon led him to appreciate the relationships between the plankton of a lake and the physical and chemical properties of its water. Birge continued his research for almost seventy years, from 1873 to 1941, and his findings are still central to the science of limnology. We are also indebted to Birge for a great deal besides his painstaking investigations of plankton; his work also took him into the realm of physical limnology, where he made important discoveries concerning light penetration, thermal properties, and currents in lake waters. In 1908 a younger zoologist, Chancey Juday, joined Birge at the University of Wisconsin. The two scientists collabo-

Figure 1.1   Lake Geneva, Switzerland, the "type locality" of limnology, where Forel's studies launched a new science. (Courtesy of Swiss National Tourist Office, New York.)

rated on their first joint publication in 1909, and for nearly thirty years thereafter they maintained a partnership that is credited with major advances in American limnology.

By 1900 limnology had already become strongly ecological in approach. To a great extent this development reflected the influence of S. A. Forbes, a naturalist and entomologist at the University of Illinois, an institution which itself earned considerable stature in the field of hydrobiology. A paper published by Forbes in 1887, entitled "The Lake as a Microcosm," had described the lake as a "small world" in which environmental features and living organisms are organized in complex interrelationships and bound together by a web of interdependence. It is basically this concept that guides the present-day study of aquatic

Figure 1.2   Dr. E. A. Birge, a founder of American limnology, at his University of Wisconsin laboratory in 1940. Birge is renowned for his fundamental work on the plankton, light penetration, heat budgets, and currents of lakes. (Courtesy of State Historical Society of Wisconsin.)

Estuarine studies are particularly important today because of the way man is manipulating estuaries and their shores. Estuaries, along with large lakes and many streams, have been customarily regarded, and treated, as virtually limitless reservoirs for dumping industrial and municipal wastes. Because they empty directly into the oceans, so that wastes seem to disappear, estuaries are especially favored as flushing basins by municipal planners and sanitary engineers: many estuaries are now no more than urban sewers. In many regions, also, estuaries are being filled in and ringed with concrete seawalls in order to create new land for development or to attract customers looking for "waterfront" homes. Hopefully, a greater understanding of the ecology of estuarine regions will stop, or at least slow, the eradication of these unique and important areas.

Figure 1.4 The rubble of stream floors is typically inhabited by many species of plant and animal. The larger individuals can be taken by loosening them from the substrate and allowing the current to carry them into a net held immediately downstream. A Surber-type net is being used by the student in the photograph.

BIOLOGICAL OCEANOGRAPHY

We have emphasized that estuaries are a border region, a kind of specialized interface, between fresh and marine waters. To understand them, therefore, we will give some special attention to the oceans themselves. Marine waters are not, of course, an obvious subject in a text on limnology, but reflection will show that many aspects of freshwater (and especially estuarine) ecology are deeply affected by events that take place in the seas and oceans: most of the earth's atmospheric oxygen is produced by the **phytoplankton,** the tiny drifting plants, of the upper layers of the oceans; marine currents and winds often determine the climate and biota of inland regions; the life cycles of many ecologically important fish, including European and American eels, steelhead and oceanic trout, and most salmon species, involve movement between fresh and marine waters. Perhaps most important, recent large-scale use of inland and offshore waters is having important effects, usually damaging, on the life and ecology of the oceans.

Ever since the earliest days of navigation, maritime peoples have been investigating the seas. Aristotle, for instance, carried out studies of marine life in the Mediterranean, and in later centuries Isaac Newton made systematic observations of the tides, and Benjamin Franklin constructed a rough chart of the Gulf Stream. Scientific oceanography, however, had its origin in the nineteenth century. Matthew F. Maury, an officer in the United States (and later in the Confederate) Navy, compiled and published the first reliable charts of marine currents and prevailing winds in 1847, and he was instrumental in establishing a pattern of international cooperation among maritime nations that still persists today; he is credited with being the founder of oceanography. At about the same time, marine biology was beginning to emerge as a distinct field of scientific inquiry. Edward Forbes, a British field naturalist, botanist, and paleontologist, was carrying out systematic surveys of life in the deep ocean regions, and encouraging others to follow his lead. Forbes died in 1854 at the age of 39, but others, including some of his students, carried on the work he had begun, and in 1872 they persuaded the British Admiralty to furnish a ship, the H. M. S. *Challenger,* to undertake a research voyage of unprecedented scope and ambition. The *Challenger* expedition lasted three and a half years. It covered nearly 70,000 nautical miles, sampled the water and bottom material at 362 positions, or **stations,** and brought to the surface specimens of 4,717 hitherto unknown species; twenty years of work were required to analyze and publish the results of the voyage, and by the end of that time marine studies had attracted the attention of many brilliant scientists. By 1900 the basic principles of marine chemistry, physics, and ecology had been recognized, and a number of permanent research stations were established for oceanographic studies. Today oceanography is a well-established science, heavily funded by universities and governments, and equipped with research facilities and ships to carry out active investigations throughout the world. Oceanographic investigations have consistently taken an ecological approach, focusing to some extent on the study of marine life, but there seems to be more emphasis recently on studies of the biota in particular, an emphasis reflected in the growing use of the terms "marine biology" and "marine ecology" as labels for distinct areas of research.

Oceanographic research often makes us think of massive and expensive equipment: not only ships, but diving gear, winches and heavy cables, electronic devices, and even submarines. It is true that marine studies can be particularly difficult, if only because of the vast size of the oceans. Researchers must often remain at sea, conducting systematic observations, for long periods, despite huge waves and violent storms; they need techniques and equipment that often seem grandiose in comparison to the tools of the limnologist. But the difference is usually only one of scale. Freshwater and marine biologists often approach their problems with virtually identical questions in mind, and even their equipment is usually similar in principle and operation.

ECOLOGICAL STUDIES OF AQUATIC SYSTEMS

We attach several labels to the different ways of studying lakes, streams, estuaries, and seas—limnology, hydrobiology, potamology, and so forth. But the same basic concepts, principles, and approaches are used to deal with all these aquatic situations. What is important is an appreciation for, and an understanding of, the conceptual framework of **ecology.**

The term "ecology" was first used in a scientific sense by the German biologist and evolutionist Ernst Haeckel, in a paper published in 1870. The word is derived from the Greek noun for "home" *(oikos),* and as it is used today the term refers to the study of the interrelationships among organisms or between groups of organisms and their environment. By "organism" we mean any living thing, and, since a living thing is always a representative of a species, we usually choose the species, or **species population,** as our fundamental unit of study. The term **environment,** as ecologists use it, refers in a broad sense to all of the physical, chemical, and biological features of the habitat in which the species lives. Since a given environment normally harbors a surprisingly large number of different plant and animal species, ecologists have to deal with such mixed populations; they therefore recognize another level of relationships, those characteristic of an assemblage of populations, or a **community.**

Thus, the study of relationships can be approached from several different directions. The study of how members of a single species affect and are affected by their environment is called **population ecology** ("autecology" in some studies). Investigating the distribution and activities of a single group of a single species, the perch in Lake Superior, for example, would be a study in population ecology. When we deal with an aggregation of several species populations, we are focusing on a community; the study of such an assemblage is termed **community ecology** (''synecology''). If we return to Lake Superior to investigate the overall ecology of the shallow shore-zone, the bottom, or even the entire lake, our work would be community ecology. Community ecology ideally sums up the autecology of each population in the system and relates each one to all the others. This demands knowledge of each of the different kinds of organisms; community ecology therefore leans heavily not only upon taxonomy, but also upon physiology, genetics, and behavioral studies. As the members of a community interact with one another and with the physical and chemical features of the environment, they form what is termed an **ecosystem. Ecosystem ecology,** the broadest approach to ecology, attempts to explain total community function in terms of energy use and transfer, and biogeochemical cycles, in a given environment. Ecosystem ecology is concerned with individual species only to the extent that their roles have an effect on the whole system, sustaining or disrupting it; at this level it is the entire mass of organisms that concerns the ecologist, much as it is the entire organism that concerns the doctor, rather than the cells of which it is composed.

Aquatic ecology today is a comprehensive area of science. It is concerned with the physics and chemistry of systems of natural water, with the forces and processes that shape and maintain these systems, and with the relationships among the water, the basin, the climate, the community of living organisms, and the movement of energy and materials through the system (Figure 1.5). It follows, therefore, that aquatic ecology must draw from several fields of science. Geology supplies principles and procedures needed to understand the origin of lake basins, drainage systems, and channels, and to explain the processes that later modify them. Geochemistry relates the composition of the substrate to the chemical nature of lake, stream, and estuarine waters. Physics offers explanations of the basic, underlying processes that drive every aquatic system: light penetration, heat dynamics, and water movements. From inorganic and organic chemistry come the techniques and principles used to describe and measure a broad range of processes involving gases, liquids, and solids. Biology contributes knowledge of plants and animals, their metabolism, life histories, and living habits, and it also helps to explain the movement of matter and energy through the living and the nonliving parts of the ecosystem.

This discussion of ecology should make our approach to limnology rather clear. Throughout out discussion we will regard every body of water as a dynamic system in which many factors and processes are constantly at work, most of them interrelated and many of them completely dependent on one another. In order to understand the system as a whole we will first have to study the factors and processes one at a time, and then see how they work together in different combinations. The rest of this book has been designed to present the material in this way. First, we describe the major physical and chemical features of natural waters and aquatic systems. Then we turn to the major biological principles that affect aquatic life. Finally, after examining the variety of living

Figure 1.5   Natural streams show wide variations in their physics, chemistry, and biology. In the upper photograph on the opposite page, a clear, turbulent, rock-bottom stream flows through the tundra of western Iceland in late summer; below, a silt-bottom stream in the heavily wooded hills of central California during winter flood conditions. The biota and ecology of the two streams are fundamentally different. (Courtesy of P. E. Salwen.)

things found in fresh and marine waters, we explore the ecosystem ecology of aquatic communities.

WORKING PRINCIPLES

This ecological appraoch can be reduced to four "working principles," which we have taken as our major themes throughout this introduction to inland and marine waters. These principles pertain to the nature of the environment, the nature of the inhabitants, and the relationships between the two:

The fitness of the environment The phrase "the fitness of the environment" was used as the title of a very stimulating book published by L. J. Henderson in 1913. Henderson was concerned, as we are, with the evolution of the environment itself; our task is to understand the ways in which physical and chemical factors are variously combined to present distinctive sets of conditions in different environments. It is these conditions that determine, in large measure,

Figure 1.6   Examination of bottom deposits often yields valuable data about the biology and ecology of the aquatic community. A Petersen dredge (background) is used here to obtain samples of bottom sediment. (Courtesy of University of Michigan News Bureau.)

the composition of biotic communities. Environmental "fitness" helps explain, for example, why an estuarine community is structurally different from that of a lake or stream. Chapters 2 through 11 should be read with this principle in mind.

Adaptation of the organisms The plants and animals that successfully occupy the various environments found in natural waters do so by virtue of adaptation. Through its long evolutionary history, each population has acquired a set of physical and biological traits that aid in its survival. For example, life in stream rapids typically demands streamlined bodies, holdfasts, or other devices which prevent the organism from being swept downstream; plants or animals of this environment characteristically possess such structures, while lake species characteristically do not. Chapters 12 and 13 describe how species distribution and other characteristics are influenced by environmental factors, and how genetic change and evolution relate to population dynamics; Chapters 14 and 15 describe typical plant and animal species of aquatic environments.

Limiting factors Every plant and animal has a certain limited physiological tolerance for each condition, or for each combination and sequence of conditions, in its immediate environment. Those conditions that approach or exceed the limit of tolerance, such as extremes of heat or cold, are termed **limiting factors.** Organisms may have a wide tolerance for some factors and a narrow range for others. Species with wide ranges of tolerance are most likely to be found in a wide variety of habitats; in this way limiting factors account for the differences in community structure of such habitats as stream riffles, pools, shallow ponds, and deep lakes. Adaptive success—whether of individual organisms or of the entire community of populations—is regulated directly or indirectly by a great complex of environmental variables. The limiting factors principle appears repeatedly throughout the text, most particularly in Chapters 12 through 16.

Traffic in energy and matter Where plants and animals have become adapted to the "fitness" of a particular environment, an ecosystem develops. Within this system plants capture solar energy, transform it to chemical energy in the form of organic compounds. These compounds serve as the basis of **food webs** of varying complexity: herbivores graze upon the plants, small carnivores prey upon the herbivores, and are in turn consumed by larger carnivores. Thus, energy is moved through the system by the consumption of organic material at various levels.

Plants and animals also need numerous chemical substances to synthesize their own tissues and to maintain their particualr life processes. These are often obtained by plants from a nonliving reservoir of water and soil, passed through the various food webs of the ecosystem, and subsequently re-released by death and decomposition. These transfers constitute **biogeochemical cycles.** In systems that are largely "closed," such as lakes and ponds lacking stream inflow, these cycles are tremendously important in maintaining the communities, for there is little input of nutrients from outside. The movement of energy and the biogeochemical cycles are the basic subject of all of Part IV, but they are discussed most fully in Chapters 12, 13 and 16.

SUGGESTED FOR FURTHER READING

Braun, E., and D. Cavagnaro. 1971. *Living water*. American West Publishing Company, Palo Alto, California. A well-done collection of photographs, interspersed with knowledgable, reflective short essays. Taken as a whole, the book strongly conveys the interest and importance of freshwater ecosystems, from the mountain headwaters to the edge of the sea.

Frey, D. G., editor. 1966. *Limnology in North America*. University of Wisconsin Press, Madison. A collection of 26 papers, by more than 30 North American limnologists, suggesting the range of freshwater studies on this continent. The volume combines a survey of the literature and history of North American limnology with a description of the major features of freshwater environments from Alaska to Central America; special chapters deal with reservoirs, farm ponds, paleolimnology (see Chapter 3 of the present text), and sanitational limnology.

Leopold, L. B., and K. S. Davis, with the editors of LIFE. 1966. *Water*. Time, Inc., New York. A volume in the "LIFE Science Library." The book uses text and illustrations—many of them remarkably imaginative—to introduce the general reader to the physicochemical nature of water and to the functions of water in the natural ecosystem and in man's economy.

Macan, T. T. 1974. *Freshwater ecology,* second edition. John Wiley & Sons, New York. An excellent introductory text, by a leading British limnologist, with a rather detailed discussion of every important area of research in freshwater ecology. The American reader might have difficulty identifying species referred to, which are predominantly European, and there is little reference to problems of freshwater management. Nevertheless the book offers a clear and vigorous presentation of the essentials of limnology.

The Physical and Chemical Properties of Water

2

If all the compounds found on earth, none is so essential as water to the maintenance of life. Water is the fluid constituent of all living matter: plants use water for photosynthesis, both plants and animals use it in metabolic processes, and, of course, water is the medium in which the members of the aquatic community live and reproduce. This means that aquatic organisms must be able to find shelter and food, to make use of gases dissolved in water, to reproduce, and sometimes to care for their young, usually without ever leaving the water: each species must be adapted to its particular niche in lake, stream, estuary, or sea.

As the environmental medium, water enters into and maintains the integrity of the entire ecosystem. The effect of water is so all-pervasive, in fact, that from one point of view it makes more sense to think of aquatic species and the water surrounding them as elements of a single, essentially continuous system rather than as separate entities. The student, however, must proceed one step at a time. Before we consider the natural communities that occur in aquatic habitats, it is appropriate, therefore, to give some attention to the properties of water as such. An understanding of these properties will be extremely valuable in later chapters, when we turn to the environmental and ecological processes that are characteristic of the various aquatic ecosystems.

THE WATER MOLECULE

A molecule of water is constructed by the joining of two atoms of hydrogen (H) with one atom of oxygen (O); everyone is familiar with the chemical formula of the resulting compound, H_2O. The bonding of atoms to form a water molecule is **covalent.** It depends, in other words, upon the sharing of electrons between the linked atoms. The nucleus of the hydrogen atom consists of a single positively charged particle, a proton. The rest of the atom consists of an electron, a particle of equal but negative charge, which lies outside the nucleus. It was convenient until recently to visualize the atom as a microscopic analogue of the solar system, with electrons circling the nucleus in orbits of various diameter—much in the way the earth and other planets are in orbit around the sun. Today scientists prefer to speak in terms of the **energy levels** of the electrons, and to visualize the electrons as occupying **atomic orbitals** (rather than actual orbits), each orbital capable of holding electrons at a particular energy level. The single electron of the hydrogen atom occupies the orbital of lowest energy, which is designated the **1s orbital.** The larger oxygen atom has eight electrons: two in the 1s orbital, two in the orbital of next highest energy (designated the **2s orbital**), and four in orbitals of still higher energy. The bonds between the hydrogen and oxygen atoms are formed by the overlapping of orbitals in such a fashion that each atom shares electrons with the other; the angle of the bonds from the oxygen atom to each of the hydrogen atoms is 105° (Figure 2.1). The oxygen atom tends to pull the shared electrons toward itself; thus the electrons tend to cluster near the nucleus of the oxygen atom, and away from the hydrogen nuclei, as shown in Figure 2.6. As a result the oxygen is considered to be negative in charge and the hydrogens positive.

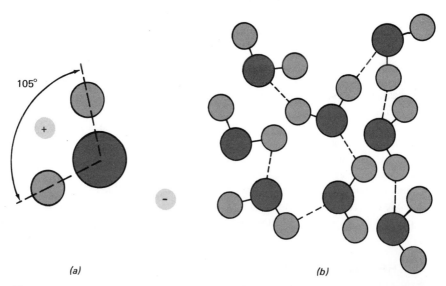

(a)   (b)

Figure 2.1   The water molecule. *(a)* Diagram of a single molecule, showing two hydrogen atoms (light circles) joined to an oxygen atom at an angle of 105°; note the polarity of the molecule. *(b)* Joining of several molecules by hydrogen bonds.

This arrangement of the hydrogen atoms, and of the resulting electric charges, imparts to water a unique physical and chemical versatility. Because of the distribution of charges, the water molecule is **electrically polar:** with a concentration of positive charge on one side (H^+ and H^+ from the hydrogen atoms) and a concentration of negative charge on the other (O^- from the oxygen atom), the molecule will respond to an electric field by turning its positive end toward the negative field source and its negative end toward the positive field source. The molecule is thus said to be a **permanent dipole,** containing a pair of equal and opposite electric charges. The strength of the polarity, which is determined by the distance between the centers of the opposite charges, is termed the **dipole moment.** For water the dipole moment is rather high—on the order of 1.8×10^{-18} electrostatic units. The importance of the high dipole moment of the water molecule is that it makes water an excellent solvent for **ionic** (electrically charged) compounds, and also makes it highly reactive with such compounds. These qualities of water are of crucial importance to the ways in which it functions in biological systems.

In its solid and liquid forms, water is a highly organized substance. As a solid, ice, water assumes the form of a highly ordered ''ice lattice,'' with its molecules held together by hydrogen bonds in a distinct pattern and at definite distances. Each molecule is linked to four neighbors in a three-dimensional tetrahedral structure (Figure 2.2). This extremely open, porous structure gives ice its unusually low density, permitting it to float on liquid water.

As ice is warmed, its hydrogen bonds begin to fracture, permitting the molecules to move more freely. At 0.0°C, the exact melting point of water, most of the hydrogen bonds remain intact, keeping the structure open. Thus the density of melting ice at this temperature is still less than that of water. At higher temperatures more hydrogen bonds are broken; when about 15 percent of the bonds are broken, complete melting occurs (Pauling, 1960). Below about 4°C, continued warming increases the density of liquid water; further warming results in its thermal expansion.

Although water molecules are arranged in a definite configuration, there remains a great deal to be investigated. Most researchers agree, however, that the breakdown of the ice structure leaves the water molecules associated in some way through the remaining hydrogen bonds. Because water is less highly organized than ice, however, there is a great deal of random movement by free molecules, and many of them seem to move into and out of the hydrogen-bonded chains. Molecular activity naturally increases with rise in temperature. The physical processes of melting and evaporation occur when a large number of hydrogen bonds are broken but the structures of the individual molecules remain unchanged.

Ordinary water consists for the most part of molecules made of hydrogen (atomic weight 1) and oxygen (atomic weight 16). There are a few exceptions, since **isotopes** of these elements—atoms with the same atomic numbers but different atomic masses and somewhat different properties—occur in pure water

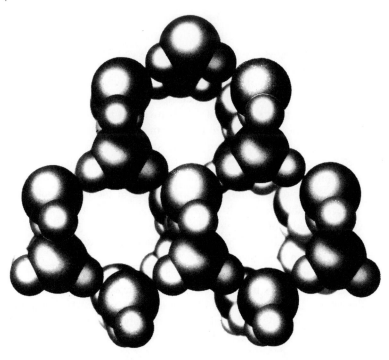

Figure 2.2   The structure of ice. The "ice lattice" is a tetrahedral arrangement of water molecules with a relatively rigid, open structure. The large spheres represent oxygen atoms, the smaller spheres represent hydrogen.

in small amounts. The commonest are deuterium (^{2}H) and tritium (^{3}H), hydrogen isotopes with two and three times the mass of ordinary hydrogen, respectively. The combination of these isotopes with oxygen produces deuterium oxide ("heavy water") or tritiated water, each of which has important scientific and industrial uses. Similarly, isotopes of oxygen (^{17}O and ^{18}O) are also present in water. These isotopes are, however, extremely rare. In natural waters, 99.985 percent of all hydrogen atoms and 99.759 percent of all oxygen atoms are the ordinary ^{1}H and ^{16}O isotopes.

PHYSICAL AND CHEMICAL PROPERTIES OF WATER

SPECIFIC HEAT

The specific heat of a substance refers to its capacity to absorb thermal energy in relation to temperature change at constant volume. Water holds a great amount of heat with a relatively small change in temperature. The unit of measure of specific heat is the gram-calorie, or small calorie, which is defined as the amount of heat required to raise the temperature of 1 g of water from 14.5° to 15.5°C. Thus, water is the basis of this parameter and is said to have a specific heat of 1; the specific heat of other compounds is calculated as the ratio of their heat capacity to that of water. Only a few substances, including lithium at high temperatures and ammonia, have a specific heat higher than that of water.

This heat capacity of water has far-reaching implications. For one thing, it permits a body of water to act as a buffer against wide fluctuations in temperature, thus modifying the terrestrial climate in regions adjacent to large bodies of water. Furthermore, an aquatic organism is subjected to much narrower ranges of temperatures than a land form: whereas land areas may reach 38°C or more, lake temperatures seldom exceed 27°C. Furthermore, the slow rate of seasonal cooling and warming of lakes and streams is attributable to the high specific heat of water. For the same reason, changes in lake temperatures lag behind atmospheric fluctuations. In natural waters, dissolved substances lower the specific heat.

Much heat is necessary to bring water to boiling because a considerable amount of heat is dissipated in energy of vaporization before the boiling point is reached. The "reluctance" of water molecules to separate and emerge as vapor is due, as we have seen, to the hydrogen bonding of the molecules. For example, methane (CH_4) has nearly the same molecular weight as water (16 as compared with 18), but it boils at -161°C as compared with 100°C for water. The explanation for the lower boiling point of methane is that the outer electrons of the methane molecule are bonded chemically rather than covalently, there being no free electron pairs to establish the hydrogen bonds found in water.

LATENT HEAT OF VAPORIZATION AND FUSION

Compared with other liquid compounds of similar, simple molecular composition, water vaporizes (evaporates) very slowly when heated, i.e., water has a very high **latent heat of vaporization.** Vaporization is a process in which thermal energy is supplied to overcome the attractive forces between water molecules (mainly hydrogen bonds in liquid water). A given volume of liquid water contains a larger number of hydrogen bonds per unit volume than most other common liquids.

Since there are two hydrogen bonds linking each water molecule to its neighbors, the heat required for evaporation is approximately twice the hydrogen bond energy (4.85 kcal per molecular number of H bonds, or 9.7 kcal per gram molecular weight). Depending upon temperature, the amount of heat required for the vaporization of 1 g of water ranges from about 500 to 600 cal (9700 cal per gram molecular weight). In other words, as much heat is used to vaporize 1 g of water as to raise the temperature of 540 g by 1°C.

The heat taken up in the change of water from a solid state to a liquid phase with no change in temperature is termed the **latent heat of fusion.** Specifically, 79.7 cal of heat are required to melt 1 g of ice at 0°C. This is about 15 percent of the heat necessary to separate the hydrogen bonds in the vaporization process. As heat is applied to ice, the molecules are set into motion. Increased motion parts the hydrogen bonds until finally the molecular lattice collapses and brings about melting. There is evidence that some few bonds may be broken before melting begins, but even so, only about 15 percent of the remainder are broken during complete melting. The cohesion of the remaining 85 percent—the bonds which remain intact in liquid water—accounts for water's high heat of fusion and heat of vaporization.

The quantities we have been discussing are referred to as "latent" heats because as water is altered from one state to another there is a period during which two states (ice/water or water/vapor) exist together. This period lasts until one state or another predominates, and during this time heat is absorbed. Because of the law of conservation of energy, the amount of energy in an aquatic system remains constant: the amount of heat liberated during the freezing process is equal to the amount taken up in melting, and equal amounts of heat are transferred to and from a given mass of water during vaporization and condensation. It is this quality that causes large bodies of water to modify the climate (as mentioned above) in adjacent areas.

DENSITY QUALITIES

Most liquids, including water, contract and become heavier with cooling: water is less dense, or lighter, at high temperatures, and becomes more dense as temperature is decreased. Fresh water is unique, however, in that it reaches its maximum density not, as with other liquids, just at the freezing point, but rather at 3.98°C. As it cools below this point the density decreases, that is, the water again becomes lighter (Figure 2.3) as the crystalline structure develops.[1] As it freezes, water increases in volume by about 11 percent. Thus water pipes burst and ice floats. Water is one of very few substances whose solid state is lighter than the liquid. This quality is important in limnology, for it means that lakes freeze from the surface downward and that only small bodies of water, and lakes in the coldest regions, freeze solidly in winter; thus in most lakes and streams life goes on underneath the ice—albeit at a reduced rate.

Density is a function not only of temperature, but also of pressure and of the concentration of substances dissolved or suspended in the water. High

1. In this respect, the characteristics of seawater are quite different. The freezing point of seawater is about −2°C, and contraction as well as increase in density associated with decrease in temperature are uniform to that point. Thus seawater is heaviest just before freezing and at that time is also some 6° cooler than fresh water. The sinking of this cold heavy water is a factor in circulation in the marine environment.

concentrations of dissolved salts account for increased density of ocean waters and the waters of such unusual inland waters as Great Salt Lake. Organic material such as finely divided detritus and organic silt, being heavier than water, will, in sufficient quantities, increase the density. Similarly, clays and fine muds increase the density of waters in lakes and streams.

Viscosity is a measure of the internal, molecular friction of a liquid and is concerned with mobility and flow. For flow to occur, the bonding forces which link the molecules of a liquid in a lattice pattern must be overcome, and there must be a void in the lattice, into which a loosened molecule can move. Compared to many liquids, gasoline for example, water is highly resistant to flow. This resistance is due to the relatively great amount of energy contained in the hydrogen bonds of water molecules. The high viscosity of water limits stream discharge, and also affects the living habits, morphology, and energy expenditure of aquatic animals, since they occupy a dense environment which offers considerable hindrance to movement. From what has been shown of the activity of molecules in relation to temperature, we should expect viscosity to diminish with increase in temperature. In fact, viscosity does change about 3.5 percent with each degree of temperature, although the relationship is not uniform.

The **mass,** or weight, of water is also related to its density. By general standards, water has considerable weight. Although it varies with temperature, the weight of one cubic meter of water is generally given as 998.4 kg (62.4 lb/ft^3). If determinations are made below the surface, atmospheric pressure must also be considered. At a depth of 30.5 m the pressure is approximately 4 kg per square centimeter or about 4 **atmospheres** (4 atms). Water pressure increases at a rate of approximately 1 atm per 10 m of depth.

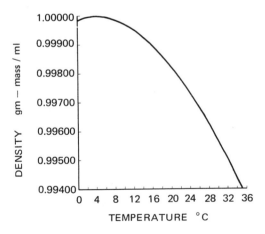

Figure 2.3   The effect of temperature on the density of pure water. Maximum density occurs near 4°C; above and below this temperature water becomes progressively lighter.

ADHESION, COHESION, AND SURFACE TENSION

Adhesion is the tendency of a liquid to cling to the surfaces of some materials by means of hydrogen bonds established between the hydrogen atoms of the water molecule and oxygen atoms of the other substance. Cellulose in wood, for example, contains a great number of oxygen atoms and is, therefore, very wettable.

Cohesion is the property of liquids which offers resistance to being pulled apart or to the formation of new surfaces. Mercury illustrates a fluid with high cohesion and low adhesion, for it forms a large spheroid drop on paper without wetting it. It has been calculated (Hendricks, 1955) that a force of 95,250 kg (210,000 lb) would be required to break a column of pure, perfectly formed water with a cross section of 6.45 cm^2 (1 in^2). At the surface of a water mass, the force mainly responsible for this resistance is termed **surface tension** and results from the unsymmetrical activity of water molecules at and below the surface. Whereas the internal molecules are bonded in all directions, those at the surface are attached only to those below and to the side of them, since there are none above. An inward force is thus imparted to a body of water which gives rise to surface tension, a phenomenon analogous to a sheet stretched tightly over the water. For many small plant and animal species, such as duckweed, snails, and water fleas, the **surface film** of a body of standing water is a firm surface, and it serves as a temporary or permanent habitat. The species adapted to such a life at the surface film are collectively referred to as the **neuston.**

The force of surface tension is measured in **dynes.** At the zone of contact between water and air surface tension is 73.5 dynes at 15°C and varies inversely with temperature. The addition of inorganic salts to water tends to increase its surface tension, and the presence of organic compounds tends to cause surface tension to decrease.

SOLVENT ACTION

Water is far from being a "universal solvent," as it is sometimes called, but it will dissolve more different substances, and in greater amounts, than any other naturally occurring liquid. Over 50 percent of the known chemical elements have been found in natural waters, and it is probable that traces of most of the others could be found in lakes, streams, estuaries, or oceans. The great capacity of water as a solvent, coupled with its tendency to ionize dissolved substances (a function—as mentioned above—of its high dielectric constant) are of major importance; these properties facilitate chemical reactions, both in the aquatic environment and, more importantly, in the internal "environment," where plants and animals carry on their physiological and reproductive functions. Basically two types of process—both involving hydrogen bonds—account for the extremely efficient solvent action of water.

One type of solvent process may be described as *inert*, for, by means of hydrogen bonding, the dissolved substances are relatively unaffected by the solvent. Many compounds of ammonia, nitrate, and phosphate, as well as sugars, alcohols, and various organic acids, are bonded to the hydrogen atom of the water molecule through oxygen atoms, hydroxyl groups, or through nitrogen bound to hydrogen. It should be pointed out that these compounds are some of the important ones involved in energy transfer and storage in biological systems. Therefore, through hydrogen bonding the substances are delivered unmodified and readily accessible, not only from environment to plant or animal, but within the physiological systems of the organisms as well.

The second type of solvent process involves the separation of the electric charge between the hydrogen and oxygen atoms in the H–O–H molecule. Water is characterized by a very high charge separation; thus various salts such as sodium chloride and the salts of potassium are retained in solution.

In still another activity, water has a remarkable ability to ionize dissolved compounds through the separation of the bonds in the H–O–H molecule. This process is due to the high dielectric constant of water. The molecule shows polarity, and total separation of one H from OH results in two charged ions, H^+ and OH^-. The H^+ becomes the hydrogen ion of acids and the OH^- becomes the hydroxyl radical of bases. A simple, reversible ionization reaction is that involving the solution of carbon dioxide in water:

$$CO_2 + H_2O \rightleftharpoons H_2CO_3$$

or more completely:

$$O=C=O + HOH \rightarrow O=C\overset{\displaystyle OH}{\underset{\displaystyle OH}{\big<}} + HOH \rightarrow O=C\overset{\displaystyle OH}{\underset{\displaystyle O^-}{\big<}} + H_3O^+$$

TRANSPARENCY

Pure water is quite transparent, allowing light to travel considerable distances underwater with relatively little distortion. The extent to which light of different wavelengths is able to penetrate into a body of water, which is usually expressed in terms of absorption of light-wave components, is a function of transparency. The transparency of water is of great importance in plant and animal relationships, and is therefore of great interest to the ecologist. Pure water absorbs light selectively, with the major absorption being at the red end of the spectrum. Minimum absorption is in the blue range at about 470 nm or 4700 Å (1 nm = 10 Ångstrom units). When sunlight enters a body of water, about 90 percent of the radiation above 750 nm is absorbed by the time the light has penetrated to a depth of 1 m. The relations between solar radiation and natural waters are complex; they will be considered more fully in chapter 7.

WATER AS A LIQUID

One of the most remarkable qualities of water is the simple fact that it occurs at most temperatures in liquid form, for liquids are found only rarely in nature. Aside from mercury, water is the only inorganic substance known to occur in liquid form at the earth's crust. (It has been suggested, but never verified, that liquid carbon dioxide might be found in quartz crystals.) Petroleum and tars occur in liquid form, but these are substances of organic origin.

Besides the familiar forms of liquid, ice, and water vapor, water is also found in somewhat different states. **Adsorbed** water may be tightly held to the surface of another substance by mechanical means; such adsorbed water is found in the soil and in the near aquatic habitats, where it may play an important role in maintaining populations of bacteria, algae, and fungi. Water is capable of forming **complexes** with the larger molecules of other inorganic and organic compounds; in this form water is an important component of soil. Many important organic substances have molecules too large to be dissolved in water. They may, however, form **colloidal suspensions,** which may have rather complex and stable structures; the cytoplasm which makes up the living component of every organism is primarily colloidal in structure. Thus water permeates every aspect of life throughout the ecosphere.

WATER IN THE ECOSPHERE

Virtually all of the earth's water may be thought of as constituting a closed system. There is a finite amount of water, which is constantly recycled in a pattern determined by solar radiation, condensation, and gravity. The recycling pattern, which is called the **hydrologic cycle,** is represented in Figure 2.4. The hydrologic cycle moves enormous masses of water from the earth's surface into the atmosphere by evaporation, back to the earth by precipitation, and thence back into the atmosphere via many complex paths.

It has been calculated that there is a mass of about $13,967 \times 10^{20}$ g of water on the accessible areas of the earth's surface as liquid and ice and in the atmosphere as vapor. Approximately 99 percent of the total is in the oceans and seas, and most of the remainder is locked in glaciers, snow, and ice. The mass of inland waters constitutes only about 0.25×10^{20} g. Water vapor in the atmosphere amounts to only a minor fraction of 1 percent of the total. The greatest mass of water, estimated at $250,000 \times 10^{20}$ g, occurs in the rock mantle of the earth.

The cycle is extremely complex if we take into account the different forms of precipitation (snow, rain, etc.), the path of water upon the earth, and the mode of loss of surface water back to the atmosphere. Studies by the United States Geological Survey throughout North America have shown that in the general region from Connecticut to North Dakota evaporation and plant transpiration return from about 50 to 97 percent of the experienced precipitation back into the atmosphere. From 2 to 27 percent of the precipitation finds its way to streams and the sea. Infiltration into the ground accounts for 1 to 20 percent of the water falling upon the surface, and this **subsurface** water ultimately finds its way back into the cycle also, via lakes, streams, and the bodies of living organisms.

In this chapter we have for the most part described the properties of pure water. It should be clear, however, that genuinely pure water is almost never found in nature, because "contamination" of some sort is inevitable at every stage of the hydrologic cycle. All natural waters are impure to some extent; even rainwater, distilled as it is by evaporation and condensation, contains a number of solutes (although these occur in very low concentrations). Thus all natural waters vary in composition. Rainwater and seawater vary the least, but water from one pond or stream may be enormously different in composition from the water in a similar environment just a few miles away. Even the water in a single stream may change dramatically as it courses over varied substrates. Thus we cannot speak of "fresh water" as if it were a single, uniform substance; the ecologist must always be aware of the extent to which natural waters are affected by local conditions.

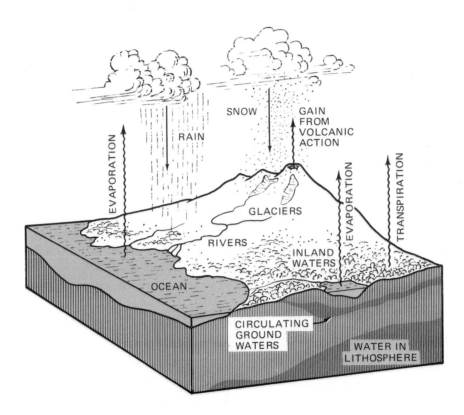

Figure 2.4   The hydrologic cycle.

SUGGESTED FOR FURTHER READING

Leopold, L. B. 1974. *Water: a primer*. W. H. Freeman, San Francisco. An excellent introduction, by a leading hydrologist, to some of the details of the hydrologic cycle in nature, with special attention to problems of runoff, flood control, uses of floodplains, and other current problems. A non-mathematical treatment, well illustrated.

Todd, D. K., editor. 1970. *The water encyclopedia*. Water Information Center, Port Washington, New York. Subtitled "a compendium of useful information on water resources," this volume presents, mostly in tabular form, a wealth of information on the hydrology, movement, physicochemical nature, use, and protection of natural waters. Chapters 7 and 8, covering water resources management and the agencies and organizations involved in planning and administering water projects, will be of tremendous value to citizens concerned with water conservation.

PART TWO

The Origins and Features Of Basins and Channels

3| Lake Basins and Lakes

s water moves over the surface of the land, its progress is often stopped or delayed. The waters then form the environments that the limnologist refers to as "standing waters," or **lentic habitats.** The technical term is derived from the Latin *lentus,* meaning "sluggish." Standing waters are generally classified as lakes, ponds, swamps, and marshes. It is relatively easy to distinguish these last two on the basis of their vegetation and soil conditions: **swamps** are wet lowlands which support mosses and shrubs, together with relatively large trees such as cypress and gum; **marshes** are broad, treeless wetland areas, occupied by abundant grasses, rushes, and sedges. Obviously, neither is a completely aquatic habitat.

It is harder to distinguish **lakes** from **ponds,** because neither term is necessarily restricted to any one kind of environment. Lakes and ponds are formed in a number of different ways, both natural and artificial. With the passage of time, they change in character, ultimately silting up and becoming transformed to wetland or meadow. Bodies of standing water also change in character as you travel from north to south. For all these reasons there can be no hard and fast difference between a "lake" and a "pond." Moreover, local usage further complicates the issue: the large lakes of Maine and the Adirondacks are referred to locally as "ponds," whereas certain stretches of Florida's slow-moving St. Johns River are called "lakes," although they are flowing, rather than standing, waters.

For our purposes we can rely on a basic distinction between large expanses of open water, on the one hand, and small bodies of water (often thickly filled with plant growth) on the other. These represent two extreme conditions, but they will serve as a basis for our working definition. For laymen and scientists alike the term **pond** generally suggests a small, quiet body of standing water, usually shallow enough to permit the growth of rooted plants from one shore to the other. Larger bodies of standing water, occupying distinctive **basins,** we will refer to as **lakes,** and these will occupy our major attention. The shape and origin of a lake basin determine its essential ecological character, so we will first consider the processes through which lakes are formed.

THE ORIGINS OF LAKE BASINS

Lakes and ponds can be created by a wide variety of natural events, but about a dozen basic processes account for the formation of most standing water basins. The ways in which basins have been formed suggest a useful system for their classification, because in any particular region the natural lakes and ponds are usually formed in a similar way. In central North America, for example, most lakes were formed by the advance and retreat of continental glaciers. Many Florida lakes occupy basins dissolved from a local form of porous limestone. Virtually any kind of geological process or disturbance can produce some form of lake basin: an excellent example is found in East Africa, where the great lakes of the Rift Valley—Nyasa, Tanganyika, and Victoria—occupy basins created by large-scale shifting of the earth's crust.

TECTONIC BASINS

Lake basins formed by folding, faulting, or movement of the earth's bedrock itself are referred to as **tectonic** basins. Tectonics (from the Greek *tektōn,* "builder") is the branch of geology concerned with the structure of the earth and, especially, of the earth's solid outer surface, or **mantle.** In the past ten years or so geological studies of many kinds have confirmed the theory of **plate tectonics.** This once seemingly incredible theory—which we shall return to in a later chapter—holds that the entire surface of the earth consists of immense granitic "plates" floating on a sea of much heavier, molten, basaltic rock, called **magma.** From time to time the magma erupts through the outer surface rock in the form of a volcano, and at all times the plates are slowly moving—upward and downward, towards and past one another—seemingly in response to currents within the plastic magma. This is "continental drift." In this view of the earth, the ocean basins are simply the lower regions of the tectonic plates and the continental masses the higher regions. Obviously, the effects of such a dynamic system include the development of many kinds of surface irregularities as the plates of the mantle, over hundreds of millions of years, move about in relation to one another.

Crustal movements

Movements of bedrock may result directly in the formation of lake basins. Reelfoot Lake in northwest Tennessee, and others in Arkansas and Missouri, were formed by the New Madrid earthquakes of 1811-1813 (Fuller, 1912; McGee, 1893). The basin of Reelfoot is about 32 km long and approximately 6 m deep. Geologic faulting, the large-scale shifting of great blocks of rock, often forms lake basins by tilting the surface of a block against the edge of an adjacent one, forming a basin in which water later collects; in Oregon, Lake Abert occupies a basin formed by such faulting and tilting (Russell, 1895). Lake Tahoe, California, was formed by the accumulation of water in a **graben,** or trough, created by the shifting of crustal masses to form steep walls (Hutchinson, 1957). Lake Ohrid in the western Balkans is a similar graben lake, in this case being found in a limestone region (Hutchinson, 1957). Fault-formed lakes typically have steep sides and a rectangular bottom profile. The deepest known lakes occupy graben basins: Lake Baikal, in Siberia, fills three basins and has a maximum depth of 1741 m; Lake Tanganyika, in East Africa, has a maximum depth of 1470 m. Tanganyika is one of many lakes found in a belt of rifts— displacements of crustal blocks along fractures—which extends some 4800 km from Rhodesia north to Israel. In these rift valleys are found such famous bodies of water as the River Jordan, the Sea of Galilee, and the Dead Sea.

Lesser deformations of the crust account for the creation of Utah's Great Salt Lake (Gilbert, 1890). Here the surrounding mountains have been thrust up to form a basin with no drainage; inflowing waters carry dissolved minerals, and rapid evaporation concentrates the salts. The resulting waters form a harsh environment in which only a few plant and animal species can live.

More subtle warping of the earth's crust may affect the course of a stream in two basic ways. If the crust bends with the direction of stream flow, the steeper gradient will increase the velocity of the stream. If, on the other hand, the crust bends upward, against the direction of flow, the stream is blocked, forming a naturally dammed lake. Lake Victoria, in Africa, was formed by such a natural impoundment. Upwarping of the land impeded the flow of the Kagera and Katonga Rivers, forming a lake basin some 68,422 km² in area.

Changes in sea level

Tectonic movements bring with them **emergence** and **subsidence**—rise and fall—of tremendous land areas, with corresponding changes in sea level. Through this process a body of water may become isolated from the sea of which it once was a part, forming a new lake. There is evidnece, for example, that Lake Okeechobee, in southern Florida, is a remnant depression in the Pliocene sea floor of some 10 million years ago, which retained its form when the region was uplifted to become land. Lake Okeechobee, with an area of 1880 km² and a maxium width of some 40 km, is the largest body of fresh water in the United States south of the Great Lakes. It is also quite shallow, and in periods of drought it may be no deeper than 4.5 m; the only important tributary to Lake Okeechobee is the Kissimmee River, which drains a watershed of about 13,000 km² and empties into the lake on the north. Other central Florida lakes, including Apopka and Weir, may have been formed by the same uplifting process: biological evidence to support this theory is the presence in these lakes of a killifish *(Cyprinodon hubbsi)* whose nearest and closest relative is found in what are now salt or mildly saline waters along the nearby coasts.

Coastal valleys which have been scoured by glaciers can be transformed into **fjord lakes** by a relatively small drop in sea level. A typical fjord is partially bounded at its mouth by a transverse ridge of bedrock, a **sill,** rising toward sea level from the floor of the embayment. A slight drop in sea level may be great enough to expose the sill, forming a lake between the sill and the headwater (source) of the fjord. A number of such lakes have been described in British Columbia, Norway, New Zealand, and the British Isles.

BASINS OF VOLCANIC ORIGIN

Volcanic disturbances may often form lake basins. Crater Lake, in Oregon, is a striking example of a lake whose basin was formed directly by an eruption (Figure 3.1). Crater Lake occupies a **caldera** (a large volcanic basin) formed by the internal collapse of the summit of Mount Mazama, a long extinct volcano. The surface of the lake is nearly 610 m below the caldera rim, and the lake is about another 610 m in depth. No streams enter Crater Lake; its water is derived entirely from rain and snow melt.

Lava issuing from an active volcano may flow across a stream valley and dam it. Various lakes in the formerly volcanic regions of Oregon and Washington were formed in this way. Lava may also issue from fissures rather than from volcanoes, in the form of a flow rather than an eruption. The sheet lava thus formed may contain depressions when it hardens, which can fill with surface water to become lake basins.

BASINS RESULTING FROM GLACIAL ACTION

Glaciation has been and is a major process of lake basin formation in many regions. Glaciers are of two principal types: **valley,** or **alpine,** glaciers, such as those shown in Figure 3-2, and **continental,** or **ice-sheet,** glaciers, such as the Greenland and Antarctic ice caps, which cover large areas of land to depths reaching thousands of meters (Strahler, 1966). The present topography of much of North America—and of other regions distant from the equator—is the result of repeated advances and recessions of continental glaciers during the late Pleistocene epoch, from about 400,000 to 10,000 years ago. The greatest concentrations of lakes in North America and Europe occur in regions once covered by the Pleistocene glaciers. Besides scouring out basins in the native rock, the glaciers also dammed streams by deposition of glacial **moraine** (soil and rocky debris).

Valley glaciers, found in most of the world's high mountain areas, commonly form lake basins. At the heads of glaciated valleys the movement of ice and captured debris often erodes out amphitheater-shaped basins called **cirques.** If water later fills the basin the result is a typical **cirque lake,** surrounded on all sides except the outlet by steep walls (Figure 3.2). Many classic examples of the cirque lake are found in Glacier National Park. Often the axis of the valley draining the cirque will be marked by a chain of lakes occupying a descending series of basins; the linked series of smaller lakes, which may be likened to the beads of a rosary, are referred to as **paternoster** lakes.

Throughout Canada and Alaska, much of the northern United States, and northern Europe, great numbers of small lakes have been formed by processes associated with continental glaciation. In the far northern regions of permafrost and tundra there are thousands of **thaw lakes,** often occurring on coastal plains and in river valleys (Vallentine, 1963). Thaw lakes have been investigated only

Figure 3.1   Crater Lake, Oregon, looking southward. The lake occupies the caldera, 10 km in diameter, formed by interior collapse of Mount Mazama, an extinct volcano. Wizard Island, at the caldera's western edge, was formed by later volcanic activity. (Courtesy of Oregon State Highway Department.)

Figure 3.2   Mountain lakes of British Columbia. The elongate lake behind the peak to the right of center occupies a cirque; drainage from the cirque lake enters a lower lake. Such valley lakes are called paternoster lakes. A braided stream pattern, such as will be described in Chapter 4, extends the length of the broad valley on the left. (Courtesy of U.S. Air Force.)

recently, but their basins seem to form when a layer of the permanently frozen soil thaws partially during the summer. Thawing causes the formation of a lens-shaped, slightly domed mass of silt, which loosens and is moved about with repeated freezing and thawing. Eventually the surface "sags" to form a small lake basin. The coastal plain of New Jersey is dotted with hundreds of similar shallow basins, less than 3 m deep, ranging in area from a few hundred square meters to over one and half km². These basins, which lie in an area just beyond the southernmost reach of the Pleistocene ice sheets, are believed to have been formed in the same way as the thaw lakes of the far north; they are referred to as **periglacial frost-thaw basins** (Wolfe, 1953).

In all the regions once covered by continental glaciers, many small lakes and ponds occupy basins called **kettles.** Kettles are formed by the melting of massive isolated chunks of buried ice, left behind as the glaciers retreated northward. Kettle basins are common in Minnesota, Wisconsin, and much of central Canada.

In many regions where the bedrock was relatively soft the continental glaciers could have essentially the same effect as a valley galcier, scouring out long, narrow basins. Often the retreating glacier would leave behind a **terminal moraine** (the load of rock and other debris pushed before it by the advancing glacier) massive enough to dam the newly formed valley. The Finger Lakes of central New York were formed in this way, as were many major lakes (such as Lucerne, Como, and Constance) of the Swiss Alps.

Evolution of the Great Lakes

In central North America, the Great Lakes—Superior, Michigan, Huron, Erie, and Ontario—offer the major example of basins created by continental glaciation. The original basins, very different from those of today, were formed in the late Pleistocene by glaciers scouring the rock of older stream valleys. The basins were later shaped by depression of the land under the weight of the massive ice sheets and by continued erosion of softer rock.

The history of the Great Lakes is rather well known. It began with the retreat of the last great Pleistocene glacier—the **Wisconsin ice sheet.** At its greatest, the Wisconsin glaciation reached south to the junction of the Ohio and Missouri Rivers. As it retreated, the ice left behind several now-vanished but very large lakes, formed between the melting ice on the north and the higher elevations to the south. This process began probably no longer than 20,000 years ago (Hough, 1962). Lake Chicago occupied roughly the basin of the present Lake Michigan; Lake Whittlesey formed from the older, smaller Lake Maumee in the present Lake Erie region; Lake Saginaw occupied an area between and to the north of the other lakes. The extent of the glacier and the original lakes is shown in Figure 3.3a. For a time, the Finger Lakes, in the east, drained into the Susquehanna River. Drainage from Lake Whittlesey was through Lake Saginaw and Lake Chicago, and then by way of the Illinois River to the Mississippi River.

As the glacier melted and withdrew farther a single lake, Lake Warren, was formed by the joining of Saginaw and Whittlesey. As shown in Figure 3.3b. the Finger Lakes probably began to drain increasingly into Lake Warren.

Later still, with the continued retreat of the glacier front, Lake Duluth (Figure 3.3c.) was formed to the northwest, in the region of what is now Lake Superior. Lake Lundy formed from the glacier and joined the waters of Whittlesey and the Finger Lakes. Lundy was large, extending from Lake Saginaw into the Basin of Lake Huron and north of Lake Ontario; it drained to the east, through the Mohawk River to the Hudson. Lake Chicago continued to expand, and at that time was nearly as large as today's Lake Michigan.

Later recession of the glacier left a large lake, Lake Algonquin, in the regions now occupied by Lakes Huron, Michigan, and Superior (Figure 3.3d);

Lakes Erie and Ontario had essentially the same form as they do today. Another dramatic change also marked this period: the St. Lawrence Valley was exposed by retreating ice, and sea water flooded the exposed valley, forming a large body of water called the Champlain Sea. One arm of the sea reached into Lake Ontario, and the basin proper extended from New York Bay as far north as the St. Lawrence. Traces of the Champlain Sea shoreline can be seen today near Montreal.

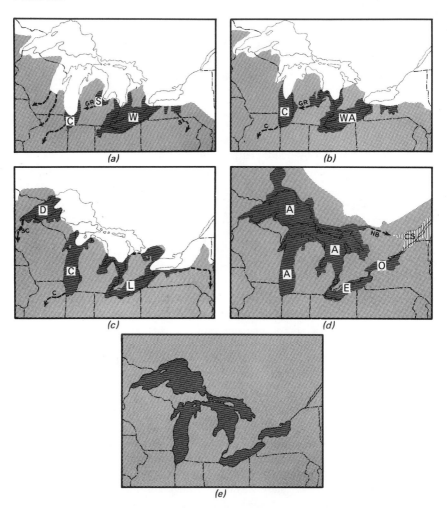

Figure 3.3   Developmental stages in the development of the Great Lakes, after the Pleistocene continental glaciation. The stages are described in detail in the text. A = Lake Algonquin; C = Lake Chicago; CS = Champlain Sea; D = Lake Duluth; E = Lake Erie; L = Lake Lundy; NB = North Bay Outlet; O = Lake Ontario; S = Lake Saginaw; W = Lake Whittlesey; Wa = Lake Warren; c = Chicago River; gr = Grand River; s = Susquehanna River; sc = Saint Croix River. (After Thwaites, 1959, and Flint, 1957.)

Immediately before the Great Lakes reached their present form, the basins went through what is called the Nipissing Great Lakes Stage, when Lakes Erie and Ontario were separated from the other three lakes. Later the ice sheet retreated far to the north, a connection developed between the Nipissing and the other Great Lakes, the lakes decreased in size, and the present configuration was established (Figures 3.3*e* and 3.4). The Great Lakes began to drain toward the east rather than, as earlier, to the west, forming the present-day St. Lawrence drainage system.

For a relatively short time, about 1000 years, the greatest of all glacier lakes covered much of North Dakota, Minnesota, Ontario, Manitoba, and Saskatchewan. Called Lake Agassiz, it was over 210 m deep and occupied a basin 1120 by 400 km at its maximum. Later, the ice dam which impounded the waters melted, the land became elevated, and most of the lake drained to the north. Today the basin of Lake Agassiz is occupied by Lake Winnipeg, Lake Manitoba, Lake of the Woods, Lake Winnipegosis, a few smaller lakes, and the fertile valley of the Red River of the North.

SOLUTION BASINS

Lake basins dissolved out of the bedrock are found in various regions throughout the world. They are commonest, however, in the Balkan Peninsula,

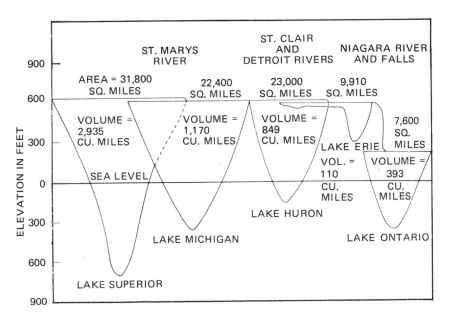

Figure 3.4 Profiles and basic dimensions of the Laurentian Great Lakes at the present time (depths greatly exaggerated). (After Chandler, 1962.)

the Calcareous Alps, Mexico's Yucatan Peninsula, and (in the United States) in Indiana, Kentucky, and Florida. All of these are extensive limestone regions, and the funnel-shaped **sinks** or (from the Slavonic) **dolines,** which form the basins, have developed through the solution and erosion of the soluble rock.

The rolling **karst** topography of central Florida (named for the geologically similar Karst plateau of Yugoslavia's Dinaric Alps) contains thousands of solution lakes formed in light-colored, usually highly fossiliferous Tertiary limestone. (Figure 3.5). The soil here is quite sandy, and the solution depressions can appear suddenly wherever the underlying, porous limestone has been dissolved either by percolating surface water or by flowing ground water. In the percolation process surface waters may start to dissolve the limestone at points of fracture or other weakness. The inflowing water moves easily through the stone,

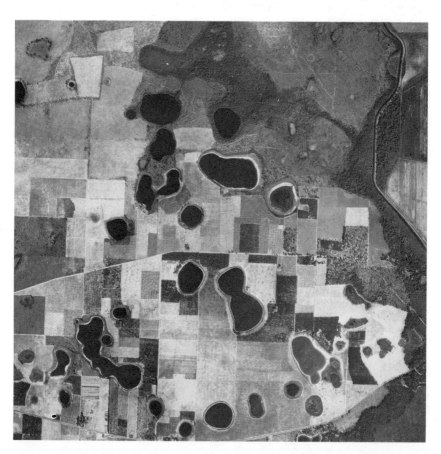

Figure 3.5   Sinkhole lakes in the karst topography of Marion County, Florida. The extensive shore zones visible in the photograph resulted from a temporary drought, which also caused almost complete drying in some of the nearby basins. (Courtesy of Commodity Stabilization Service, U.S. Department of Agriculture.)

and with continued flow the doline develops. The limestone may also be dissolved from below, when water moving in a subsurface aquifer weakens the roof of an underground chamber. This may cause the roof of the chamber to collapse, leaving a fairly regular, cone-like doline.

In the Calcareous Alps of western Switzerland many solution lakes are formed at high altitudes, the basins sometimes taking the form of cirques. In a drainage pattern that is unusual except in such limestone regions, the outflow from these basins may flow into caverns and disappear underground.

Sinkholes do not necessarily contain water. If the bottom of the doline does not reach down to the local water table, the sink will be essentially dry. If the bottom of the sink extends down to the range of fluctuation of the water table, a temporary lake appears in the rainy season but dries up when the water table drops. A deeper sink will be the basin of a permanent lake. Occasionally a newly formed sink may suddenly alter the shape of a lake bottom—or a clogged sink may suddenly be opened—with dramatic effects on a preexisting lake. This happened to Orange Lake in northern Florida in the late 1950s; a new sinkhole suddenly appeared and began to drain the waters of this large, shallow lake. The original lake was preserved only by construction of a dike around the sinkhole.

BASINS OF WIND ORIGIN

In arid regions the movement of fine, light materials such as clay and sand may result in the development of lake basins. The movement of the soil may be such as to block an existing stream, thus creating a dammed lake. Moses Lake, in Washington, was formed when wind action shifted sand dunes across Crab Creek. Removal of loose materials from an area may also form eroded **deflation basins,** and many lakes in Nebraska, New Mexico, and other regions east of the Rocky Mountains lie in basins scoured and shaped by winds. Obviously, the development of permanent lakes in deflation basins will depend upon adequate rainfall and suitable rock and soil conditions. Such conditions are not always found in deflation regions, however, and many of the basins do not contain water. On the other hand, lakes do occur in some arid regions, and it has been suggested that in these cases the basins were formed earlier under arid conditions, becoming filled later after repeated changes in climate. In some areas, depressions, created by shifting dunes and hollowing of troughs between the dunes, contain small ponds and shallow swamps. Cherry County, Nebraska, contains such waters.

BASINS FORMED BY STREAM ACTION

Lake basins can be formed by stream movements. Flowing waters may release their load of materials held in suspension, such as soil and organic debris, and they may cut away at their beds, forming lake basins through erosion. Streams with steep gradients and relatively high velocities lose their carrying power and drop their loads of sediment when they join larger streams. Lake Pepin, on the Minnesota-Wisconsin state line, was formed by the deposition of alluvium carried by the Chippewa River. At its junction with the Mississippi, the Chippewa deposited its load across the main stream, thereby creating a **fluviatile dam** in the Mississippi; Lake Pepin was formed upstream. Similarly, Tulare Lake, California, had its origin in an alluvial fan built by the King's River across the valley of the San Joaquin.

The reverse process may occur when a large stream builds its banks with sediments so quickly that the junction with a tributary is closed off. The result is a lake impounded by the action of the main river. Such **lateral lakes** are found along the Sacramento River of California and the Red River in Texas and Louisiana.

A meandering stream may erode the outside shores of its broad bends, and in time the loops may become cut off, leaving basins. The resulting shallow, crescent-shaped lakes are called **oxbow lakes** (Figure 3.6). The mature flood plain of the Mississippi and some of its tributaries in Louisiana and neighboring states contain hundreds of oxbow lakes occupying abandoned sections of river

Figure 3.6 Oxbow lake formed by the cutting off of a loop in the Connecticut River. (Courtesy of Soil Conservation Service, U.S. Department of Agriculture.)

channel. The lower ends of many of the oxbows are blocked by sediments deposited by the parent stream as the meanders were cut off.

The force of a stream flowing over a cliff as a waterfall will erode a pool at the foot of the fall. This **plunge pool** may become enlarged and the original stream may eventually become extinct, leaving a basin which may later fill to become a lake. The lake at Dry Falls in the region of the Grand Coulee, Washington, occupy such former **plunge basins** (Hutchinson, 1957).

BASINS RESULTING FROM MASS WASTING

Large masses of rock, soil, mud, or other unconsolidated material may fall away from an adjacent slope and block a steam valley. Lakes in basins formed in this manner are found in the Alps, in western North America, and in other regions. In 1925, a landslide on the Gros Ventre Range in Wyoming dammed the Gros Ventre River and formed a lake some 5 km long. Two years later the lake waters reached the top of the dam, overflowed, and caused a disastrous flood. The dam was not completely destroyed, however, and a smaller lake remains today. In this instance the materials in the slide were not tightly consolidated, so eventual erosion and destruction of the dam are rather to be expected.

If the slide material is compacted, and the impounded stream is small and fills the basin slowly, long-lived lakes may result. Deep Lake, Wyoming, was formed by a landslide which dammed Clarks Fork of the Yellowstone River behind some 800 feet of rubble; a mudflow was responsible for the formation of Lake San Cristobal in Colorado.

PLUVIAL LAKES

The climate of the Pleistocene epoch was strongly influenced by the extent of continental ice sheets, and in most regions the climate was quite unlike that of today. In western North America, for example, the environment was periodically much more moist than it is at present. The glacial climate produced **pluvial** periods, times of increased rainfall and reduced evaporation, leading to extensive lake formation. Most of these Pleistocene lakes disappeared with larger changes in climate, and with increased aridity, evaporation, and lack of outlets, most of the few that remain are saline. During the Pleistocene, Lake Bonneville, in northwestern Utah, covered some 51,000 km^2 to a depth sometimes exceeding 300 m, and drained northward to the Pacific through the Snake and Columbia Rivers (Gilbert, 1890). After repeated episodes of rising and falling waters (Bissell, 1963), all that remains of Lake Bonneville today are three saline lakes, the largest of which is Great Salt Lake. Most of the original lake bottom and shore zones now make up the Bonneville Salt Flats. Thus, although the original basin was of tectonic origin, the basins of the present saline lakes might be thought of as "fossil" basins, formed ultimately by the drying of the earlier lake.

LAKES DEVELOPED BY SHORELINE FORCES

In large bodies of water the shifting of sediment by nearshore currents can form lake basins. If, for example, sediments are deposited in quantity across the mouth of an embayment or estuary, the blockage may result in the formation of a new pond or lake. Along seacoasts, sand spits formed by longshore currents (those moving parallel to the coast) may effectively block a lagoon or bay. Wave activity, too, may form bars of sand or sediment, which can likewise close off the mouth of an embayment. Many impressive shoreline lakes, or "ponds," are found along the New England coast, especially in the area of Martha's Vineyard and Nantucket (Figure 3.7).

These processes are by no means limited to seacoast areas; they may operate in any sufficiently large freshwater lake. In Minnesota, Buck Lake was cut off from its parent body, Cass Lake, by the formation of a spit which eventually closed off the smaller lake. For an example of this process in action on a smaller scale, see Figure 3.11.

Figure 3.7   Shoreline lakes, or "ponds," along the New England Coast. Material eroded from headlands has been distributed by wave action and longshore currents across the mouths of estuaries, closing them off to form new bodies of fresh water. (Courtesy of Lowry Aerial Photo Service.)

LAKES OF ORGANIC ORIGIN

Those lakes that are formed entirely or partly by the action of plants or animals may be said to be of **organic** origin. The activities of the beaver, *Castor canadensis,* have created innumerable ponds throughout the northern United States and Canada. In northern regions, at least, the beaver instinctively cuts logs and sticks and arranges them into a latticework dam, impounding upstream waters to form a protective pond for the den. Beaver dams may reach impressive sizes: one in Montana was reported (Mills, 1913) to be 600 m long. Most, however, are only some 25 m long and seldom more than 2 m high.

Of all animals, man himself is of course the one which creates the greatest artificial lakes and ponds. In recent years tremendous bodies of water have been impounded for flood control, water storage, to create hydroelectric power, and for other reasons; there have even been serious proposals to dam Arizona's Grand Canyon. Agricultural policies have led to the creation of many thousands of farm ponds throughout the United States and the world. Industrial ponds, called "lagoons," are often constructed as reservoirs for various unwanted byproducts. Because these artificial bodies of water are created for many purposes and are managed in countless different ways, their ecological features are sharply different from those of natural impoundments. The limnology of artificial ponds and lakes promises to be a rich and illuminating area of research, but it will not be dealt with in this text except where it touches on the limnology of natural surface waters.

Plant material, living or dead, may contribute to the formation of a lake or pond by blocking the normal flow of water. Dense growths of aquatic and marsh plants may impound waters effectively, especially in the more southern regions where there is a long period of vigorous plant growth. In Florida, Lake Okeechobee is partially impounded by vegetation. In Louisiana, dense mats of alligator weeds, *Alternanthera philoxeroides,* may choke watercourses (locally called "bayous") and form ponds. A log jam—or any accumulation of debris and detritus—may also block a stream and thus create a pond or small lake.

BASINS OF METEORIC ORIGIN

Very rarely, a meteorite may strike the surface of the earth with an impact great enough to blast out a large depression, in which water may later accumulate to form a lake. Several lakes are known to have been formed in this way, of which Lake Chubb in Quebec Province is a classic example.

LAKES OF UNKNOWN ORIGIN

Most lake basins can be shown to be formed by one or more of the processes discussed above, but there remain a few lakes whose origins are unknown. This is true of a number of lakes found in the Atlantic coastal plain. Of these, the most studied have been a cluster of small, shallow lakes in eastern North Carolina. Referred to locally as the **Carolina Bays** (although they are not "bays" in the technical sense), these lakes all occupy elliptical basins with the long axis oriented roughly northwest–southeast (Figure 3.8). The shores and bottoms are generally sandy, and the water, which reaches a maximum depth of about 6 m, is relatively clear. They are particularly unusual in that some of them seem to be increasing in size, whereas most freshwater lakes and ponds tend to become smaller as a result of sedimentation and encroaching shorelines. Many theories have been offered to explain the origin of the Bays, including meteoric impact (Melton and Schriever, 1933) or meteoric shock waves (Prouty, 1952); wind action leading to the development of deflation basins (Shand, 1946; Odum, 1952); solution of the underlying rock by groundwater (LeGrand, 1953); erosion by currents set in motion by schooling fishes; and burning of peat accumulations. Many geologists and biologists have studied the Bays, but so far no theory has proven acceptable to all of them.

SURFACE FORMS OF LAKE BASINS

Lakes vary widely in form and size, and the problem arises of finding a way to classify lakes of different types. Our discussion of the origins of lake basins suggests a general classification of lakes on the basis of the forces that created them. In a general way, the form of a lake will be a function of its origin: unless the basin is later reshaped by geological or organic forces, the size and morphology of a lake will usually reflect the nature of the process by which it was originally formed.

GEOMETRICALLY SHAPED BASINS

Basins of approximately geometric shape are particularly well described in terms of their formative processes. A descriptive terminology developed by Hutchinson (1957) distinguishes seven principle forms:

Circular As we noted earlier, typical solution sinks such as the dolines of Florida are primarily circular in shape. The Florida dolines often overflow a portion of the basin on the downslope side of the sink, forming shallow marshes, but these usually show gently rounded outlines. Other often circular basins are those of meteoric or volcanic origin, especially craters and calderas.

Subcircular Cirque lakes and tarns in mountainous areas are mostly of subcircular form, reflecting the action of valley glaciation.

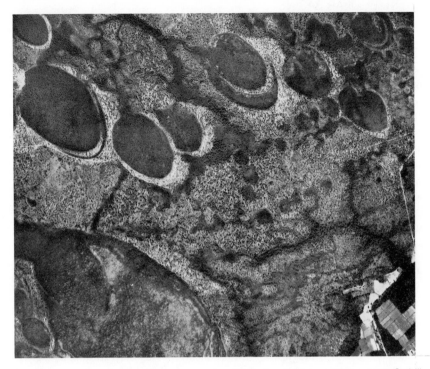

Figure 3.8   Carolina Bays, Horry County, South Carolina. (Courtesy of Commodity Stabilization Service, U.S. Department of Agriculture.)

Elliptical The Carolina Bays, of unknown origin, are examples of elliptical lakes. Such basins are quite unusual.

Subrectangular Lakes of tectonic origin situated in grabens (Lake Tahoe is an example) usually show an elongate, angular surface form of roughly rectangular shape.

Dendritic When a highly branched stream valley is impounded by the development of a dam, the subsequent flooding of the main channel and its tributaries results in a lake with many arms, or embayments.

Lunate Oxbow lakes in mature river valleys typically occupy lunate, or crescent-shaped, basins.

Triangular Coastal lakes formed by the development of a barrier beach or sand spit across the mouth of a stream valley are usually triangular.

IRREGULAR BASINS

Glacial action by continental ice sheets often results in basins of irregular shape. This may occur when glacial erosion wears away the upland regions between several basins, linking them in a single new basin. Irregular basins may also be formed by glaciers scouring rocks of different hardness to different depths, or by deposition of debris by a retreating glacier, with subsequent damming of streams.

Many lakes, especially older ones, do not have shapes regular enough to fit them into any of the geometric categories listed above. Once a lake forms, intrinsic processes—currents, sedimentation, and freezing, to name a few—continually work to change the original form of the basin. External factors, human activities being the most conspicuous, can also modify the original basin. Construction of dams and waterways, and diversion of water for human use, have far-reaching effects on lake basin morphology.

ISLANDS

Islands are often conspicuous features of lakes and large ponds, and the presence of an island can have strong and direct effects upon the currents, shore-building processes, and biological characteristics of a lake. Because most natural islands are related geologically to the islands in which they are found, we will consider here a few generalities concerning lacustrine islands.

The action of longshore currents in moving sediments has been discussed above in relation to the origin of lakes. It has been pointed out that sand spits projecting at various angles from a shore could be formed through the deposition of loose materials carried by the currents. The near-shore portion of a spit may later be eroded by changes in currents and/or wind direction and intensity, thereby leaving an island developed in the lake. Long Point, on the Canadian side of Lake Erie, apparently developed in such a manner. Wizard Island in famous Crater Lake, Oregon, represents the remnant of a secondary volcanic cone formed in the older caldera now occupied by the lake (Figure 3.1). Lakes in regions subjected to Pleistocene continental glaciation typically contain islands formed of materials resistant to the scouring action of the glacier or of debris dropped or pushed by the moving glacier.

Islands that bear little relationship to the origin of the basin are common in certain lakes in various parts of the world; these are **floating islands.** They have been described for lakes in Germany, the English Lake District, Chile, and other localities. In North America, floating islands have been reported to occur in Minnesota bog lakes and, in Florida, in the upper St. Johns River and Orange Lake. In Orange Lake, the islands are free-floating structures composed of a substratum of densely compacted peat-like material supporting shrubs and small trees; they vary in size up to several hectares.

SOURCES OF LAKE WATERS

After a basin is formed, and indeed, often while it is taking shape, water enters and reaches a given level, and, subject to fluctuations of varying magnitudes, it remains at that level. The sources of lake waters, the processes by which waters fill lake basins, and the maintenance of lake volume, all relate directly to the operation of the hydrologic cycle, described at the end of Chapter 2. In the cycle water is lifted from the surface waters and plants as vapor, and ultimately returned as precipitation. Studies by the United States Geological Survey in various regions of North America have shown that in the general region from Connecticut to North Dakota, evaporation and plant transpiration return from about 50 to 97 percent of the experienced precipitation back into the atmosphere; from 2 to 27 percent of the precipitation finds its way to streams and the sea; and infiltration into the ground accounts for 1 to 20 percent of the water falling upon the surface.

Water in lake basins, therefore, may come from a variety of sources. Although lakes do receive water directly from precipitation, land surface drainage in some form contributes most of the water. Water of infiltration arriving in the basin through seepage or springs is another source. In some instances all three sources may contribute.

Lakes have been described as "temporary stopping places for water on its way to sea." This suggests, accurately, that lakes and lake districts are most numerous in those regions of the world where streams arise and find their way to ocean basins. There are other regions in which streams are common, but end in dry valleys or in lakes without outlets; lakes are less common in these regions. In desert and semi-arid localities streams are rare, and, as would be expected, lakes are also. In such areas there are many lakes that are "landlocked" and water escapes only through evaporation or seepage.

Lakes often may be classified on the basis of the presence or absence of outlets, or **effluents.** Most freshwater lakes possess some form of outlet and are termed **open** lakes. The effluent may be a stream which drains the lake, or drainage may be by seepage into the basin substrate in the form of ground water. Whatever the influent and effluent of open lakes, there is a relationship between the two systems that determines the water level of the lake. Open lakes serve as "settling basins" for sediments introduced by inflowing streams because the tributary loses its carrying capacity as it enters the calm water of the lake. The water in the effluent stream usually is clearer than that in the influent, attesting to the settling action of the lake. Salts are removed with the stream outflow or seepage and the lake remains fresh.

Lake Geneva (Forel's "Lac Leman") is an open lake in the western Alps. Its major tributary is the Rhone River, which annually contributes great amounts of sediments to the basin of the lake. Originally about 75 km long, Lake Geneva is now nearer 64 km as a result of delta formation in its upper region. It is about 300 m deep in places, and the delta is quite thick. Continued sedimentation brings about the encroachment of the delta toward the lower part of the lake and will eventually fill the basin. At the same time, the effluent reduces the water level by erosion of the stream channel. It has been calculated that the Rhone waters take over 11 years to travel the length of the lake.

Surface streams are not the only means of moving water from lakes. In some instances water may be removed through shallow seepage—no surface outlet will be apparent. Many such lakes are found in glaciated areas, and are called **seepage** lakes because the water filters out through the basin substrate. Lakes in karst regions often drain into subsurface channels through openings in the lake bottom. Indeed, a distinctive characteristic of most sinkhole lakes in limestone areas is frequently an extreme drop of the water level and occasionally complete emptying. Either of these phenomena may result from lowering of the water table, solution of a portion of the basin permitting drainage into underground channels, or the opening of a previously clogged channel to subsurface drainage.

Lakes which lack an outlet and lose water mainly by evaporation are called **closed** lakes, and are found under extremely diverse conditions. Seepage lakes are one form of closed lake. In arid or semiarid regions streams often flow into lakes without effluents. Under conditions of high evaporation and low precipitation such lakes are usually saline. The salt lakes in the Great Basin of the western United States are closed and occupy sealed basins. Evaporation removes only pure water and leaves behind the accumulated dissolved salts. Thus the lakes tend to become increasingly saline. The salinity of Great Salt Lake, for instance, was 138 parts per thousand in 1877 when the surface area was 5720 km². In 1932, when the lake's area had been reduced to 3380 km², the salinity had increased to 276 parts per thousand (Langbein, 1961).

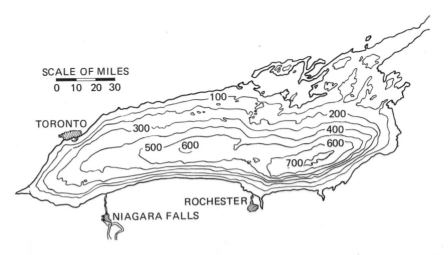

Figure 3.9 Hydrographic chart, showing bottom topography of Lake Ontario. (After Hough, 1958.)

All fresh waters contain some quantity of dissolved salts of various kinds; the proportions depend upon the nature of the soil and the chemical composition of underlying geologic formations which may contribute to the content of the stream water through seepage. In some regions, precipitation may influence the composition of surface waters. The tributary streams of a closed lake introduce into it varying amounts of dissolved salts, which, in turn, become further concentrated in the lake through evaporation. Thus the nature of the salts in a lake is dependent upon the geochemistry of the drainage basin. Great Salt Lake, in Utah, derives its salinity mainly from sodium chloride and sodium sulfate. The Dead Sea salt is almost wholly chlorides. Several lakes in Oregon are characterized by high concentrations of carbonates, chiefly of sodium. The "bitter lakes" of the arid lands between the Sierra Nevada and the Rockies are high in sulfates. The chemical nature of lake waters will be discussed further in Chapters 10 and 11.

SOME IMPORTANT PARAMETERS OF LAKES

Now that we have considered the nature and origins of lake basins, the sources of the waters that fill them, and the maintenance of the water in the basin, we can address ourselves to the major physical features relating to the form of the lake. We have seen some of the great diversity in form exhibited by lakes, and of course lakes also vary greatly in size. Going further, we see that such **morphometric** features (measurements) as depth, mean depth, length, breadth, area, volume, extent and development of shoreline, water level, and elevation above sea level are basic data for any limnological investigations of a lake. These features of a basin are also essential to an understanding of the dynamics of a lake. In comparative limnology these measurements are the starting point for any quantitative study; the usual procedure is to use them to construct a **bathymetric map** (one giving depth measurements) of the body of water that is being studied. Such a map is given in Figure 3.9; the features of a bathymetric map are discussed below. The details and methods of obtaining morphometric data and constructing a map may be found in any standard work on limnological methods.

MAXIMUM DEPTH

Maximum depth is best determined by field measurement (sounding the lake) or from existing maps. It must be remembered, however, that sedimentation, erosion, and water-level fluctuation can rapidly alter the depths in a lake. Thus old data must always be used with caution.

Lakes may range in depth from a few meters to nearly 1800 m. Lakes of great depth are not necessarily large in area. Lake Baikal, in Siberia, is the deepest lake known. While its maximum depth is 1741 m, its surface area, 31,500 km², is not remarkably great. Lake Tanganyika, in Africa, is the second deepest lake on record, with a maxium recorded depth of 1470 m. The Caspian Sea is 946 m deep, but its waters are saline. Lake Nyasa, also in Africa, has a maximum depth of 706 m. In the U.S.S.R., Lake Issyk-kul is 702 m deep. All of these spectacularly deep lakes are of tectonic origin and, with the exception of the Caspian Sea, all occupy basins in grabens. North American deep lakes include Crater Lake, Oregon, which has a maximum depth of 608 m (recall that Crater Lake occupies a volcanic basin); Lake Tahoe, a lake in a graben between California and Nevada, is about 500 m deep; and Lake Chelan, Washington, formed by glacial activity, has a maximum depth of 458 m. Morphometric data for the Great Lakes are given in Table 3.1.

MEAN DEPTH

Mean depth $(\bar{z})$ is the relationship between the volume (V) and the area (A) of a lake. The formula is $\bar{z} = \frac{V}{A}$. This datum is of particular value because it renders a more accurate description of depth–area proportions than does maximum depth. For example, the Caspian Sea, with a maximum depth of 946 m, has a mean depth of only 182 m when its area of 436,400 km² is considered in relation to the volume of 79,316 km³. On the other hand, Lake Baikal, with a volume of 23,000 km³ and covering an area of 31,500 km² to a maximum depth of 1741 m, has a mean depth of 730 m.

LENGTH

Length (l) is simply the distance between the farthest points on the shore of a lake. It can be determined from field mapping or from an accurate map or aerial photograph of known scale.

BREADTH

Breadth is the distance from shore to shore measured at right angles to the longitudinal axis. In view, of the often irregular nature of lake shorelines this dimension may not be highly meaningful. **Mean breadth** $(\bar{b})$ is a somewhat more useful measure than breadth, and is defined as the area (A) divided by the length (l), or $\bar{b} = \frac{A}{l}$

AREA

A lake's **area** is the extent of its surface. As a result of seasonal or other variation in volume, the surface area may fluctuate considerably. In artificial

TABLE 3.1 Morphometric data for the Great Lakes[a]

	Lake Superior	Lake Michigan	Lake Huron	Lake St. Clair	Lake Erie	Lake Ontario
Length[b]	560	490	330	42	385	309
Breadth	256	188	292	38	91	85
Length of coastline (including islands)	4,768	2,656	5,088	270	1,369	1,161
Area (km²)						
Water surface, United States	53,618	58,016	23,569	512	12,898	9,324
Water surface, Canada	28,749	—	36,001	756	12,768	10,360
Drainage basin land, United States	43,253	117,845	41,958	7,380	46,620	39,370
Drainage basin land, Canada	81,585	—	86,500	10,570	12,224	31,080
Drainage basin land, total	124,838	117,845	128,464	17,948	58,793	70,448
Drainage basin (land and water), total	207,200	175,860	188,034	19,217	87,434	90,132
Maximum depth (m)	460	281	228	6	60	244
Average depth (m)	148	84	53	3	17	86
Volume of water (km³)	12,221	4,871	3,535	4	458	1,636
Mean elevation above sea level (m)	182.99	176.42	176.42	174.67	173.86	74.61
Average annual precipitation (mm)	736	787	787		863	863
Approximate length of outflow river	112	—	43	51	59	808

[a] Data from Chandler, 1962.
[b] Data in kilometers except as otherwise indicated.

impoundments, for example, ''drawdown'' can significantly reduce the area as well as other aspects of a lake. Area is determined most accurately and easily by the use of a planimeter run on a well-drawn map or aerial photograph of known scale.

Of the world's most extensive lakes, the Caspian Sea has the greatest area, covering 436,400 km². Lake Superior, at 83,300 km², is the second largest lake in the world. Lake Victoria is third, with an area of 68,800 km², and the brackish Aral sea, in the U.S.S.R., is fourth, with an area of 62,000 km². The two last-mentioned lakes are of tectonic origin. Lakes Huron and Michigan, glacial lakes of North America, are also among the largest of the world. In the United States there are about 250 lakes with surface areas of 259 km² or more; about 100 of these are in Alaska and another 100 in Maine, New York, Minnesota, Michigan, and Wisconsin (Bue, 1963).

VOLUME

Measurement of total volume of a lake is made by deriving the amount of water contained in each of the strata bounded by depth contours. For this a bathymetric map is needed, such as that in Figure 3.9. Because of the slope of the bottom, it is necessary to consider the area of both the upper and the lower surface of each contour stratum. The volume of the stratum (V_s) may be calculated from various formulas, one of which is:

$$V_s = \frac{1}{3}\left(A_1 + A_2 + \sqrt{A_1 A_2}\right)h$$

where A_1 is the area of the upper surface of a contour stratum, A_2 is the area of the lower surface of the same stratum, and h is the height of the stratum. The volume for each stratum is computed from the formula, and the sum of all the volumes gives a figure for the total lake volume.

Of all the inland waters, the relatively shallow Caspian Sea has the greatest known volume, 79,319 km³ (Hutchinson, 1957). Its volume is a function of the great area of the lake. Lake Baikal, by virtue of its considerable depth, has a volume of 23,000 km³, and is the second most ''massive'' lake in the world. Lake Tanganyika contains 18,940 km³ of water. Table 3.1 gives volume of each of the Great Lakes.

EXTENT AND DEVELOPMENT OF SHORELINE

Extent of shoreline is a simple statement of the length of a shore, the distance around the lake perimeter. This figure can be obtained from a detailed map by using a mechanical map measurer such as a rotomer or chartometer. Shoreline measurement is necessary for calculating the area of a lake, the extent of its shallow shore zone, and (as below) in deriving an important index called **shore development,** or **development of shoreline.**

Shore development is a quantitative expression derived from the shape of a lake. It is defined as the ratio of the shoreline length to the length of the circumference of a circle of the same area as the lake. Shoreline development D_L can be calculated from the formula:

$$SD = \frac{S}{2\sqrt{\pi A}}$$

in which L is the length of the shore and A is the lake area. Since the ratio is related to a circle, it is seen that a perfectly round basin would have an index of 1. Increasing irregularity of shoreline development in the form of embayments and projections of the shore is shown by deviations from the value of 1.

Small doline lakes of Florida are often nearly circular and consequently have shoreline development indexes approaching 1. Fjord lakes with numerous embayments in valleys, and lakes formed by drowned river mouths, would be expected to yield high values. Lake Salsvatn, a fjord lake in Norway, has a shoreline development index of 5.5 reflecting its very irregular shape (Hutchinson, 1957).

MEAN SLOPE AND AREA OF SHOAL

The extent of shallow water in a lake has important effects on the magnitude of wave action, turbidity, and total biological activity. Where sunlight penetrates to the bottom, photosynthesis and the development of bottom organisms contribute to a generally rich pond or lake area. Thus slope characteristics and the amount of shoal receiving light are valuable morphometric data for describing a body of water. A lake with a gently sloping basin or with broad shoals could normally be expected to be more productive biologically than a deep lake with steep sides.

The **mean slope** is an expression of the proximity of bathymetric contours to one another. This may be expressed quantitatively (Welch, 1948) as percent slope of the basin from the formula:

$$\overline{S} = 1/n(\frac{1}{2}L_0 + L_1 + L_2 + L_3 \ldots + L_{m-1} + \frac{1}{2}L_n)\frac{D_m}{A}$$

where $\overline{S}$ is the mean slope, L is the length of each contour, n is the number of contours on the bathymetric map, D_m is the maximum depth, and A is the area of the lake surface. Note that in Figure 3.9 the contour lines along the north side are far apart, indicating a gradual slope; the closely spaced contours on the south side of the basin indicate a steep wall.

Shoal area indicates the extent of shallow water. The depth to which the shallow area is said to extend is arbitrary; the shallow area is given as a percentage of the total bottom area.

ELEVATION OF LAKES RELATIVE TO SEA LEVEL

Most existing lake basins are located above sea level. A few, however, occupy depressions. The deepest portion of such a depression may lie below sea level, and that portion of the depression is known as a **cryptodepression.** The depth of that portion of the basin below sea level is termed the depth of the cryptodepression. For example, we have seen that Lake Baikal has a maximum depth of 1741 m. Of this total depth, some 73 percent (1279 m) occupies a cryptodepression. Lake Chelan, in Washington, is contained in a cryptodepression nearly 129 m in depth. As Figure 3.4 indicates, the basins of four of the Great Lakes (Superior, Michigan, Huron, and Ontario) extend below sea level.

The surfaces of some lakes are below sea level. The surface of the Dead Sea, for example, lies 392 m below sea level; the surface of the Caspian Sea is 25 m below sea level.

At the other extreme, many lake basins are found high in mountainous regions. Lake Titicaca, in South America, is situated at an elevation of 3842 m. Several lakes in California's Inyo National Forest have surface elevations slightly over 3500 m, and a few Colorado lakes lie above 3800 m. The importance of elevation to lake processes, both physical and biological, are many. Lakes at higher elevations are exposed to longer periods of cold and ice cover than those of lower elevation; they may also receive more intense sunlight, since the thinner elevated air absorbs less radiation. As a result the biota of higher lakes is usually different from that of lakes lying at lower elevations.

LAKE SHORES AND LAKE SHORE ZONES

The morphology of a lake basin is by no means permanent. Modification of shore outline and depth are inevitable in lakes, and are brought about by a variety of processes, some internal and some external.

A newly filled lake is a dynamic system. Internal forces such as waves and currents work against the shores. The shores offer resistance to these forces to varying degrees, depending upon the nature of the rock or soil and the slope of the basin. A basin formed tectonically in an area of igneous geologic formations often has steep sides, and these sides are only slowly affected by the work of waves. Conversely, gradually sloping basin rims in regions of less resistant rock, such as limestone, are much more subject to erosion by hydromechanical processes and they change rather rapidly. The rate and extent of erosion are dependent upon the size of the lake, the magnitude of waves, the depth of the water near shore as it influences currents and the breaking of waves, and, as just mentioned, the composition of the shore material.

The shores and lake bottoms near shore of many typical older lakes in the Rocky Mountains are composed of coarse gravel and rocks. Mud and fine materials occur in the deeper regions of the lakes. Unconsolidated or soluble materials such as sands and limestone are rather quickly eroded to develop gently sloping lake shores of sand and nearshore bottoms of fine sedimentary particles. The erosional processes result in the formation of a terrace on the periphery of the lake. The exposed portion of the terrace is termed the **beach.** In the cutting of the terrace by water action some of the lighter, more easily transportable materials are deposited along the shore as a lakeward extension of the beach; this submerged portion is called the **littoral shelf** (Figure 3.10). As a result of the differences in size and mass of the shore materials, there will be graduation in texture from the water's edge to the high-water mark. Fine sands occur in the low-water shallows, and progressively larger sands and gravels are distributed further up the slope of the beach.

The external factors that contribute to the modification of lake shorelines are, for the most part, those associated with the processes of erosion and transport of sediment from the land areas. As a producer of currents and waves in lakes, wind might be included in this category. Longshore currents created by winds may carry wave-stirred sediments to some point in the shallow zone and there drop the sediments. As this transport and dumping continues, a **spit** is built from the shore out into the lake, usually at a point of some indentation or embayment of the shoreline. If the spit develops across the embayment to separate the bay and lake the sedimentary structure is termed a **bar.** Figure 3.11 illustrates a late stage in the development of spits and bars.

Streams are often very important external factors, shaping the morphology of a lake through alteration of the shoreline and bottom. Depending upon the velocity of the stream and the material through which it is flowing, even a small stream may contribute sediments to a lake. Heavier particles will usually settle out near the mouth of the stream and form a delta; lighter sediments may be carried farther into the lake where they sink to the bottom. The formation and growth of a delta is dependent upon the stream contributing sediments faster than

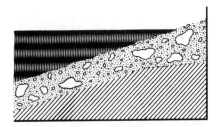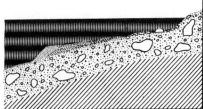

Figure 3.10   Erosion and sorting of lakeshore sediments, leading to the formation (right) of a littoral shelf. (After Macan and Worthington, 1951.)

Figure 3.11   Bay-mouth bar and lagoon in St. Mary's Lake, Glacier National Park. The bar is formed by growth of a sand spit originating at the shore. (Courtesy of U.S. Geological Survey.)

lake currents and waves can distribute them within the basin. Obviously a delta cannot be formed if the stream sediments are spread generally over the lake floor. Ideally, a delta is triangular in shape, with the apex projecting against the river current (Figure 3.12). The deposited materials will ideally be stratified in the pattern shown in Figure 3.13. As the coarser sediments are deposited they form a series of reposing layers known as the **foreset beds.** The lighter sediments are carried lakeward and settle out on the bottom of the lake to form the **bottomset beds.** Continued transport and deposition of materials results in the lakeward growth of the delta, and consequently the stream must extend its mouth to the base of the delta. The extension of the river channel deposits material over the foreset beds; this material is termed the **topset beds.** Various factors, including wind intensity and direction, wave action, and lake currents, act to modify the form of deltas. Thus few natural deltas closely resemble this theoretical ideal.

SEDIMENTS AND LAKE BOTTOMS

From its beginning and throughout its existence the composition of the lake bottom is influenced by numerous factors, and for this reason great variation exists. Even within a small area two lakes may differ significantly in bottom type and in bottom associated features. In Florida, Orange Lake is connected with Lake Lochloosa by narrow Cross Creek, which is about 3 km long. Orange Lake is surrounded by extensive marsh, and its bottom is a layer of organic detritus a meter or more thick in places; Lake Lochloosa has no such broad marsh areas and much of its bottom is sandy.

Four major factors largely determine the nature of lake bottoms:

Age Young lakes are more apt to have rocky or sandy bottoms with little deposition of sediments and organic materials. As a lake matures, sediments accumulate more deeply.

Size Increased surface area of a lake permits greater wave action, eroding shores and adding the eroded materials to the bottom sediments. Similarly, subsurface currents operate to a greater extent in large, deep lakes than in small ones. These currents serve to distribute sediments.

Figure 3.12   Delta of the Missisquoi River, Vermont, as it enters Lake Champlain. (Courtesy of Soil Conservation Service, U.S. Department of Agriculture.)

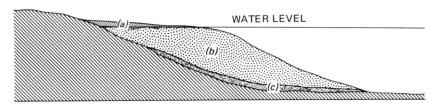

Figure 3.13   Idealized longitudinal cross-section of a delta, showing the arrangement of sediments *(a)* topset beds; *(b)* foreset beds; *(c)* bottomset beds.

Latitude and climate The latitude of a lake and its climate are often complexly related to a number of other features, such as the chemical nature and size of the basin. Together they affect the nature of the bottom in many ways. Seasonal temperatures and day length are, of course, correlated with latitude. Thus, chemically rich lakes in low latitudes with long growing seasons often contain large amounts of organic material on the bottom. However, similar bottom conditions are often found in higher latitudes, where many species are adapted to rapid growth during the shorter growing season. Chemically poor or geologically "young" lakes in any region are often rocky or sandy on the bottom, with only a meager layer of organic material. The local annual rainfall, and its intensity and seasonal variation, may alter the nature of lake bottoms by causing the water level to fluctuate and by bringing in new sediment.

Soils and underlying rock formation Lakes occupying basins in sandy soils or in geologic formations which are easily fractured may have thick deposits of sediments over the bottom. This condition stands in contrast to lakes in basins of highly resistant rocks in which little erosion occurs.

Generally, lake sediments are made up of variously sized rock fragments and soils ranging from clay and silt to sand, gravel, and boulders; chemical precipitations and compounds including marl, tufa, ferric hydroxide, ferric carbonate, and silicon dioxide; and organic deposits such as peat or ooze.

Coarse sediments are commonly found in lakes. As a rule the larger sized particles are found in the shallow zones, while clay and silt may occur over any part of the bottom at any depth. The color of the clay and silt often grades from white through shades of blue and green to black and red. Sands may range from white to gray to black. The color of any type of lake sediment is primarily determined by the mineralogical nature of the basin.

Lake **marl** is essentially calcium carbonate precipitated by certain bacteria and algae, and it is most characteristic of small lakes, ponds, and some streams. Marl forms in alkaline waters as a result of photosynthesis; algae of the genus *Chara,* through their physiological deposition of lime, are a major contributor to marl formation. The typical color of marl is white to grayish, although blue and black marls are not unusual. **Tufa** is a porous lime carbonate formed primarily by some of the algae. Ferric hydroxide, particularly in the form of the mineral limonite, is known from bottom deposits in lakes in northern Europe, Canada, and other regions.

In some organically poor lakes, and in some that are relatively productive (but usually clear), the sediments are composed of a gray or reddish-gray, highly viscous material referred to as **gyttja** (from the Swedish; pronounced "yit-cha"). Gyttja is a fine-textured, largely organic sediment, basically a mixture of plant and animal remains, chemical precipitations, and mineral substances (Hansen, 1959). In terms of acidity gyttja is circumneutral, and the humus in the sediment usually contains less than 50 percent organic carbon. Decomposition of gyttja under aerobic conditions contributes much elemental nutrient material to organic production within a lake. Fossilization of gyttja leads to the formation of anthracite, a form of hard coal.

Where anaerobic conditions prevail in regions of peat formation, lake sediments may be brown or blackish-brown due to the addition of brown humus colloids to gyttja-like material. Such sediments, called **dy** ("dee"—also from the Swedish), are deposited in lakes with brown water, and are typically acid. The organic carbon content of the humus in dy is usually greater than 50 percent.

Under anaerobic conditions, particularly in summer, organically rich sediments may decompose to form **sapropel,** a blue-black substance containing hydrogen sulfide and methane. It has been suggested that fossilization of sapropel results in the formation of oil.

Partially carbonized plant material in the form of **peat** occurs rather commonly in lakes in various stages of compaction and decomposition. Peat is normally thickest near the shore in the regions of extensive plant growth. Lignite and bituminous (soft) coal are formed from the decomposition of peat.

Whatever the factors and materials concerned in the composition of lake bottoms, there usually exists a mixture of substances. The bottom of Lake Providence, Louisiana, for example, is composed almost entirely of mud, the shallow shore zone being the only region of sand (Moore, 1950). The lake shore is heavily wooded, and thus considerable organic detritus is present along the border. Cultus Lake, British Columbia, is described as having a bottom predominantly of gray ooze with a high content of carbonaceous plant remains. Only a few small areas of sand are found and the steep shores are littered with boulders and large rock fragments (Ricker, 1952).

SEDIMENTS, LAKE HISTORY, AND PALEOLIMNOLOGY

The rate, quality, and amount of sediment deposited on lake bottoms vary with seasonal and climatic changes. In most of North America spring rains and melting snow increase the carrying capacity of streams, so that they bring greater amounts of sedimentary materials into lake basins. The coarser particles settle to the bottom quickly, while the finer ones remain suspended. When the streams freeze in winter—or during the dry season in arid climates—lake currents are less vigorous and the lighter particles settle out, covering the spring-summer deposit. Thus two layers (or **laminae**) are formed each year. In lakes of glacial origin and in mountainous regions the summer layer is usually composed of coarse silt and is light in color. The thinner winter layer is darker because it contains organic material and finer-grained sediments. The layers that have accumulated during a single year are called a **varve,** and examination of the successive layers from a borehole in a lake bottom is a valuable means of estimating the age of a lake or basin. The number of varves in a now extinct lake basin in the Connecticut Valley, for example, indicates that the lake existed for about 4000 years. More recently, however, investigators have realized that the number of layers deposited yearly will not always be exactly two, because of seasonal fluctuations in the rate and extent of sedimentation. Counting varves will give the age of a basin only in regions with consistent periglacial climates; this method has been very useful, for instance, in the Scandanavian countries.

Studies of lake deposits have yielded data on many aspects of lake basins in addition to simple estimates of age. This is possible because plant and animal remains are often found in the depositional strata; since the strata are arranged chronologically—with the most recent sediments closest to the surface—studies of bottom deposits can tell us a great deal about the history of the lake and the surrounding watershed.

In a number of lakes it has been possible to piece together a rather detailed account of the development and evolution of the basin, and of the different associations of plant and animal species that have occupied the lake, by examining the microfossils embedded in the sediments. The fossils most often used for such studies are those of the plankton—in particular those members of the plankton that had hard shells or similar exoskeletons. As these often microscopic organisms died, their remains continually "rained" down upon the lake bottom and became embedded in sediment. Over the centuries, as different species flourished and were later replaced by newcomers, the changes in the lake biota would be reflected in the changing makeup of the bottom sediments.

Some of the pertinent details in the history of Lake Windermere, the largest lake in England, are summarized in Figure 3.14. The full history of the lake is lengthy and complicated, but we can quickly note some of the more conspicuous features and processes. First, the column at the extreme left of the figure represents a core from the bottom showing the several sediments underlying the lake. Lowermost are clay varves, showing one fine and one coarse layer per year, and indicating a periglacial climate (see above) until about 8000 years **B.P.** ("before the present"). Above the clay laminae a transition zone leads to a relatively uniform zone of brown mud and finally to a surface layer of ooze.

The histogram figures in Figure 3.14 indicate the depth and relative abundance of certain plants in the sediments. The absence of plant and animal remains in the clay certainly means that there was little life in the lake at the time the deposits were formed, suggesting that the clays were deposited during glacial times of little life. Observe that certain forms appeared rather abruptly in the lake during the transition period; later changes diminished the numbers of some while *Gomphonema* has persisted and increased. Other forms invaded the lake while the brown mud was being deposited. Note that the alga *Asterionella* becomes most abundant in the ooze. The ooze is attributed to sewage derived from increased human habitation of the lake country beginning about 100 years ago. Variations in pollen figures correlate reasonably well with changing agricultural, forestry, and cultural habits of the early people. The time scale in the figure is based on this latter information.

Linsley Pond is a small kettle lake formed during late glacial times in southern Connecticut. Its bottom strata have been dated by analysis of animal remains, permitting investigators to reconstruct the major events in the lake's history (Deevey, 1942). By recognizing taxonomic differences between two genera of midges found in the sediments, and comparing them with species found in lakes today, it has been possible to determine that shortly after it was formed Linsley Pond was organically unproductive **(oligotrophic).** Later it became generally productive and rich in organic nutrients **(eutrophic).** The sediments from the oligotrophic stage show an abundance of the genus *Tanytarsus;* when the lake became eutrophic *Tanytarsus* was replaced by *Chironomus,* a better-adapted form. The remains of crustacean exoskeletons can often be

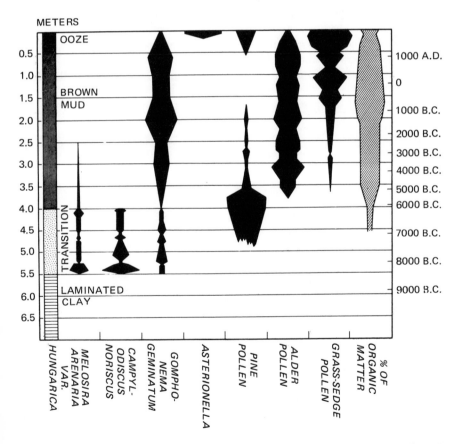

Figure 3.14   Summary of the biotic history of Lake Windermere, England, as indicated in mud cores from deep water. Relative abundance of the materials is represented by the varying thickness of the histograms. (After Macan and Worthington, 1951.)

identified precisely, providing useful data for studying lake history. In Linsley Pond the older sediments contain remains of the water flea *Bosmina coregoni,* a species adapted to the nutrient-poor waters. Later, as the lake became eutrophic, the related species *B. longirostris* replaced *B. coregoni* as the dominant microcrustacean (Figure 3.15).

Many species of diatoms—algae which secrete distinctive species-specific siliceous shells—are highly sensitive to changes in environment. Since their shells and fragments are often well preserved in sediment, diatom remains are useful in reconstructing both the long-term and the short-term history of a lake. Examination of diatom remains, for example, can yield valuable data on the effects of human use of lake waters. Lake Washington is a large, organically rich lake adjacent to Seattle. During the past 80 years different species of the diatom flora have fluctuated considerably in number (Stockner and Benson, 1967). The population fluctuations seem to have been responses to increased human habitation and the subsequent introduction of sewage—first raw, and later treated sewage—into the lake water.

The study of lake history through such examination of sediments is termed **paleolimnology.** Paleolimnology has made it possible to reconstruct not only lake history, but also many aspects of regional climate and land use. Pollen from

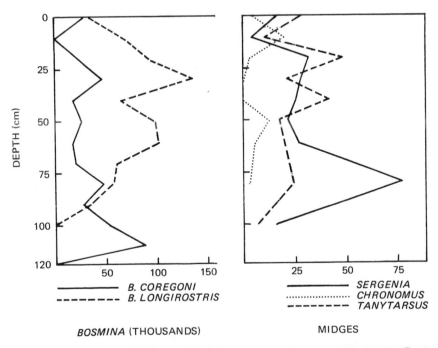

Figure 3.15   Microcrustacean component in sediment cores from Linsley Pond. Changes in the relative frequency of different species indicate changes in climate during the history of the lake. (After Goulden, 1964.)

lake basin sediments is often closely examined and analyzed, for example, and can tell an investigator just what plant species were abundant at any past time. On the basis of pollen profiles researchers have been able to determine the climate of the Pacific Northwest for thousands of years in the past (Heusser, 1960). Pollen profiles from a lake in the Petén region of Guatemala indicate that, contrary to the usual pattern, the peoples of that region converted large areas of grassland to forest (Cowgill *et al.,* 1966).

What we have said about the dynamic processes that are characteristic of lakes should make it clear that no lake is a permanent feature of the landscape. Once a lake has formed, and indeed even while the basin is first being filled, a wide array of natural forces immediately begin to level the basin rim, to fill the basin with sediments, and to erode an outlet. It has been said that lakes are "born to die;" lakes are indeed ephemeral, and their impermanence is a basic

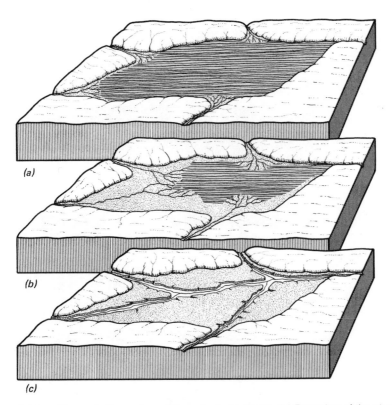

Figure 3.16   Stages in the sedimentation of a typical lake. *(a)* Damming of the stream course at the upper right forms a lake, covering the junctions of the earlier streams and parts of the original stream channels. Streams entering the lake drop their loads of sediment and delta building begins at the stream mouths. *(b)* Continued sedimentation gradually fills the lake basin. The deltas join and become continuous around the lake perimeter. *(c)* The effluent stream erodes a new outlet. The basin is drained and stream-cutting of the deltaic sediments leaves terreced deposits. (After Longwell *et al.,* 1939.)

(a)

(b)

(c)

principle of limnology. As we have seen, sedimentation alone plays a major role in filling lakes and ultimately replacing them with dry land (Figure 3.16). Living plants and animals, however, also have a conspicuous role in this process (Figure 3.17). Because aquatic organisms are so important in **lake succession,** as the process is called, we will defer a more detailed discussion of this phenomenon until after we have surveyed the more important inhabitants of lakes and marshes.

SUGGESTED FOR FURTHER READING

Amos, W. H. 1967. *The life of the pond*. McGraw-Hili, New York. A carefully written, extremely readable introduction to the biota and ecology of temperate-zone ponds and lakes. Illustrated throughout with color plates and drawings, the book is designed for the general reader; using nontechnical language, for the most part, the author nevertheless manages to impart a great deal of detailed information.

Coker, R. E. 1954. *Streams, lakes, ponds*. University of North Carolina Press, Chapel Hill. Available as a paperback in the Harper Torchbook series. One of the first, and still one of the best, introductions to limnology for general readers.

Flint, R. F. 1957. *Glacial and Pleistocene geology*. John Wiley & Sons, New York. For students interested in learning more about the formation of lake basins by continental glaciation, this book is a reliable, standard text on ice-age geology. Lakes and lake basins are discussed briefly but in detail, as is nearly every other aspect of glacial morphology.

Lind, O. T. 1974. *Handbook of common methods in limnology*. C. V. Mosby, St. Louis. A brief but up-to-date handbook for use in introductory level limnological field studies.

Niering, W. A. 1966. *The life of the marsh*. McGraw-Hill, New York. A companion to the book by Amos, above, and with the latter a volume in the publisher's ''Our living world of nature'' series. An extremely clear and

---

Figure 3.17   An actual example of lake sedimentation. *(a)* Lake Como, an artificial impoundment near Hokah, Minnesota, as it appeared in 1926. *(b)* The same locality in 1935. The watershed has been altered by heavy cutting of timber and intensive cultivation; with little vegetation remaining to retain surface water and slow down runoff, erosion and sedimentation have increased dramatically. Heavy fish kills were frequent in the lakes, and most of the basin has been filled in. *(c)* By 1950 the former lake bed was almost filled with sediment washed down from three upstream valleys. Little remained of the lake but a narrow stream and a few ponds. Some gardening was carried out in the valley, but periodic damage was still caused by runoff flooding. (Courtesy of Soil Conservation Service, U.S. Department of Agriculture.)

enjoyable introduction to wetland ecology, although perhaps not quite so strong on hard data as the Amos book. It is strongly recommended as a supplement to this chapter and also to Chapter 5, "Estuaries."

Welch, P. S. 1948. *Limnological methods*. McGraw-Hill, New York. A major, standard handbook in field limnology, frequently cited throughout this text. Gives a straightforward, detailed account of the basic techniques for mapping and measuring lakes and streams, and for measuring the major physical, chemical, and biological parameters of freshwater environments. Although many of the techniques and items of equipment described have been improved upon by more modern (especially electronic) techniques, this book is still an important tool for the field researcher.

4 | Stream Channels and Streams

We are all familiar, to some extent, with most of the flowing water habitats—rivers, streams, brooks, and channels of other sorts. The ecologist groups these together under the term **lotic habitats** (from the Latin *lautus,* a form of the verb meaning "to wash"). The water in these habitats is derived, of course, from the hydrologic cycle. We have seen that the cycle returns water to the earth's surface in the form of precipitation, and although the amount of rain and snow that falls varies widely from one region to another, all regions probably receive some amount of water from precipitation. Even Death Valley, in California, receives an average of 50 mm per year, while parts of eastern North America and western Europe receive between 110 and 250 mm of precipitation per year. Of the water that falls upon the earth, some evaporates, some is absorbed by the soil and sinks in to form subsurface water, and some flows over the ground as **surface runoff** in the form of streams. That precipitation which enters the ground may later issue forth in the form of **groundwater runoff.**

It has been estimated that approximately one third of the precipitation received on the land eventually finds its way into runoff. In local regions the proportion may be considerably higher; it has been shown, for example, that 92 percent of the winter rainfall in Vermont moves as runoff. Local variation in runoff is dependent upon (1) nature of the soil (porosity and solubility); (2) degree of slope of the surface; (3) development and type of vegetation; (4) local climatic conditions such as temperature, wind, and humidity; and, of course, (5) volume and intensity of experienced precipitation.

Because of inequalities in the nature of surface rock formation and soils, and because land areas are not flat but rather possess some degree of slope, runoff accumulates quickly in small gullies; these coalesce to form larger channels for rivulets, which finally converge and give rise to brooks and eventually to larger streams. As the water moves, it carries materials which have been picked up along the course of the stream. A **stream** then may be defined as a mass of water with its load moving in a more or less definite pattern and following the course of least resistance toward a lower elevation. The great mass of water in rivers is best emphasized by considering the estimate of some 27,000 km³ of water carried yearly to the sea by streams. The load of a stream relates to both dissolved and suspended materials; it has been estimated that the Mississippi River annually transports about 136 million tons of dissolved matter and over 340 million tons of suspended material.

The nature of a stream is essentially a reflection of the fluvial processes concerned in the transport of water and materials. As they relate to these processes, certain outstanding qualities differentiate streams and lakes. As suggested previously, the movement (flow) of water in streams is unidirectional and, in larger streams, continuous. Since precipitation varies seasonally in volume and frequency, it follows that wide fluctuations in water volume, rate of flow, size of channel, and rate of land erosion are to be expected in streams, depending upon local climatic conditions and the size of the stream. Because such fluctuations are characteristic of streams, it also follows that bottom and shoreline areas are relatively unstable. The inhabitants of streams exhibit characteristic forms and ways of life as adaptations to their environment. The water of smaller streams is subjected to a greater variety of movements and to more thorough mixing than that of lakes. Streams are more apt to be highly turbid than lakes, at least seasonally. Oxygen content in unpolluted streams is usually higher than in lakes. There is considerable interchange between land and the water of streams. Throughout its linear extent, which may dissect many soil types as well as more than one climatic zone, a stream exhibits a great variety of physical, chemical, and biological conditions. Because of these characteristics, especially that of land–water interchange, a stream is said to constitute an **open ecosystem** in contrast to a lake as a **closed ecosystem.**

The foregoing general aspects are important to our understanding of the dynamics of streams. These features are related to physicochemical and biological fundamentals which must form the basis for stream studies. There are additional fundamentals underlying the appreciation of the stream as a natural community. Each will be considered in more detail. Unfortunately, streams have not received as intensive study as have lakes. Admittedly, much is yet to be learned about the relationships and processes within lakes, but even more remains to be discovered concerning streams.

ORIGIN OF STREAM BEDS AND VALLEYS

The original bed of a stream may be a depression previously developed by an agency other than the stream itself. In areas of volcanism and lava deposition streams occupy folds or surfaces formed by irregular lava flow. Glaciers may scour valleys or trenches, which may later be occupied by streams. Even within such "preformed" valleys, however, stream forces early begin deepening and widening the valley through erosive processes.

Most streams, however, originate on newly formed land surfaces and excavate their own channels and valleys. As stated above, precipitation falling upon the ground flows into depressions and seeks the least resistant route to lower levels. Gullies are soon excavated and join to form a shallow entrenchment containing the contribution of a number of tributaries. The greater the volume of water and the slope, the faster the erosion (its rate depending upon the type of substrate). With increased erosion the channels become larger and transport more runoff. As erosion continues at the head of the gully, it grows by extending the upper reaches. With headward extension and vertical cutting, a valley is formed. Streams developed upon the initial slope contributing the runoff are termed **consequent streams.** The classification of streams according to hydraulic and erosional processes with respect to the land area includes additional types such as **subsequent** streams, **resequent** streams, and others. Reference to a recent textbook in physical geology should give details relating to these types.

We have made free use of the term **erosion** by using it in a general sense. Actually this very significant characteristic of fluvial work is the result of two processes. **Mechanical erosion** is accomplished mainly through abrasion of stream sides and bedrock by mineral sediments carried in suspension or rolled along by the current. **Chemical solution,** or **corrosion,** of channel materials by stream waters is important in regions of soluble substrate, as, for example, in limestone areas.

The rate at which a stream erodes its channel is determined by the nature of the bedrock, composition of the water, climate, and the grade or slope. The greater the slope the greater the capacity to transport abrasive materials through increased velocity. Some rocks are more susceptible than others to erosion, and the channel is more rapidly deepened in a zone of soft material than in an area of hard bedrock. Hard rock may be cut rather quickly, however, if angular, sharp fragments are moved rapidly by stream flow. It would seem that the time of most rapid erosion is during flood season, for the water level is high, and the increased volume and velocity move more debris and larger particles than during normal stages.

We have seen that in the earliest stages of development of a stream, surface runoff is probably the most important single source of water. But once the valley has been eroded and several streams unite to pool their load and work, maintenance of a stream depends upon additional supplies of water. Because the size and morphology of river valleys are direct results of the work of the waters of the streams, let us consider first the sources of stream waters and later the form of the channels and valleys.

SOURCES OF STREAM WATERS

Nearly everyone has observed the sudden and voluminous rise in stream levels following periods of heavy rainfall or spring thaw of winter snow. We have also seen that the lands in the drainage area of a stream soon dry and the flooded stream returns to "normal" flow within a short time after the rains have subsided or snow has passed. Yet, depending upon its size, the nature of the surrounding soils, and climate in the region, the stream, in most cases, continues to flow. It would appear, then, that the stream waters are being derived from sources other than the immediate surface runoff.

Recall that in the hydrologic cycle varying amounts of precipitation may be absorbed by the soil and held as **subsurface water.** The subsurface distribution of water is dependent upon the local climate, topography, and the porosity and permeability of the underlying soils and rocks. Where permeable rocks occur, water may be taken in and held. The surface of the saturated zone of permeable rocks is called the **water table.** Water in the soil above the water table is referred to as **vadose** water, and its volume with respect to the soil is subject to considerable fluctuation. **Groundwater** is the water contained in the rocks below the water table and usually is of more uniform volume than vadose water. The water table normally follows the topographic relief of the land, being high below hills and sloping into valleys (Figure 4.1). It is this subsurface water that controls to a great extent the level of lake surfaces, the flow of streams, and the extent of swamps and marshes.

The intercalation of subsurface water into cavities, channels, loose sand, and gravels provides a reservoir for water supply to streams during periods of drought. Whereas surface runoff may be the start of a stream valley and the cutting of the valley may depend upon seasonal rainfall, once the valley has eroded to the water table the stream flow is quite apt to become regular and fairly uniform. For now water can issue as seepage or as springs into the stream channel. Conversely, the stream may also contribute to the supply in the water table. As demonstrated in Figure 4.1, there may be exchange between ground water and the stream bed or lake basin depending upon the nature of the substrate and the slope of the water table. In some instances, however, the stream bed

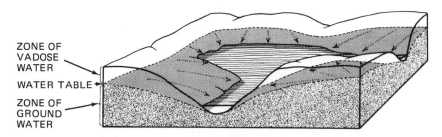

ZONE OF
VADOSE
WATER

WATER TABLE

ZONE OF
GROUND
WATER

Figure 4.1   The subsurface water zones. Arrows indicate direction of flow in a hypothetical drainage basin. Note the relations between subsurface water zones, water flow, and surface topography; the water table is typically higher in upland regions. (After Longwell et al., 1939.)

may be perched in impervious rocks above the water table, and interchange between the two waters be inhibited.

Although some streams may receive water from springs or from glacial melting, these are not, as a rule, common sources of water in streams. As we have just seen, the major supply is that from ground water, primarily as seepage. Nor should we conclude that ground water enters the stream throughout its entire length. More frequently the inflow is in the region of the source of the stream and its tributaries. The source may be in a marsh, or soggy meadow, or perhaps a periodically dry brook bed. On the basis of continuity of flow, we may distinguish three basic streams:

Permanent streams Streams which receive their waters mostly through seepage and springs from subsurface water. In the immediate drainage area the water table usually stands at a higher level than the floor of the stream.

Intermittent streams Streams which receive their waters primarily from surface runoff. Because the runoff is seasonal, stream flow occurs during the wet periods. In regions of considerable rainfall and melting snow, surface runoff may be sufficiently uniform to maintain relatively "permanent" streams.

Interrupted streams Streams which flow alternately on and below the surface. The subsurface flow is usually through coarse sands or gravels (as in portions of the Rio Grande of the southwestern United States), or in limestone; the Santa Fe River of Florida disappears into a limestone sink and follows a subterranean channel for several miles before reappearing on the surface.

FLUVIAL DYNAMICS

Having considered some fundamentals in the formation of stream channels and valleys and the sources of water in the streams, we need now to give attention to some of the dynamics of running water. Consideration of these stream processes necessarily precedes a study of stream valley morphology. For whereas the morphology of lakes is basically attributable to the mode of origin of the basin, stream-bed and valley forms are direct results of stream flow. The processes and parameters with which we shall be concerned belong essentially to the realm of hydrology, and the student of streams should refer to standard texts and references such as the "Hydrology Handbook" of the American Society of Civil Engineers Committee on Hydrology for further information. Our concern at this time is primarily with those fluvial processes responsible for the shaping of the stream channel, valleys, and ultimately the drainage basin.

TYPES OF STREAM FLOW

Stream flow may be of two basic kinds: **laminar** and **turbulent.** Laminar flow is an approximated ideal that takes place in channels with smooth sides when the rate of flow is low. Water exhibiting laminar flow moves in paths that are straight and parallel to the sides of the channel. Since the channels of streams are usually rough and the rate of flow above a certain critical velocity at which flow becomes turbulent, laminar flow is seldom found in natural currents of any significant magnitude. Laminar flow may occur in shallow streams or in sheets of runoff down smooth slopes.

In most streams the water flows in contact with irregular and rough stream beds. This results in the development of a system of eddies and circular currents giving rise to turbulent flow. Most stream flow is turbulent, and eddying effect is a major force in transporting particles in suspension. More will be said about turbulence in Chapter 7. **Shooting,** or **jet flow,** is a type of turbulence that results when stream flow develops a very high velocity. Turbulence is created and fluctuations in velocity result in spurts of water mass. Such a flow is best observed in narrow channels of considerable gradient, or over waterfalls.

VELOCITY

Velocity is the distance a mass of water moves per unit of time. The measurement is generally expressed in meters (or feet) per second. Upon this parameter depend movement of dissolved and suspended materials, rate of

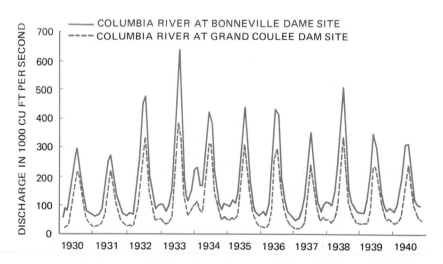

Figure 4.2 Seasonal variation in the discharge of the Columbia River. Note that the annual patterns are affected very little by tributary inflow between the Grand Coulee Dam upstream and Bonneville Dam downstream. (After Robeck et al., 1954.)

discharge, erosion (in part), the distribution of animal and plant life, and other aspects of stream ecology. Stream velocities may range from near motionless in pools and lower reaches of rivers to relatively high rates of 9 m or more per second (Coker, 1954). Silver Springs, Florida, a stream of generally uniform flow, was found to have a velocity of 0.2 m per second. The velocity of swift-water trout streams in Ontario ranges seasonally, from about 0.45 m per second in August to 1.8 m per second in April (Ferguson, 1947). The velocity of a stream is determined basically by the volume of the water in the stream, the load of suspended sediments, and the gradient. Obviously, a stream flowing on a steep slope has greater velocity than on near-level terrain. Similarly, with a given gradient, an increase in volume will increase the velocity. Friction between water and the sides and bottom of the stream bed tends to reduce velocity. Thus there may be a velocity differential from shore to midstream and from top to bottom. Turbulence develops in the zones of contact between waters of different veloci-ties. It has also been shown that velocity varies inversely as the load.

DISCHARGE

The total volume of stream water passing a point in a given period of time is termed discharge. The measurement of discharge is called **gauging** and the data describing discharge are usually expressed in cubic feet per second (cfs).[1] The Niagara River has an annual mean discharge of 219,850 cfs. At New Orleans, the Mississippi discharges at a rate of about 1,740,000 cfs. Figure 4.2 shows the cyclic nature of discharge of the Columbia River.

The rate of discharge is proportional to the rate of flow and volume of water being fed into a stream. Thus the discharge varies with season and contribution by tributaries. Discharge is determined by the shape of the channel, the cross-sectional area of the channel, and the gradient. From these factors a formula for discharge may be stated as follows:

$$Q(\text{cfs}) = WD_m V_m$$

where the discharge (Q) is related to mean channel depth (D_m) in feet, channel width (W) in feet, and mean velocity (V_m), in feet per second. The formula can be further refined to take into account other features of the channel. If the bottom is rough, for example, Q is multiplied by 0.8; if the bottom is smooth, Q is multiplied by 0.9. The figure for Q is further multiplied by 1.33 if the stream is less than 2 ft deep, or by 1.05 if it is more than 10 ft deep (Welch, 1948).

1. Other units—acre foot and gallon—are sometimes used. An acre foot is the amount of water needed to cover an acre to a depth of one foot; it is equal to 43,560 ft^3. Metric units are not given here because in the United States flow is still measured using the English system.

TRANSPORT AND LOAD

The ability to do work in the form of movement of materials is one of the conspicuous attributes of stream flow. The movement of suspended and dissolved substances eroded from the valley or brought in by other means is termed **transport.** The transport of **dissolved** materials is not dependent upon stream velocity. The movement of suspended matter is a function of **competency, capacity,** and **load,** the first two aspects being related, in part, to velocity.

Competency is a measure of the maximum particle size of sediments transported by a stream, and is dependent upon velocity and turbulence. A stream requires a high competency to raise and carry sediments. Also related to competency is the ability of stream currents to roll large particles along the bed.

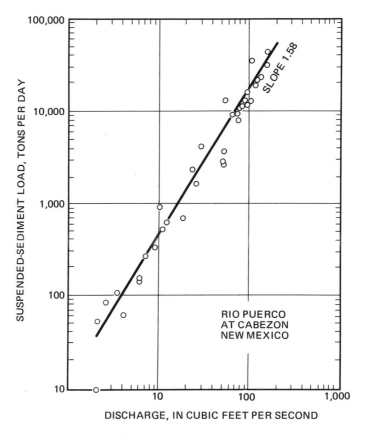

Figure 4.3   Stream load as a function of discharge, as measured in the Rio Puerco, near Cabezon, New Mexico. The data from this ephemeral arid-region stream dramatically illustrate the effect of velocity and turbulence on a stream's capacity to transport materials. (Data from Leopold and Miller, 1956.)

It has been shown that the size of particles transported varies as the square of the velocity. A current of approximately 1 mph can, under proper conditions, move a stone of ¼-in. diameter along a normally smooth bottom.

Capacity of a stream refers to the quantity of material that can be transported under certain optimum circumstances. Capacity is also a function of velocity and turbulence. The capacity of a stream is not constant throughout its course. The result is uneven beds, formed by scouring in one portion of the stream and deposition in another.

The volume of sediments actually held by a current at a given time is termed the **suspended load.** The load seldom equals the capacity. As seen in Figure 4.3, load increases with discharge, although an increased load tends to reduce the capacity of the stream. As the velocity is diminished, the load is deposited on the bottom and along the sides of the stream bed. Deposition, as will be seen later, plays an important role in the form, or morphology, of stream channels and valleys.

STREAM ADJUSTMENT

In conformance with natural laws of equilibrium and "orderliness," fluvial processes tend toward establishing balance among the several forces operating within the stream proper, and between the stream and its surroundings. In most streams there is, as described above, an apparent increase in discharge from the upper reaches to the mouth of the stream. There is also a decrease in gradient over the course of the stream from source to mouth. This adjustment to opposing phenomena is due primarily to correlations between depth, width, and velocity as one set of factors and discharge as another factor. As shown in Figure 4.4, the correlation is logarithmic; as discharge is increased the dimensions of the stream are increased. Or, conversely, the stream bed becomes enlarged to accommodate the greater discharge. The hydrodynamic factors involved in discharge and channel adjustments include erosion, transport, and deposition of the eroded materials along the stream bed.

MORPHOLOGY OF STREAM VALLEYS

In the preceding section we have considered briefly some of the basic dynamics operating in flowing waters. All of these hydraulic processes are of concern, for they relate ultimately to form of stream valleys at a given time as well as to modifications of valley and basin morphology. Bearing in mind the forces responsible for shaping the basins, let us now consider the valley in its linear and transverse aspects, and then the depositional features shaping the valleys.

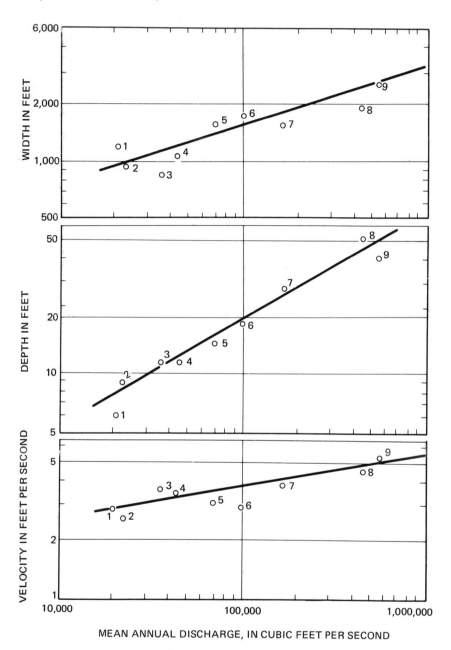

Figure 4.4 Relation of width, depth, and velocity to mean annual discharge as it increases downstream. Stations 1 through 5 are sites on the Missouri River between Bismarck, N.D., and Hermann, Mo.; stations 6 through 9 are on the Mississippi River between Alton, Ill., and Vicksburg, Miss. (After Leopold and Maddock, 1953.)

STREAM LENGTH

Streams, especially mountain brooks, creeks, and rivers, show great variations in length. Some mountain streams, important locally to fishermen and to water supply, may be less than a mile in extent. Important rivers may range from relatively short ones such as the 61-km-long Mobile River of Alabama, to the great Nile of Africa, over 6400 km long. The Mississippi-Missouri System of North America is said to be 6233 km in length.

LONGITUDINAL GRADIENT

As defined, the gradient of a stream is the slope of its longitudinal course. This characteristic is expressed in vertical descent per unit of horizontal distance. Generally the region of steeper slope is near the headwaters, while the more gentle gradient is in the vicinity of the mouth of the stream. In the upper reaches of the Yuba River of California, the gradient averages 305 m/km for about 20 km. By contrast, the gradient in the Mississippi River from Cairo, Illinois, to its confluence with the Red River of Arkansas is no greater than 0.93 m/km.

In its early evolutionary stages, the gradient of the stream is essentially the degree of slope of the land. This gradient enters into adjustment processes tending toward equilibrium with channel characteristics. Through erosion and deposition the longitudinal profile of the gradient attains the form of a shallow, irregular arc with an upward concavity flattening toward the mouth of the stream. This curve is termed the **long profile.** At points along the stream local equilibria are established through adjustments between channel form, velocity, gradient, and discharge. Erosion of the stream bed is reduced as transport of sediments increases, that is to say, much of the energy formerly directed toward scouring (degrading) the channel is diverted toward deposition, or refilling, of the bed (aggrading). In time, irregularities along the stream slope are more or less leveled. When this stage has been reached the stream is described as **graded,** and the long profile has developed a **profile of equilibrium.**

Stream profiles may be shown graphically by plotting elevation from source to mouth against horizontal distance as in Figure 4.5. Since the gradient of most streams is relatively slight, the vertical distance is considerably exaggerated on the graph. An irregular curve, such as that for the Snake River in Figure 4.5, suggests that the stream is still in the process of grading. A smooth curve depicts a stream in equilibrium, one in which local inequalities have been minimized. Ideally, continued erosion and lowering of the land surface result in the curve becoming progressively more flattened. However, various geologic forces such as volcanic activity, uplift, faulting, and change in sea level complicate the ultimate developmental pattern.

There is a limit, although often temporary, to which a stream may erode its channel. This limit is called the **base level.** In the case of streams entering

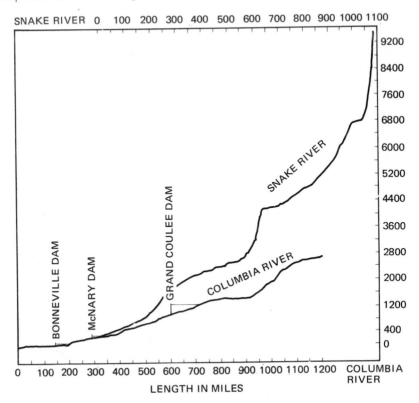

Figure 4.5 Condensed profiles of the Snake and Columbia Rivers. The longitudinal profile of a stream (from its upper reaches to its mouth) is typically concave to the sky. Variations in erosion rates along the profile account for the irregularities. (After Sylvester, 1958.)

the sea, the sea level is the deterrent to further downcutting, for generally streams do not erode below sea level. Similarly, a stream discharging into a lake is eroded no deeper than the lake level. Since lakes are essentially impermanent, the stream base level controlled by a lake is called **temporary base level.** The base level of a stream can be manipulated by artificially draining lakes or erecting dams in valleys.

TRANSVERSE PROFILES

The form of the stream valley in cross-section, or **transverse profile,** may reflect the nature of the area geology, local climatic conditions, and the period of time the stream has been cutting its valley. Ideally, stream erosion should result in a channel having nearly vertical sides. However, due to variations in discharge and water level of the stream proper, and because of surface runoff from the surrounding terrain, the valley in its early stages acquires (in humid regions) a V-shaped transverse profile. Such valleys, characterized by narrowness and steep sides, are termed **young,** and are usually found in regions recently uplifted or otherwise newly formed, geologically speaking. Young streams in narrow valleys are typically fast-flowing, the water often encountering falls and rapids in its course. Vertical down-cutting is proceeding at a faster rate than horizontal erosion. Consequently the stream bed may occupy nearly the entire width of the valley floor. Yellowstone Canyon, in Yellowstone National Park, Wyoming, is an example of a young stream and valley.

In time, sediments carried by young streams are deposited as velocity and discharge fluctuate, the valley sides are eroded to more gentle slopes, and irregularities in the channel cause the stream to vary its channel, or **meander,** within the valley (see below). The valley floor becomes widened, and a plain resulting from stream flooding develops. The longitudinal gradient decreases. When these conditions obtain, the valley is said to be **mature.**

A stream valley is **old** when, as a result of continued erosion, the valley sides are reduced to nearly level surfaces and a broad flood plain has been built by deposition resulting from decreased ability of the stream to carry its load. Broad meanders and oxbow lakes occupy the valley floor. The lower Mississippi River is a classic example of a stream and valley in old age.

The Rhone River of Europe and the Colorado River of North America are examples of **rejuvenated streams.** We have seen that streams in old age typically meander over a broad valley floor. Geostrophic processes such as upwarping or bending of the land surface may act to increase the slope of the area, thereby increasing the gradient and erosional power of the stream. Thus the stream is "rejuvenated," and as a result of increased cutting ability comes to occupy deeply cut, winding gorges known as **incised meanders.**

VALLEY FEATURES RESULTING FROM STREAM DYNAMICS

We have considered some of the major forces of flowing water and how these forces may operate to scour a channel and to modify the form of the valley. Similarly, various hydrodynamic factors, acting singly or together, often result in the formation of valley features which may bear greatly on the characteristics of local segments of the stream as well as the stream generally.

Reference has been made to the cutoff of stream meanders as a method of lake formation (Chapter 3) and in connection with mature and old streams. Meandering is a mechanism typical of streams that have eroded to base level. As a result of decreased velocity the stream is deflected from one side of its valley to the other, thereby widening the valley in the process. The nature of the substrate also relates to this process. As erosion of the valley banks continues, acute bends develop in the channel. Because of increased turbulence and velocity on the outside of the bend, erosion of the shore imparts a load of sediment which is carried downstream to the next bend. Since the velocity is reduced on the inside of the lower bend, the load is dropped and the channel becomes partially filled, thereby pushing the current farther toward the eroding outer side of the bend (Figure 4.6). The bends become more sinuous, and as they approach and exceed 180°, they are termed meanders. The downstream arc of the bend in the meander system migrates downstream, leaving a layer of deposition across the valley. Oxbow lakes and meander scars develop when the downstream migration of one bend proceeds at a faster rate than the one below. The result is the formation of a new channel, or cutoff, between the bends.

Another feature often found in broad valleys develops when scattered masses of debris in the flood plain cause the stream to flow in many small, intermeshed channels. The debris is left when the stream velocity is reduced by seepage or excessive evaporation. A stream exhibiting such a form is termed a **braided stream** (Figure 3.2). Streams occupying glacial outwash plains and alluvial fans often show braiding.

An **alluvial fan** is an expanse of sediment deposited in the form of a broad fan by a rapid stream flowing onto a level plain. Although more characteristic of regions of arid climate, for example Death Valley, California, fans may form

Figure 4.6   Meandering stream in Big Horn County, Wyoming. Extensive sand bars on the inside of each meander result from decreased load-carrying ability of the stream. Meander scars representing former channels are visible to the left of the bend at the lower edge of the figure. Note the dendritic drainage pattern formed by small streams toward the left edge of the photograph. (Courtesy of Commodity Stabilization Service, U.S. Department of Agriculture.)

wherever conducive conditions prevail. Sudden floods in mountainside gullies transport a great load. The load is dropped and the waters usually soak into the fan as the stream flows onto the more level surface. Fans might be thought of as the land form of a delta.

A delta is, of course, another result of fluvial deposition and is formed when the velocity and carrying capacity of a stream is reduced. In this instance the reduction occurs as the stream flows into a standing body of water such as lake or sea. The morphology and structure of a delta were considered earlier as they relate to lake form. Seaward deltas are typical of the great streams such as the Nile, Rhine, and Ganges. The Mississippi delta covers an area of over 31,000 km² on the Gulf Coast of the United States. The depth at New Orleans is said to be nearly 300 m and the advance of the delta is estimated at some 90 m per year. Many streams lack deltas because the streams may not carry sufficient sediments or because currents at their mouths remove the sediments rapidly.

Two additional depositional features of streams in broad valleys are worthy of consideration here because they are often significant factors in determining the nature of the shores of mature and old streams, especially in relation to

Figure 4.7   Floodplain of the Rio Grande, north of Española, New Mexico. The older floodplain, now under cultivation, is clearly differentiated from the adjacent desert-like areas (upper corners). The river is at a high-water stage and follows a meandering course with cutoff streams. (Courtesy of Soil Conservation Service, U.S. Department of Agriculture.)

vegetation and other biological aspects of the streams; these are **natural levees** and **floodplain deposits.** Natural levees are formed by deposition of sediments along the normal channel of a stream as it overflows its channel. These levees take the form of ridges of fine materials which stand higher than the flood plain proper. The areas behind the levees are frequently maintained as swamps or marshes. Along the Mississippi in its delta region natural levees rise to heights of nearly 6 m above the swamps. A floodplain (Figure 4.7) is a flat expanse of land bordering an old river; it is usually built up of materials left by streams during flood times or as the stream meanders over the flood plain. A profile through the flood plain will often give an indication of channel movements, floods, and other events in the history of the valley (Figure 4.8). Often the flood plain may take the form of a very level plain occupied by the present stream channel, and it may never, or only occasionally, be flooded. Where several steplike landforms are found along the valley wall, these are called **floodplain terraces.** Several terraces may be found in a single valley, representing alternating episodes of flooding and channel downcutting.

Upstream towards the headwaters, and in the tributaries of a broad valley, we encounter a great variety of streams, grading from creeks to brooks. These often contain water only during the wet season. But when flowing, the dynamics of running water are operating as in the larger streams; currents erode exposed surfaces and land areas, sediments are transported and dropped, and the discharge from the smaller streams contributes to the maintenance of the major stream of the system. As we follow a stream toward its source through the

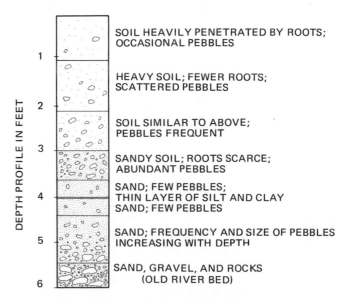

Figure 4.8 Soil profile of the ancient floodplain of the Raritan River near New Brunswick, New Jersey. Variations in the nature of the sedimentary strata are attributed to shifts in the course of the river over its bed. (After Wistendahl, 1958.)

headwaters, we are aware of gradations in the structure and morphology of the valley and the stream channel. Along the main stream these gradations may be subtle and we observe that, according to size, velocity, bottom nature, and the like, we usually proceed from brook, to creek, to river. However, in another instance, for example where a small brook joins a larger stream, a more abrupt transition will be observed.

STREAM-CHANNEL PARAMETERS

Because of the role of running water in creating and modifying the stream channel, and since the morphology of the channel is a direct consequence of hydraulic factors, certain of the more important stream parameters have been previously considered. It should be evident that certain morphological features described for lakes are also applicable in principle and method to stream studies. On the other hand, the very fact that the water in streams is flowing introduces a number of features not encountered in lakes.

Of the factors common to lakes and streams, let us consider depth, mean depth, length, breadth, area, volume, extent of shoreline, and mean slope with reference to streams.

DEPTH

The depth of a stream usually refers to the maximum depth within a given segment of the stream. For most streams the parameter is determined by sounding, although navigation charts for major rivers indicate depths. **Mean depth** $(\bar{d})$ takes into consideration the transverse bottom slope and is a relationship between cross-sectional area (A) and width (w) of the flowing stream:

$$\bar{d} = \frac{A}{w}$$

LENGTH

Stream length may refer to the total extent of the stream or to a segment under study. Stream length should describe the "true" length including bends. A straight-line measurement should be so stated.

BREADTH

The breadth, or width, of a stream channel may be defined in two ways. It may be (1) the actual cross-channel measurement of water at a given time, or (2) the width at near bank-full stage. The second parameter is evident in regions where well-developed flood plains and V-shaped valleys exist; it is less conspicuous in dry, flat areas.

AREA

Stream surface area is simply the measure of exposed water surface. This parameter varies considerably with seasons of drought and flooding. Another expression of areal extent often used in stream studies is that of **cross-section area;** this dimension is obtained roughly by multiplying the mean depth by the actual surface breath.

VOLUME

As in lakes, the volume of a stream is the amount of water held in the basin, or more preferably, the channel. Also as in lakes, stream volume is the sum of the volume in each of the strata bounded by bottom contours. Since the water mass is in motion, the term **discharge** is applied.

EXTENT OF SHORELINE

The extent of shoreline is simply a measurement of the length of the shore with respect to the length of the stream or a segment thereof. Features such as main-channel pools, side pools, and embayments increase the shoreline length relative to stream length.

MEAN SLOPE AND AREA OF SHOAL

Stream cross sections may be somewhat trapezoidal, semielliptical, or triangular. Depending upon the shape of the channel, the bottom may slope gently or drop off abruptly from shore. The mean slope describes as a percent the bottom grade with respect to horizontal distance transversely across the stream channel. The area of shoal describes the extent of shallow water as a proportion of the total area of a stream segment. The depth used to designate a shoal is arbitrary.

The major stream-channel parameter not shared with lakes is longitudinal gradient. Because such important hydraulic features as velocity, discharge, load, and erosion relate directly to gradient, it becomes of prime interest in stream studies. This measurement may be obtained from topographic quadrangles where available and of sufficient scale. Field determinations can be taken with stadia rod and transit along the shore, in the channel, or in the stream bed when the channel is dry or the water low.

Thus far, those parameters associated more directly with morphology of stream channels have been discussed. The characteristics of channels and dynamics of running water combine to maintain a stream and drainage basin. Certain relationships among the stream features are established and can be expressed in quantitative terms, as we have seen. In Table 4.1 some of the characteristics of channel-shape and flow are given for selected North American rivers. Note especially, with reference to the Mississippi River data, that the velocity increases downstream, seemingly in contradiction to the usual pattern of higher velocities occurring upstream. This will sometimes be the case when discharge is uniform along a stream but increased depth compensates for a decrease in stream gradient. In the case of the lower Mississippi, however, the velocity increases downstream because tributaries cause the discharge to increase faster than the cross-sectional area.

STREAM BEDS AND SHORES

The nature and composition of the channel bed and shores of streams are as varied as the terrain through which the streams flow and the internal dynamics of flowing water. From the general survey of fluvial processes just concluded, it would appear that gradient and types of substrata are the major factors in determining the composition of shore and bottom of streams. Rocky to gravelly substrates occur where the stream runs rapidly over a steep gradient, the coarse materials being derived from the mountainous zones (or at least the higher regions) where the stream slope would naturally be greater. Sands and muds may be deposited in the upper reaches but usually occur only in wider or deeper regions where the velocity is reduced. Finer sediments are transported to lower regions, and light organic materials do not settle out until the stream reaches near-base level. Thus, a gradation in the nature of the bed and shores exists from the fast-water, rough-bed zone to the slow-water, mud-bed region. Associated with this gradation (but related to a number of ecological factors) is a similar gradual change in the distribution of animals and plants, which we shall consider in more detail later.

LOWER STREAM COURSE

As a stream nears its base level and approaches its mouth, the bottom typically is composed of loose muds, silt, and organic detritus. Because the stream is operating in a broad flood plain of low relief, the shores are generally

TABLE 4.1 Parameters of stream-channel morphology at stage corresponding to mean annual discharge[a]

	Years of record	Mean annual discharge (CFS)	Width (ft)	Area of cross-section (sq ft)	Mean velocity (FPS)	Mean depth (ft)	Drainage area (sq miles)
Scioto River at Chillicothe, Ohio	26	3,289	200	1,840	1.8	9.2	3,847
Tennessee River at Knoxville, Tenn.	48	12,820	933	13,450	1.0	14.4	8,934
Yellowstone River at Billings, Mont.	16	6,331	253	1,343	4.8	5.3	11,180
Kansas River at Ogden, Kansas	30	2,514	342	1,300	1.9	3.8	45,240
Kansas River at Wamego, Kansas	28	4,114	525	2,150	1.9	4.1	55,240
Mississippi River at Alton, Ill.	12	96,670	1,750	32,500	3.0	18.6	171,500
Mississippi River at St. Louis, Mo.	14	166,700	1,586	44,440	3.8	28.0	701,000
Mississippi River at Memphis, Tenn.	13	454,900	1,950	99,400	4.6	51.0	932,800
Mississippi River near Vicksburg, Miss.	16	554,600	2,610	105,650	5.3	40.1	1,144,500

[a] Data from Leopold and Maddock, 1953.

poorly defined, the stream being bordered by marshes or swamps. Common features are backwaters, sloughs, and bayous, such as those of the lower Mississippi.

MIDDLE STREAM COURSE

The middle stream course typically exhibits a more inclined gradient and greater velocity than are found in the lower course. In this middle zone the bottom is usually characterized by coarser materials, with muds and lighter sediments being found only in pools or sidewaters. The flood plain is, as a rule, less extensive, leading to more discrete shores standing considerably higher than the stream. Wooded stretches rather than marshes and swamps occur along the banks.

UPPER STREAM COURSE

Mountain streams and brooks typify the stream in the upper reaches. The current in high velocity rushes and tumbles down through a steep gradient, occasionally slowing in pools. In the fast stretches, the channel and often the shores are strewn with boulders and rubble of various sizes. Where a pool interrupts the course, or the stream flows briefly out onto a meadow, the bottom may contain light deposits of sands and organic detritus.

These impressions of stream bottom and shores are, of course, very generalized. Local climatic conditions in terms of humidity and rainfall, soil types, land elevations and configurations, and age of the area and streams all serve to greatly modify the stream from source to mouth and to contribute to the countless variations exhibited by streams in nature. The topic is introduced at this point to emphasize the importance of the nature of bottom and shore in stream studies as well as in the ecology of a stream or stream segment. In defining a lotic environment and distinguishing it from standing waters, the considerable interchange between a stream and its basin was noted. This relationship becomes increasingly important in terms of contributions of water and nutrients from the valley and shore to the several courses, or regions, of the stream. Contrast the relative productivity of a stream flowing over highly insoluble rock substrates with that of a stream moving through a zone of soluble material rich in chemical nutrients.

DRAINAGE BASINS

The total land surface from which a system of streams receives its waters is termed the drainage basin. We have seen how a system may develop from continued erosion of a land area by water carried first in small trenches and eventually in stream valleys. As a rule, definite patterns of confluence of the various tributaries develop. Regardless of its size, each stream possesses its own drainage basin and the water received usually follows a particular course. The drainage basin of the Mississippi extends over some 3 million km². Such coverage would demand an impressive number of rivers such as the Tennessee, Ohio, Missouri, and Red with their countless tributaries. The Nile River system serves to drain over 2.6 million km².

The form of the drainage basin exerts considerable effect upon the regularity of flow in streams within the basin. Two methods of measuring the relationship between stream dynamics and form of the basin have been suggested; these are (1) the **form factor,** and (2) the **coefficient of compactness.** The form factor expresses the ratio of the mean width of the basin to the length from mouth to uppermost part, or

$$\text{form factor} = \frac{\text{mean width}}{\text{axial length}}$$

The lower the form factor, the less likely a widespread rainfall over the entire basin. The compactness coefficient relates shape of the basin to a circle of the same area. A high value indicates a less circular form for the basin, and therefore a smaller chance of heavy rainfall over the entire region.

Within a basin the pattern of streams with respect to orientation and branching usually reflects the composition of the substrate and surrounding terrain. It also indicates the geologic history of the stream and the region.

For the ecologist investigating streams, the most important consideration of the drainage basin relates to the present nature of the underlying geologic formations and the surface soils. It is from these sources that the materials carried in solution by the stream are obtained. Thus the physicochemical nature of the stream as well as the richness of plant and animal life can be attributed to a great extent to the drainage basin. Some examples of the accepted designations of

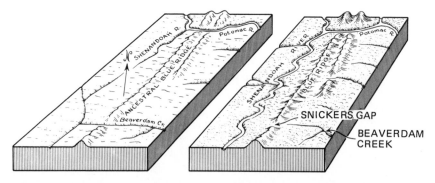

Figure 4.9   Stream piracy in the Blue Ridge of Virginia. See text for discussion. (After Longwell, *et al.,* 1939.)

basin patterns are given below. There are others in evidence on topographic maps of various regions.

Trellis pattern Just west of the old Appalachian Mountains is found a belt of rolling, uniformly crested mountains. These parallel ridges extend northeast–southwest and represent folds in the earth's crust. Streams in this region are remarkably parallel, with tributaries joining at right angels to the main stream. Some of the streams join through water gaps eroded across the long ridges. This distinctive pattern is given the name **trellis.**

Dendritic pattern A **dendritic** pattern, so-called because of its resemblance to tree roots, develops on a relatively smooth, gently sloped surface. Although inequalities in the surface determine the channels, the nature of the surface permits random and rather uniform coverage of the land by stream drainage (Figure 4.6).

Anastomosing pattern On flood plains and other flat depositional features, streams deposit their loads according to flood conditions and meanders. These deposits, together with meandering channels, may divide the usual stream channel, forcing the branches into braided networks. The braided stream shown in Figure 3.2 forms an **anastomosing** pattern.

Radial pattern A **radial** pattern occurs as streams flow from a symmetrical high region. Drainage on the sides of a volcanic cone, such as Mt. Hood, Oregon, typically occurs in a radial pattern.

STREAM PIRACY

Before closing this section on the origin, dynamics, and morphology of streams, we should give attention to one additional aspect of stream erosion; this is **stream piracy,** or **stream capture.** Stream piracy is a process by which one stream, through more rapid headward erosion, captures the headwaters of a neighboring stream. In the process, the second stream is beheaded.

Examples of stream piracy are common, and one classic example will suffice. Figure 4.9 illustrates the manner in which an ancestral tributary of the Potomac River captured the upper reaches of Beaverdam Creek, Virginia. As shown in the figure, the Potomac and Beaverdam flowed through separate water gaps in the Blue Ridge Mountains. Because of hard rock in Snickers Gap, the Beaverdam was unable to erode its channel as fast as the Potomac tributary. The result was the beheading of Beaverdam by the tributary to form the present Shenandoah River. In losing its stream, Snickers Gap became a wind gap.

SUGGESTED FOR FURTHER READING

Amos, W. H. 1970. *The infinite river: a biologist's vision of the world of water.* Random House, New York. A popularly-written account of the major aquatic habitats, with special attention to freshwater environments and their susceptibility to disruption by human agencies. The author's principal concern is with the plant and animal species to be found—at present or in the recent past—in the major habitats. However, his discussion, especially in Chapters 4 through 6, also includes a graceful and accurate account of stream bed formation.

Gregory, K. J., and D. E. Walling. 1973. *Drainage basin form and process: a geomorphological approach.* John Wiley & Sons, New York. A comprehensive summary of recent developments in the geomorphology of rivers. Fluvial systems are discussed first in terms of the major parameters—geology, vegetation, runoff, and erosion—of drainage basins. The rest of the book examines the nature and measurement of the effect of stream action on channel (basin) morphology. The discussion is detailed, but is well within the grasp of the beginning student with some background in geology.

Strahler, A. N., and A. H. Strahler. 1973. *Environmental geoscience.* Hamilton Publishing Company, a division of John Wiley & Sons, Santa Barbara, California. A detailed and authoritative text by two of the leading writers in the field of physical geography. In particular, Chapter 13 is concerned with stream and lake hydrology, particularly the former, and Chapter 14 deals with the relationships between fluvial processes and landforms. In both chapters the authors are concerned with the effect of human activity upon the hydrology and composition of natural waters.

5 | Estuaries

Most fresh water remains in lakes and streams only temporarily. Standing waters and streams, for the most part, flow eventually into major rivers, which in turn flow down to the sea. At and near the region where the fresh water from the land meets the salt water of the sea, a distinctive aquatic environment occurs, the **estuary.** The estuary is an **ecotone**—a rather complex "buffer zone," sharing some characteristics of both types of aquatic ecosystems, but identical to neither.

A standard definition identifies an estuary as a basin in which river water mixes with and dilutes sea water (Ketchum, 1951); an estuary might also be defined as the wide mouth of a river, or an arm of the sea, where tides ebb and flow as they meet the river currents (Emery *et al.*, 1957). These definitions, incomplete as they are, make it clear that a number of factors and processes that do not occur in fresh waters will be very important in the ecology of estuaries, and that in estuaries we will find communities with highly variable environmental conditions. In the first place, two opposing current systems, the unidirectional stream currents and oscillating tidal currents, meet and exert considerable and complicated effects upon sedimentation, water mixing, and other physical features of the estuary. These also greatly influence the biota. Secondly, the mixing of salt and fresh water produces a chemical environment unlike that of the typical sea or river. Thirdly, the diurnal shifting of the more saline waters with tides may necessitate physiological adjustment of the inhabitants of the community. These and other characteristics of estuaries will be considered further under more specific headings.

Estuaries are quite varied, not so much in their origins (as are lakes and streams), but rather in form and extent. Such variation is usually attributable to local modifications subsequent to original formation. Similarly, the amount and distribution of salt water in estuaries are functions of stream inflow, tide and other currents operating within the estuary, and the topography of the area. These attributes also vary. The biota, usually a unique assemblage of organisms, is related to the various physical and chemical features of the estuary. These general considerations should be borne in mind as we study in this chapter the estuary, its origin, morphology, modifying forces, and bottom and shore features. For many years estuaries were largely disregarded by both freshwater ecologists and oceanographers, since the study of estuaries was not a vital part of either specialty. In the past ten years or so, however, attitudes toward estuaries have changed considerably. This is partly an effect of recently aroused interest in ecological issues, and of the recognition that estuarine ecosystems—including their surrounding salt marshes—are a unique and endangered type of natural habitat. Economic interest also plays a part. Many estuaries, which in earlier times had been rich sources of fish, game, and shellfish, have become stagnant and unproductive as a result of unregulated economic exploitation and pollution.

As a result, estuarine studies have benefited from a great upsurge of interest. In New York City's Jamaica Bay, for example, a 33,000-acre tract has been set aside as a preserve for wildlife preservation and study; a special laboratory for estuarine studies has been established along the New Hampshire shore at Great Pond; in San Francisco Bay, an estuary fed by two of California's major rivers, the United States Geological Survey recently carried out an intensive, year-long study of all aspects of the region's ecology. The basic data for any estuarine study is the material we shall cover in the following pages: the origin of the basin, its morphology, modifying forces, and the major features of its bottom and shore.

THE ORIGINS OF ESTUARIES

As we pointed out in our discussion of tectonic lake basins, the crust of the earth consists of vast, moving plates. Thus the surface of the earth, including the floors of the seas, is subject to bending and other deformations. Continental land masses may be raised or lowered, and where the continent is in contact with the sea, conspicuous changes in the shoreline take place. As a land area is depressed, the shores become submerged and the coastline migrates inland. Conversely, as land is uplifted the existing shores emerge and the coastline is moved seaward. The basic physiography of coastlines and shores is, therefore, due primarily to **submergence** or **emergence.** Either or both of these processes may form estuaries. Some shores show the effects of both processes acting at different times in the past. Various local activities, for example, the movement of currents, wave action, tides, stream deposition, glaciation, and wind, operate to modify the original form of shores (Shepard, 1937).

Thus the forms of coastlines may be classified under two headings. Features developed as a result of movements of sea and land in relation to each other are called **initial** forms. Tectonic factors, glaciation, and climate are examples of forces that result in coastlines of initial form; shoreline patterns resulting from the action of marine forces on land masses (initial forms) are termed **sequential forms**.

Shoreline submergence results in a very irregular coastline of peninsulas, islands, and uneven bottoms (Figure 5.1). The eastern shore of the United States is primarily one of submergence, although subsequent emergence has altered local areas. The region from the Carolinas to New Jersey gives evidence of having been subjected to periods of both emergence and submergence. It is along submerged shorelines that many well-developed estuaries occur as **drowned stream valleys** (Pritchard, 1951). Chesapeake Bay represents such an estuary.

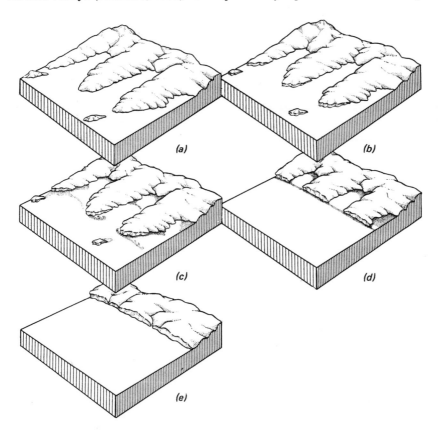

Figure 5.1   Developmental stages in a shoreline of submergence. The *initial stage (a)*, characterized by headlands projecting seaward, results from drowning of the lower reaches of stream valleys. In *early youth (b)*, sea cliffs at the ends of the headlands are typically present. Depositional features, such as bayhead beaches and sandspits, are characteristic of *late youth (c)*. In *early maturity (d)*, the water between headlands has become somewhat enclosed by the formation of bayhead bars across the entrance resulting in the development of lagoons or "lakes." Continued erosion and filling of the lagoons *(e)* results in the formation of a more regular shoreline lying back of the original headlands; this stage, essentially one of equilibrium, is termed *full maturity*. (After Johnson, 1919.)

Most streams entering the sea flow near base level across coastal plains or some other region of low relief. Since the stream lies at or very near sea level, almost any perceptible submergence of the coast results in the encroachment of the sea and the flooding, or drowning, of the stream's mouth. The extent to which the sea invades the stream valley is determined by the gradient and size of the valley, stream discharge, and the range and force of tides of the adjacent sea. Ideally, a typical estuary lies perpendicular to the shoreline.

MORPHOLOGY OF ESTUARINE BASINS

As the mouth of a stream is drowned by the sea, the young estuary assumes a rectangular or triangular shape, depending upon size, shape, and gradient of the stream channel, the volume of discharge of the stream itself, and the nature of the substrate and surrounding topography. The early form of the estuary is rapidly altered by fluvial and marine processes.

Where the drowned stream valley consists mainly of a single channel, the form of the basin is fairly regular and the estuary is said to be **simple.** The

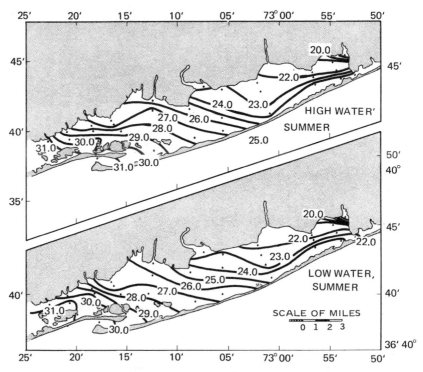

Figure 5.2   Great South Bay, Long Island. The bay is a coastal lagoon lying behind a barrier beach. Distribution of salinity in parts per thousand is shown at mid-depth for high and low water in the summer of 1950. Further discussion of salinity features will be found in Chapter 11. (After report of Woods Hole Oceanographic Institute, 1951.)

estuaries of the Delaware and the Hudson Rivers are simple. The flooding of a channel with numerous tributaries results in an **irregular estuary** such as Chesapeake Bay.

Estuaries along the Mediterranean Sea are essentially triangular, broadening from the stream valley toward the mouth of the estuary. Deltas are usually present. In this region the sea is generally calm and tidal range is slight. Along the English coast, estuaries are mostly broadened and tidal mud flats are common. This morphology is associated with moderate wave action and considerable tidal range. Stream channels are scoured by flow during the recession of tides. In South Africa, the shores experience great wave action, but slight tidal effects. Here sand bars develop in the mouths of the estuaries, causing the water to spread out broadly behind the bars.

Most of the estuaries along the coasts of temperate and subtropical North America deviate to varying degrees from the "classical." From New England southward many estuaries have been deprived of direct entrance into the sea by the formation of barrier beaches. Through the same process embayments lacking a conspicuous stream inflow have been partially enclosed, resulting in the development of a salinity gradient within the bay. Thus these bays take on something of an estuarine nature.[1] One such estuarine bay is Great South Bay (Figure 5.2), on New York's Long Island.

Delaware Bay is an example of a relatively "pure" estuary. The Delaware River flows directly into the Atlantic Ocean. The salinity of the estuary ranges from zero in the river to that of ocean water in the bay: approximately 35 parts per thousand (‰). Depending upon stream flow, tides, and winds, salt may be detected upstream as far as Wilmington, Delaware.

Apalachicola Bay, Florida, is a shallow coastal plains estuary behind a barrier beach. The estuarine waters cover some 362 km². Connections with the Gulf of Mexico are through "passes" between barrier beaches. East Bay, an arm of Galveston Bay, Texas, represents the **lagoon-like** estuary. Such an estuary is described as "lagoon-like" because, although the salinity of the water grades from fresh to moderately salty, the bay lies parallel to the shore behind a marine depositional feature, probably a spit. East Bay is nearly 37 km long and averages about 3.2 km in width. Depths range to nearly 4 m at mean low tide.

Spectacular estuaries occupy glaciated mountainous coastlines. In Norway, southern Chile, and at Puget Sound, for example, elongate, steep headlands

---

1. Such coastal bodies of water, partly separated from the sea by barrier beaches or bars of marine origin, are more properly termed **lagoons.** As a rule, lagoons are elongate and lie parallel to the shoreline. They are usually characteristic of, but not restricted to, shores of emergence. Lagoons are generally more shallow and more saline than typical estuaries.

Great South Bay, on the southern shore of Long Island, is a lagoon developed behind a barrier beach. The beach is primarily a reflection of slight tidal range with considerable wave action. The Bay averages about 4.8 km in width and is approximately 40 km long; mean depth at low tide is about 1.2 m. The deeper areas range to nearly 7 m. Fire Inlet, less than 1 km wide, is the only connection between the Bay and sea. The salinity grades generally from 20 to 30 parts per thousand (‰).

alternate with deep U-shaped valleys. Estuaries formed in such areas are called **fjords** (Figure 5.3). Sogne Fjord, in Norway, is 179 km long, averages 6.4 km in width, and has a maxium depth of 1200 m. Puget Sound extends from the Pacific Ocean some 130 km south into the state of Washington. Typically, the deepest regions of a fjord are in the upstream reaches; the seaward extension is often bounded by a **sill,** an upreaching section of the bottom, which limits free movement of water.

MODIFICATION OF THE ORIGINAL ESTUARY

When a shoreline is submerged, stream and sea processes of erosion, transport, and deposition immediately begin to modify the topography of the adjacent regions, and eventually of the entire coast. The dropping of sediments by the stream as it meets the encroached sea initiates the building of a delta in the upper reaches of the drowned mouth. Continued building by sedimentation gives rise to broad, level mud and silt deposits eventually becoming **tidal flats.** Meanwhile, currents and tidal action of the sea erode the peninsulas, or headlands, depositing the materials on the bottom of the seaward region of the estuary. These deposits may develop as bars and spits in various positions relative to the mouth of the estuary. In time, the estuary may become filled by stream and tidal deposits.

The rate and extent of sedimentation and filling of an estuary are primarily dependent upon the original size of the estuary, its age, the present rate of

Figure 5.3   A fjord on the coast of Alaska. (Courtesy of U.S. Geological Survey.)

erosion upstream and deposition by the stream at its mouth, and marine forces such as tides and longshore currents. Where stream discharge and tidal currents are relatively slight, coastal currents tend to build depositional features such as spits and baymouth bars across the mouth of the estuary, restricting the entrance. In such an instance, the process of filling is apt to be accelerated. On the other hand, if the scouring action of stream and tidal currents is sufficiently strong, the deposition of entrance barriers is inhibited, fluvial sediments are carried farther from the estuary, and the filling process is slowed.

A great portion of the volume of water, representing the difference between low and high tides, flows into an estuary and out again through the entrance in a short time. If the stream discharge is great and the entrance narrow, a deep channel is cut. If the entrance to the estuary is through easily eroded sediments, the depth is such as to be in equilibrium with the water volume moving through. Once an equilibrium is established, there follows only slight scouring or deposition. As filling of the estuary continues, the entrance is made more shallow, gradually decreasing the volume of water passing through in the tidal prism. In some regions of mountainous coastline and great tidal range, restricted entrances

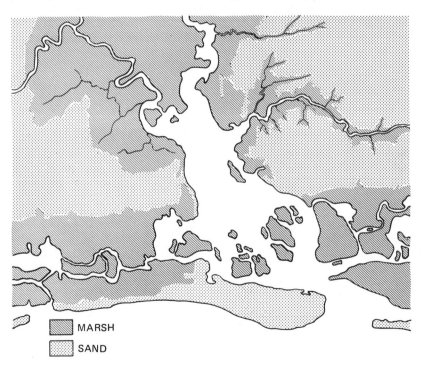

MARSH

SAND

Figure 5.4 An estuary formed by meandering marsh streams and partially blocked by a barrier beach. A marsh with tidal streams lies between the beach and the mainland. Coastal geomorphology such as this is characteristic of later stages in the development of a shoreline of emergence. Note that the seaward slope of the barrier beach is sandy, while the landward side grades toward marsh conditions.

result in tidal flow of considerable velocity. Because of the eroding ability of the current, the entrance is made deeper.

The sequence of events in the filling of an estuary and erosion of the shoreline of a region of high relief such as the New England coast is shown in Figure 5.1; the illustration is quite generalized since the manner and rate of erosion are dependent upon the nature of the materials composing the shore. Erosion of a homogeneous substrate results in steep, straight cliffs; differential cutting of a heterogeneous substrate leads to an irregular shore.

In regions of low coastal relief, erosional effects of the sea are often buffered by the inflowing stream. Deposition of sediments exceeds removal by ocean currents, and barrier beaches and other features are formed. From New Jersey southward, the Atlantic coast of North America is characterized by a sea island system of barrier beaches and other land forms. Some of these lie opposite the mouths of estuaries; others contribute to the development of brackish lagoons lying between islands and the mainland (Figure 5.4). This chain of sea islands extends the length of the Atlantic coast, southward from Long Island, and it affords protection to a continuous inland waterway, consisting of lagoons and passages, along which boats can travel in relatively calm waters from New York to Florida.

PARAMETERS OF ESTUARINE BASINS

Estuaries are, for the most part, drowned river valleys at the edge of the sea. This fact immediately suggests a great deal about the shape and functioning of estuaries. The parameters of estuaries are essentially the same parameters typical of lakes and streams, but these are complicated considerably by the confluence with seawater, and its particular physicochemical characteristics. The origin and location of an estuary determine its original form, but, as we have seen, the initial form is soon changed.

MORPHOLOGICAL PARAMETERS

The shape of an estuarine basin is of primary importance in determining the nature of the hydrologic forces operating within the estuary. These dynamics in turn often bear upon the animal and plant inhabitants and their relationships with the environment. A triangular estuary with a wide, deep mouth permits incursion of marine waters to considerable distances upstream, depending upon amplitude of tide and stream gradient. This results generally in increased mixing of waters from stream and sea, greater over-all circulation, and, often, the development of strong currents. A narrow-mouthed estuary, on the other hand, is character-ized by decreased circulation, more pronounced longitudinal and vertical salinity gradients, and more rapid development of sedimentation features such as spits and bars.

With respect to more specific morphologic features, such as depth, length, breadth, and area, it becomes immediately apparent that instability and fluctuation are inherently conspicuous in each. Daily ebb and flood of tide enlarge and contract each of the characteristics. Shorelines, normally clearly discernible in most lakes and along the upper regions of streams, are often obscured by marsh and depositional features and also fluctuate with tides.

DYNAMIC PARAMETERS

The parameters useful in describing and delimiting static and dynamic attributes of estuaries are essentially those discussed previously for lakes and streams. Depth, mean depth, areal features such as length and breadth, shoreline development, and volume are especially useful from the purely descriptive standpoint. These parameters take on an even more important function, however, in determining and describing dynamic qualities such as rate of flushing and flushing number. These attributes (to be considered later in relation to currents in estuaries) result from the conspicuous and highly important action of tides in estuaries.

The presence and action of such tides demand extended expression of the rather standard parameters listed above. It now becomes necessary to picture the body of water under various conditions of tidal range, for many coastal areas are greatly modified during successive ebb and flood of the tide. A relatively moderate range of conditions and variations is shown in the data pertaining to Biscayne Bay, Florida, a shallow, bar-built estuary, or lagoon.

The northern part of Biscayne Bay has an area of approximately 6.54×10^7 m^2. At low tide the mean depth is 2.1 m, thereby resulting in a low-water volume of 13.7×10^7 m^3. Because of the size of the bay and the nature of the entrances through which the waters flow, the bay does not experience a uniform high or low tide. Allowances for this give an ''adjusted'' tidal range of 0.56 m. Thus the volume during a mean high tide is estimated at 57.1×10^7 m^3. The difference between the high-tide volume and that at low tide suggests that about 12.2×10^7 m^3 of water is introduced into the bay during a given flood tide. The volume brought in is termed the **tidal prism volume** and, in the case of Biscayne Bay, increases the bay volume by approximately 27 percent.

It is interesting to note that the area of Biscayne Bay is not greatly affected by the tide range. This is doubtless due to the fact that a major portion of the ''shoreline'' is walled with concrete or otherwise retained. In an estuary of low relief and shores the area would fluctuate with tidal prism volume. The relationship between area, volume, and depth in Narragansett Bay is shown in Figure 5.5.

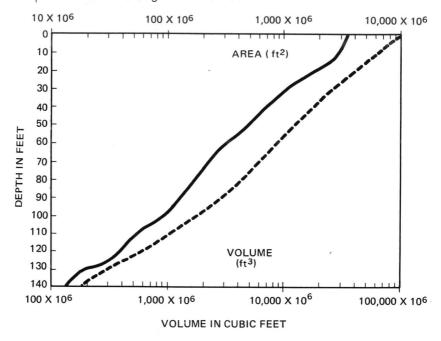

Figure 5.5 Narragansett Bay, Rhode Island. Chart gives area at any given depth (solid line) and volume below any given depth (broken line). All depths are corrected to mean low water. The high correlation indicated by the curves reflects the close interdependence of estuarine parameters. (After Hicks, 1959.)

The movement, or **discharge,** of water (from the river, via the estuary, to the ocean) is rather complicated. Fresh water moves outward from the river, often in complex flow patterns, resulting in a decided difference in temperature and salinity between the surface and the bottom waters. The greater the river discharge, the further will fresh water extend down the estuary. With rising tides, however, an inflow of seawater forces the fresher water back upstream; thus there is an oscillating movement of salt water, as the upstream and the downstream currents alternately predominate. The overall effect will be a net flow downstream, which may, however, be rather slow.

The time it takes for all the water in an estuary to be moved out to sea is called the **flushing time** (see Chapter 9). Complete flushing, which removes a mass of fairly uniform water, along with its associated plants and animals, may take from several days to many months. This fact is of special importance to those concerned with the removal of toxic pollutants from estuaries. In the past, great amounts of pollutant materials have been dumped into estuaries, on the assumption that they would "go out with the tide," presumably at the tidal speed of 1 to 3 knots. Unfortunately, pollutants may remain in an estuary for an extremely long time.

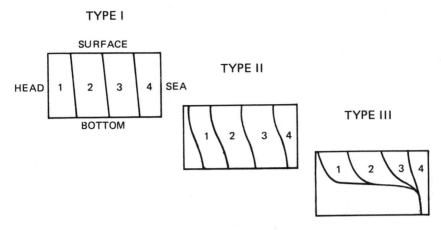

Figure 5.6   A classification of estuaries, based upon the shape of the vertical salinity gradients throughout the length of the estuary. (After Stommel and Farmer, 1952.)

CHEMICAL PARAMETERS

The salinity of an estuary is of particular ecological importance. In fact, one system of classifying estuaries (Stommel *et al.*, 1952) is based on hydrodynamic characteristics, as reflected in the salinity pattern of the waters (Figure 5.6). By analyzing water samples, taken at fixed depths throughout the tide cycle, it is possible to plot lines (isohalines) connecting points with the same salinity values. The resulting charts reveal the pattern of salinity changes. The estuary may then be classified on the basis of the bends or slopes in the isohaline lines. Type I shows little or no slope, type II shows a distinctly curved gradient, and type III shows the region of salinity gradation to be restricted to an upper stratum (probably as a result of a sill across the mouth of the estuary).

Other dissolved chemicals are also important in estuaries. Oxygen concentration in water is largely due to absorption from the atmosphere, but it is also strongly affected by biological activity. In general, O_2 concentration rises during sunny days and declines at night. Oxygen depletion may be an important indicator of pollution by excessive microbial action. Changes in CO_2 content are also affected by biological conditions, with CO_2 concentration dropping during the day and rising at night. The pH values of estuarine waters are also of interest. The pH will usually be determined by the mixing of various seawater solutes with the solutes in the fresh water; this in turn is affected by biological factors. In the upper reaches of an estuary, pH values between 7.5 and 9.0 are not unknown, but in the seaward regions pH values will fall within the normal marine range (8.0 to 8.2) because of the buffering action of seawater.

ESTUARINE SHORES AND SUBSTRATES

The shores and substrates of estuaries attest to the vigorous, rapid, and complex sedimentation processes characteristic of most coastal regions of low relief. These sediments are derived through the hydrologic processes of erosion, transport, and deposition described for other aquatic environments. In estuaries, however, these processes are being carried on by both the sea and the stream, thereby complicating the nature of the estuary.

The shores of a coastal-plains estuary are composed, in the main, of mixtures of silt and sand in varying proportions and degrees of compaction. Near the mouth of the estuary where predominating forces of the sea build spits or other depositional features, the shores and substrate of the estuary are conspicuously sandy. Just inside the entrance, the sand contains considerable quantities of fine sand and mud. Indeed, marked zonation, from the coarse sand shores of the seaward slope to the tidal flats or marshes of the inner slope of bay-mouth bars and spits, is typical of most estuaries. This is especially noticeable in estuaries lying parallel to the coastline and in lagoons formed behind barrier beaches.

From the estuary mouth with its generally coarse bottom sediments, there usually exists a gradation toward finer materials in the head of the estuary. This sorting is associated with current action. In the head region and other zones of reduced flow, fine silty sands are deposited, while in the main channel and in the mouth where stream and tidal flow carry greater loads, coarser sediments make up the bottom.

The nature of the substrate is known to exert considerable influence on the plant and animal inhabitants of the estuary floor and shore. Both pure sand and pure mud present problems in the maintenance of living organisms. Some studies have indicated that mixtures of sand and silt support the richer faunas. The silty sands of estuary bottoms tend to hold the more saline waters as the tides ebb. Thus bottom-dwelling plants and animals that require higher salinities are able to exist farther into the estuary than ecologically similar forms in the fluctuating water above the bottom. These and other relationships will be discussed further in later chapters.

Tidal, or mud, flats are commonly built up in estuarine basins. These depositional features, composed of loose, coarse and fine sand and silt, often develop in the estuary and divide, or **braid,** the original channel. Depending upon the composition of the substrate and tidal action, vegetation may eventually occupy the flats. Otherwise, the broad, flat, or slightly arched areas remain as barren features of the estuarine basin. Barring pollution, these tidal flats provide a habitat for an abundant fauna that feeds upon materials brought in by the tide or upon organic detritus of the substrate. Similar flats characterize the low-tide shoreline of estuaries.

One of the most significant features of estuaries throughout a major portion of the world is the oyster reef. This assemblage of organisms is usually found near the mouth of the estuary in a zone of moderate wave action, salt content,

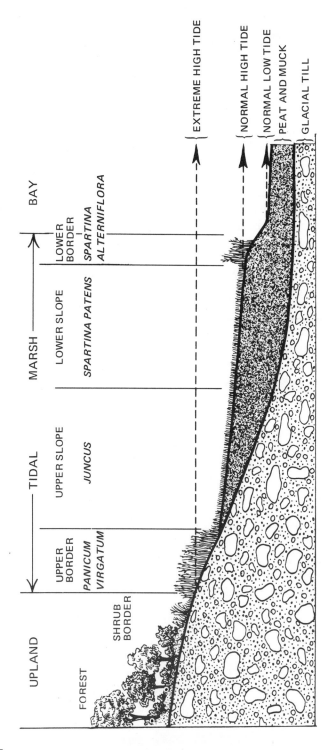

Figure 5.7 New England tidal marsh. A diagrammatic cross-section shows the major features of plant zonation between upland and the sea. Vertical scale is considerably exaggerated. (After Miller and Egler, 1950.)

and turbidity. Because of its location, depth, and areal dimensions, an oyster reef is often a salient factor in modifying estuarine current systems and sedimentation. The form and position of the reefs vary depending upon the nature of the substrate, currents, and salinity. The reef may occur as an elongate island or peninsula oriented across the main current, or it may develop parallel to the direction of the current. In shallow coastal areas, reefs may grow as islands, often exposed for considerable periods. The oyster reef is similar ecologically to a number of other aquatic communities. Similar reefs, for example, occur in a number of New England estuaries, where they are formed by such other shellfish as the mussel *Mytilus edulis*. Each of these communities is, however, biologically unique. The composition and functioning of the oyster reef will be considered in more detail in our discussion of aquatic communities.

At the approximate line of average high tide, we find the lower margin of the tidal marsh; at this point there may be an abrupt growth of salt-tolerant marsh grasses, such as *Spartina alterniflora*, which is nearly ubiquitous in temperate tidal marshes. Seen from the seaward, the *Spartina* growth resembles a solid wall. From this point back, the marsh extends as a flat plain, tilted almost imperceptibly to seaward, frequently covered by extreme high tides, and cut by tidal creeks and other channels. Tidal salt marshes support a characteristic, abundant flora of grasses, sedges, and other aquatic plants. As a result of the rich plant growth, the marsh substrate is often peat-like in composition. From the seaward margin of the marsh to the higher-land elevations, zonation of plants is a marked feature of the coastal marsh and estuary shore (Figure 5.7). The plant zonation appears to be related to moisture and salt tolerance. Along the estuary-stream shore from the saltwater zone inland, the salt marsh grades into a freshwater marsh or swamp. Thus the subtle transition between fresh water and the brackish estuary is re-emphasized.

RECENT TRENDS IN ESTUARINE STUDIES

We have mentioned that widespread interest in estuarine studies has developed only in recent years. Present evidence indicates that estuaries are among the most important coastal life-support systems. A productive estuary may be indispensible for a healthy marine biota in nearby regions. Thus it becomes increasingly important to assess properly the ways in which estuarine systems are manipulated. Up to now, common practices have included filling in large areas of estuarine shoreline to make new seaside real estate, constructing protective bulkheads around the edges of existing shores, and, often, cutting drainage channels through marshes as a mosquito control measure. Recent ecological studies (Mock, 1966; Odum, 1970) suggest that these practices are generally detrimental to the estuary and the surrounding areas. In many cases, such studies reveal ecological damage of almost "disaster" proportions, and they may provide valuable guidelines for new public policy.

SUGGESTED FOR FURTHER READING

Lauff, G. H. 1967. *Estuaries*. American Association for the Advancement of Science, Washington, D.C. A collection of papers, largely technical, presented at a conference sponsored by the University of Georgia Marine Institute and several other organizations. Somewhat dated, but still of value for more advanced undergraduates.

McLusky, D. S. 1971. *Ecology of estuaries*. Heinemann Educational Books, London. A straightforward introduction to estuarine ecology, useful for college students and requiring no extensive background in the subject. The author describes the principle physicochemical and biotic factors, the major plant and animal communities, and the special areas of brackish seas and hypersaline lagoons. Also included are a discussion of man's effects on the estuarine ecosystem and a brief introduction to field methods in estuarine studies.

Teal, J., and M. Teal. 1969. *Life and death of the salt marsh*. Little-Brown, Boston. An excellent popular account of estuarine ecology, with emphasis on the animal species of tidal marshes and on increasing threats to estuarine ecosystems.

6 Oceans

Most of the earth, some 71 percent, is covered by vast interconnected expanses of water, the oceans and seas. These have been a primary route of exploration throughout the world since early times; they have served as vital routes of transportation, and as an important source of food for maritime civilizations. Unlike lakes and estuaries, whose limits are more or less clearly circumscribed, the oceans are largely uninterrupted masses of water. The Atlantic and Pacific Oceans, for example, extend virtually unbroken from Antarctica north to the Arctic Circle. Thus, in the oceans the distribution of plants and animals does not follow the limits of such landforms as mountains, plains, and deserts. Instead, chemical-physical factors such as temperature, salinity, depth, and currents largely determine the nature, distribution, and abundance of marine life.

Oceans and seas are much alike; they are distinguished principally by size. The term **ocean** is generally reserved for the largest masses of water. The **seas** are the others—generally of secondary size, more or less landlocked, like the Caribbean, or forming part of (or connected with) an ocean, like the Bering or Arabian Seas. Some seas (so-called) are inland bodies of salt or brackish water, while the ''Sea'' of Galilee, in contrast, is fresh water and is considered a lake.

In many ways the oceans and seas are much the same as lakes and ponds; but in other ways—such as their vast size, their tremendous depth, their uniform salinity, and their massive tides—marine waters are decidedly different from any inland waters. One difference is overwhelmingly important: only a minute region of the oceans and seas is shallow enough to support attached plant life. Therefore, while many rooted plants—pondweeds, water willows, and so forth—are often found in inland waters, it is characteristic of marine waters that the great mass of their biota—plant as well as animal—consists of motile or permanently suspended organisms. Only in the narrow region of the coastal waters do attached living things become important in marine ecology. However, despite other differences between the coastal, or maritime, and the oceanic forms, these two ocean realms are closely related. Many fish and other animals of the open seas (salmon, green turtles, and seals are well-known examples) return to shallow inland waters during spawning or other stages in their life cycles. In this chapter, we will note many similarities between the basic principles of marine and freshwater ecology; this will help us to recognize more clearly those other features that make the oceanic ecosystem unique.

Most of the world's marine waters occupy the large ocean basins, those of the Pacific, Atlantic, Indian, Arctic, and—by some reckonings—Antarctic Oceans (Stembridge, 1958). Lesser basins include those of the seas: the Caribbean, Mediterranean, Arabian, Phillippine, Coral, North Sea, Bering Sea, and others (Figure 6.1). Other recognizable basins, **gulfs** and **bays,** are formed by indentations in a coastline; thus the Gulf of Mexico, the Gulf of Guinea, the Bay of Bengal, and Hudson's Bay. Marine waters occupying narrows between landmasses are identified as **straits** or **channels;** familiar narrows include the English Channel, Mozambique Channel, Bering Strait, and the Strait of Gibraltar. The largest body of marine water is the Pacific Ocean, with an area of 165,246,000 km^2; the Baltic Sea, with an area of 422,000 km^2, is the smallest. Basic hydrographic data for the major oceans and seas are given in Table 6.1.

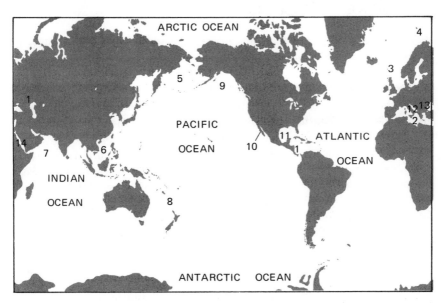

Figure 6.1   Major divisions of the world's marine waters. Besides the oceans (shown in figure), the major seas include the Caribbean (1), the Mediterranean (2), the North Sea (3), the Baltic Sea (4), the Bering Sea (5), the South China Sea (6), the Arabian Sea (7), and the Coral Sea (8). Other important subdivisions include the Gulfs of Alaska (9), California (10), and Mexico (11), the Adriatic (12), Black (13), and Red (14) Seas, and the inland Caspian Sea (15). Many additional regions are also generally recognized.

TABLE 6.1 Hydrographic data for the major oceans and seas[a]

	Area (km²)	Volume (km³)	Mean depth (m)
Pacific Ocean	165,246,000	707,555,000	4282
Atlantic Ocean	82,441,000	323,613,000	3926
Indian Ocean	73,443,000	291,030,000	3963
Mediterranean and Black Seas	2,966,000	4,238,000	1429
North Sea	575,000	54,000	94
Baltic Sea	422,000	23,000	55
Caspian Sea[b]	436,400	79,319	182

[a] Data from Sverdrup et al., 1942, and Hutchinson, 1957.
[b] Inland; surface is approximately 26 m below sea level.

No such data are given for bays or channels, because the latter are not usually treated as comparably independent systems.

ORIGIN OF THE OCEANS

The oceans were first formed as water condensed out of the atmosphere of the primeval earth and settled in the lowermost parts of the newly formed crust. According to current geological theory all the land on earth—that portion of the earth's crust above sea level—formed a single "supercontinent," **Pangaea,** during the first 3 or 4 billion years of earth history. The continent was surrounded by a single ocean, **Panthalassa.** Some 200 million years ago this single landmass began breaking apart, ultimately to form the continents we know today (Figure 6.2). From this process of continental drift, which was explained in Chapter 3, the ocean floors have spread and enlarged, and thus taken on their basic morphology. The Pacific Ocean of today is a remnant of the original Panthalassa, and the Atlantic Ocean fills a rift that was left when the Americas and Antartica became separated from the African and Eurasian landmasses. The Mediterranean Sea, now an arm of the Atlantic Ocean, has undergone episodes of dramatic change, having been at different times a desert and the floor of an inland sea.

The origin of the present ocean basins through **sea-floor spreading** is reflected in the distribution of marine species. A vigorous marine biota was already alive when the new basins started forming; thus animals and plants that were widespread at that time would invade the new basins as they formed. However, any species which later evolved in a new basin would ordinarily be restricted to the region where it first appeared. Thus the restricted distribution of many species of crab, coral, and fish indicates their relatively recent emergence as distinct species.

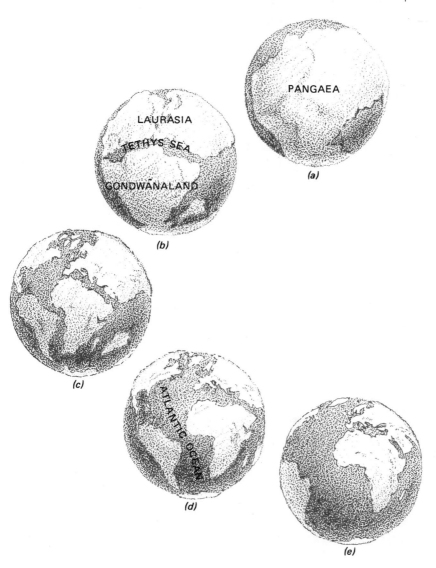

Figure 6.2   Formation of the ocean basins. During the early Mesozoic era, approximately 200 million years ago, the earth contained a single landmass, Pangaea, and a single great ocean, Panthalassa *(a)*. At about the start of the Jurassic period the northern and southern parts of Pangaea began to drift apart, the region between them filling with water to become the narrow Tethys Sea *(b)*. Further division of the landmass produced the beginnings of the present continents *(c)*, and the early Tertiary, when the Rocky Mountains were forming, the region between the eastern and western landmasses began to widen into the present Atlantic Ocean *(d and e)*.

MORPHOLOGY OF OCEAN BASINS

Most of the ocean floor is fairly level or gently undulating. The average depth of the oceans is about 3800 m, with about 50 percent lying at a depth of between 3000 and 6000 m. The deepest regions of the ocean floors are long furrows, referred to variously as **trenches, deeps,** or **troughs,** which occur usually parallel to the shoreline of a continent or to an **archipelago** (a chain of islands). The Philippine and Japan trenches reach depths exceeding 10,000 m. The greatest oceanic depth so far recorded, 11,515 m, is in the Mindanao Trench, east of the Philippine Islands (Voss, 1972); the greatest depth recorded in the Atlantic Ocean, 9219 m, is in the Puerto Rico Trough (Sverdrup *et al.*, 1942).

A conspicuous feature of the Atlantic Ocean is a submarine mountain range, the Mid-Atlantic Ridge, which extends the length of the ocean from north to south. The ridge is continuous with comparable ridges, or with zones of frequent earthquakes and volcanism, which extend around the world and are prominent in the Indian and Pacific Oceans (Figure 6.3). These ridges and zones of activity are thought to be the primary regions where new material rises from the earth's mantle to form new crust. New sea-floor appears at these areas, which are the centers of sea-floor spreading (and consequently of continental movement). As the ocean floor is moved toward the continents it is forced beneath the continental masses at some points, thus forming the trenches and deeps.

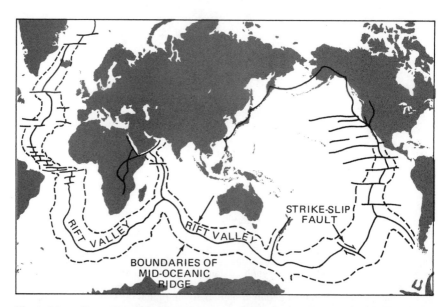

Figure 6.3   The midoceanic ridge and rift valley system. The Mid-Atlantic Ridge (left) is continuous with the East Pacific Rise (west of Mexico) and the fault system of the North American west coast. (After Strahler, 1965.)

Other major features of oceanic basins are **continental shelves,** submerged extensions of the continental masses, usually lying at depths of less than 200 m. The depths and widths of the continental shelves vary considerably (Table 6.2); in the Atlantic the continental shelf accounts for a significant part of the ocean's surface area, whereas in parts of the Pacific, as along the southern California coast, the continental shelf is so narrow as to be nonexistent.

Submarine **canyons** are common in coastal regions, where they may cut deeply through the continental shelf and extend for great distances down the continental slope. A number of spectacular canyons extend outward from the North American coasts; well-known canyons arise at the mouths of the Columbia and Hudson Rivers, and a particularly large one, Scripps Canyon, begins at La Jolla, southern California. Other canyons are not so clearly associated with river mouths, but begin instead at the edge of the continental shelf. These canyons are thought to be created by repeated mudslides and turbidity currents; submarine "avalanches" consisting of sand-flow have been observed and filmed by the Scripps Institute of Oceanography at La Jolla.

MORPHOLOGY OF OCEANIC COASTS

Oceanic shores, like those of lakes, are by no means permanent. The original form of an oceanic shoreline, especially in a region of violent weather, is strongly modified by dynamic forces. The single factor that has most strikingly altered coastlines in the recent past has been the advance and melting of the continental ice sheets. The great icecaps of the Pleistocene era trapped enormous amounts of water; the melting of the polar icecaps has caused the water level

TABLE 6.2   Depth zones of the major oceans: percentage of total area which extends to indicated depths[a]

Depth (m)	Atlantic[b]	Pacific[b]	Indian[b]
0-200	13.3	5.7	4.2
200-1000	7.1	3.1	3.1
1000-2000	5.3	3.9	3.4
2000-3000	8.8	5.2	7.4
3000-4000	18.5	18.5	24.0
4000-5000	25.8	35.2	38.1
5000-6000	20.6	26.6	19.4
6000-7000	0.6	1.6	0.4
greater than 7000	—	0.2	—

[a] Data from Kossinna, 1921.
[b] Including adjacent seas.

of the oceans to rise some 90 m in the past 10,000 years or so. Water level has also been affected locally by warping of the continental plates. In North America the eastern edge of the continental plate has moved downward in New England and upward in the southeast; thus in New England the water level has risen, while along the southeast coast miles of beaches have been newly exposed. On the basis of the relation of sea level to land, it is possible to distinguish three basic classes of coastlines: shorelines of **emergence,** shorelines of **submergence,** and **neutral** shorelines.

Figure 6.4 Major features of a sinking shoreline. Visible features include arch, narrow rocky beach, a stack (at far end of beach), and the beginning of an undercut notch at far end of beach. There is no sandy foreshore, as would be found in a rising shoreline beach. Rather, the beach extends out into the water to form a wave-cut terrace or abrasion platform. (Courtesy of P. E. Salwen.)

SHORELINES OF SUBMERGENCE

Shorelines of submergence, or sinking shorelines, are those in which the coasts are generally rocky and steep, with bays, referred to as **drowned rivers,** formed by the flooding of river mouths. Along such a coast, cliffs are often cut by the ocean (Figure 6.4), forming a narrow beach below of eroded rocks and boulders. If the cliff consists of soft rock, the swash of waves may cut away exposed areas, leaving a smooth, horizontal wave-cut terrace slightly above the water level.

Because shorelines consisting of softer rock erode relatively quickly, horseshoe beaches often develop between headlands of harder rock (Figure 5.1); the longshore currents carry sand, depositing it in beach areas, while silt is transported farther offshore. If rocks are broken up into fairly uniform, small pieces, shingle beaches or gravel beaches may be found, with little or no sand but, instead, flattened or rounded stones.

SHORELINES OF EMERGENCE

Rising shorelines, or **shorelines of emergence,** typically are gently sloping, with long beaches and sand (or sandy sediments). River mouths and bays tend to be shallow and heavily sedimented, with slow water flow. Such a rising shoreline occurs at Cape Hatteras, Virginia; there the extensive lagoons of

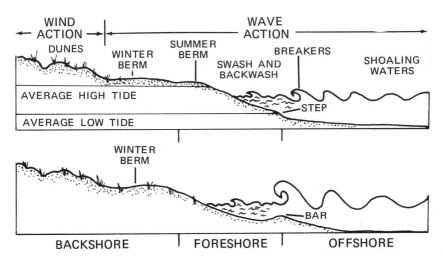

Figure 6.5   Structure of a barrier beach changes from summer (above) to winter (below). Wave action on the upper, middle, and lower beach (between average high and low tide) moves the sand so as to form a relatively broad summer berm, which is removed again in winter.

Pamlico Sound form a shallow inland water passage, with a low-tide depth of barely 0.6 m. Where bays occur, bars develop across the mouth, forming enclosed lagoons (see Figures 3.7 and 5.1). Offshore bars develop parallel to the beach, and gradually move landward, where onshore winds drive the beach sand inward to form dunes. Dunes of considerable size may become stabilized as plant roots hold down the surface sand and a thin soil develops which gradually caps the dune.

Sandy beaches are subdivided into an **offshore** zone, lying below low water, a **foreshore** zone, the region of wave activity, and a **backshore,** lying beyond the area reached by high tides. Along the foreshore the **swash** (the breaking wave) lifts and mixes particles of sediment, and the **backwash** (the retreating wave) carries them back out, spreading them to form a flat, uniformly sloping beach. The foreshore extends inland to a level strip, the **summer berm,** formed by the gentler summer waves depositing sand along the beach. The more violent waves of winter cut away much of the beach sand, possibly exposing a rocky substrate, and leaving a narrower **winter berm.** The wider beach is ordinarily rebuilt when a new berm is deposited by wave action during the following summer (Figure 6.5).

SHORELINES OF OTHER TYPES

Where forces of deposition, rather than tectonic movements, are the major determinants of coastline morphology, the coast is defined as a **neutral shoreline.** Deposits left by rivers—such as **deltas** (Page 57) and **alluvial fans** (low, cone-shaped deposits left by braided streams)—are common along neutral shorelines. Glacial outwash, lava flows, and coral reefs may also affect the form of a neutral shoreline; typical features include barrier islands and offshore bars, spits, hooks, and lagoons.

A rather unusual form of shoreline, a **fault shoreline,** may occur when a fault, or break, occurs in the earth's crust so as to form a new shoreline. A portion of crust may be raised above sea level, or a down-faulted block may sink below sea level; in either case, a new shoreline forms against the steep edge of the fault plane. In many cases a shoreline will show evidence of more than one morphological process: such a shoreline is referred to as a **compound shoreline** (Strahler, 1965). Compound shorelines occur, for example, where an episode of deposition is followed by a period of emergence or submergence.

ISLANDS AND REEFS

Independent land masses smaller than continents are referred to as **islands.** Islands are classed into two major types, continental and oceanic. **Continental islands** are exposed parts of a submerged continental land mass, usually an upward extension of the continental shelf, separated from the mainland by submergence of the connecting lowland areas. Greenland, an island of nearly 22 million km^2, is a continental island, rising on the submarine extension of eastern North America. Islands which arise independently from the floor of an ocean basin are referred to as **oceanic islands.** The Hawaiian Islands, with a total area of about 16,600 km^2, are volcanic oceanic islands, rising to a height of 9756 m above the ocean floor. Oceanic islands often occur in arcs, as in the Aleutians southwest of Alaska or the outer islands of the Caribbean; usually such arcs are associated with areas of tectonic activity.

Among oceanic islands a distinction is usually made between **high** and **low** islands. The former usually are continental or volcanic in origin and consist of igneous or metamorphic rock; the latter are usually raised coral reefs or portions of atolls, lying generally close to the water and consisting of limestone. The low islands, the reefs, and the atolls are formed by the growth of corals and coralline algae in tropical or subtropical marine waters. Corals are small polyps that resemble sea anemones and secrete a limestone casing in which the individuals live and feed. In the clear, well-oxygenated marine waters of the tropical latitudes, corals and coralline algae form immense colonies which can live at depths of up to some 60 m (Wells, 1957). On a subsiding island, or along a subsiding coastline, extensive reefs may form as new material is deposited by the growing algae and coral. Conversely, masses formed in this manner may be raised to considerable heights above sea level by subsequent uplift of an island or continent. The coral or **organic** reef forms a unique ecosystem, which will be described in further detail in Chapter 16.

OCEANIC SEDIMENTS

The floor of the ocean is largely covered with unconsolidated sediment, consisting of particles which have settled down from higher regions. The sediments reach thicknesses of thousands of meters, and the great pressure exerted by the weight of such masses of material consolidates the bottommost portion into sedimentary rock. These rocks may be classified (Sverdrup *et al.,* 1942) on the basis of the materials from which they were originally formed. Six basic types of sediment are distinguished: terrigenous deposits from eroded (often volcanic) land; volcanic discharge; skeletal remains (e.g. diatom cases) of plants and animals; inorganic precipitates from the seawater; products of organic chemical processes; and fine oceanic particles (such as red clay), which accumulate very slowly into fine sediment.

The nature of the uppermost layer of benthic sediment will be determined by whether it is organic or inorganic in origin. The nature of the bottom will be affected, too, by the types of organisms of which it is composed. Where diatoms are the predominant life form, a siliceous ooze is found; where foraminifera are most abundant, a predominantly calcium carbonate ooze develops. A number of deep-sea animal forms—small, light, and adapted to the great oceanic pressures—live on or in the oozes.

OCEANIC CURRENTS

The volume of water in the ocean basins is estimated at 1.37 billion km^3, and most of this is in motion in the major ocean currents. Near the equator, water flows generally westward within the tropics, along the north and south sides of the equator as a result of the earth's eastward rotation. This effect is termed the **Coriolis force,** and is a major determinant of oceanic currents. When they reach the major landmasses, these **equatorial currents** are deflected north or south, moving poleward along the east coasts of the continents. In subpolar latitudes the waters then move eastward, and are again deflected back toward the equator along the west continental coastlines (Figure 6.6). These currents describe great cycles, called **gyres;** such a gyre forms the limits of the calm Sargasso Sea in the North Atlantic. A different (and simpler) flow pattern is observed in the Antarctic Ocean, whose waters flow continuously eastward around the earth, being joined by eastward flows from the other ocean basins.

The constant flow of oceanic waters affects conditions where organisms live, warming the eastern continental coasts and northern portions of west coasts of continents and cooling the other western portions of continents. Currents flow in different directions at different depths. Antarctic water, for example, flows northward beneath the equator, upwelling along the west coasts of Peru and

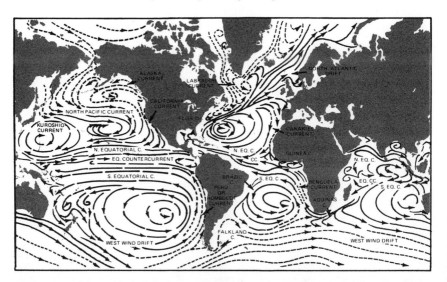

Figure 6.6 The major oceanic surface drifts and currents, showing average January movement of warm currents (solid lines) and cool (broken lines). Note the great gyres in the ocean basins and the circum-antarctic drift in the southern oceans. (After Strahler, 1965.)

southern California. Upwellings of this type help bring nutrient-rich cool water to the surface, and support enormous growths of plankton and seaweed (Dawson, 1966), which in turn support great numbers of crustaceans and fish. A vigorous whale and seal migration pathway follows the direction of these upwellings of nutrients (Sheffer, 1969).

Ocean currents flow at velocities of 0.93 to 4.3 km/hr (0.5 to 2.33 knots). The effect of oceanic currents was dramatically shown by the expedition of the raft *Kon-Tiki* (Heyerdahl, 1950), which was an effort to prove that early man might have colonized the Pacific islands from the west coast of South America. The *Kon-Tiki* drifted almost 8000 km along the South Equatorial Current, travelling from Peru to the Tuamotu Islands in the central Pacific in 101 days. The currents are of great importance in the ecology of islands and other terrestrial ecosystems, since currents transport water-resistant seeds (such as coconut and mangrove) to distant regions, and many plant and animal species have been dispersed widely by drifting on "rafts" of matted vegetation. Similar drifting determines the distribution of suspended forms of marine life. Oceanic currents also affect plant and animal life throughout the earth by distributing solar heat. As will be understood from Chapter 2, marine waters absorb tremendous amounts of heat in the equatorial regions; oceanic currents redistribute this heat, warming coastal regions and landmasses quite far poleward.

PARAMETERS OF MARINE WATERS

The oceans and seas are so large that the parameters which are important for inland waters, such as depth, mean depth, breadth, and volume, have little or no effect on the ecology of marine life. These measurements become significant only in inshore habitats, where the effects of depth and shoreline configuration may be important. In the open ocean the usual hydrographic measures are of interest mainly in physical oceanography, which explores such phenomena as wave development and dispersion of heat from warmer to colder waters. Marine ecologists are more concerned, for the most part, with environmental rather than hydrographic factors, and their interests will include water motion and mixing, light, temperature, and dissolved gases and solids in seawater. These will be set forth in detail in Chapters 7 through 11. In this chapter our discussion will be limited to a few factors whose effects upon marine ecology is of particular concern.

ECOLOGICAL VARIABLES OF THE OPEN OCEAN

Four major factors—tides, waves, circulation cells, and settling velocities —are of special importance in marine ecosystems. Two other phenomena— water pressure and osmotic pressure—are less obvious in their effects, but they are also important and will be considered briefly.

Tides

The tides are the **diurnal** or **semidiurnal** (once or twice daily) changes in water level that can be observed along seashores. The normal range of tide rise and fall is between 0.5 and 4 m, although extremes can vary from 0.0 to as much as 15 m. Tides are most noticeable along coasts, but the same rise and fall does occur at sea. Thus tides seem to have no effects on organisms of the open oceans, but shoreline species must have special adaptations to the semidiurnal change in the environment. The mean height of local tides also varies over the 28-day lunar schedule, increasing for 7 days to the **spring** (highest) **tide,** and then decreasing for 7 days to the **neap** (lowest) **tide.** These and other longer-term variations in the tides require that shoreline plants and animals be able to tolerate extended periods of immersion and exposure which might last weeks or even months.

In everyday usage, we tend to speak simply of "high" or "low" tide. It is obvious, however, that for precise work a more detailed terminology is needed. Tides are usually described in terms of their relative heights. Thus:

higher high water	HHW
lower high water	LHW
higher low water	HLW
highest higher high water	HHHW
lowest lower low water	LLLW

and so forth. The last two take into account the variation between the extremes of spring and neap tides. Another system of notation that also takes these extremes into account uses different wording. Thus

HHHW (= extreme high water spring tide)	E.H.W.S.
LLLW (= extreme low water neap tide)	E.L.W.N.

and so forth. The reader should bear in mind that tide height in the United States is reckoned on the basis of average low tide, which is referred to as **datum.** Since the metric system is not generally in use in this country, tide measurements are calculated in feet (e.g. +1.6 or −0.2) above or below datum.

Tides may also affect the ecology of coastal regions by creating **tidal currents,** through changing the water level in a restricted basin such as a narrow estuary or a coastal valley. In certain basins, such as the Bay of Fundy between Maine and New Brunswick, local morphology forces the tides into a narrow channel. The resulting **bore**—a flood tide that approaches with a strong, abrupt current—is strong enough to back up the St. John River for several miles into northern Maine.

Waves

Like the tides, **waves** affect marine organisms by alternately raising and lowering them, and the effects of waves are conspicuous in the open ocean as well as near shore. Waves are important primarily because they circulate the upper layers of water (mixing heat, dissolved nutrients, and oxygen), they cycle plankton in and out of the bright surface light, and they disperse the gametes and larval forms of many species.

Waves are of two basic kinds: progressive, or **surface,** waves, and standing, or **long,** waves. (See also Chapter 9.) Surface waves are created by wind, their size depending upon both wind velocity and **fetch** (the distance over which wind has blown without obstruction). As the wave moves ahead, masses of water rotate vertically beneath it as it advances. The water does not actually move along with the wave front, but rotates in place; large waves rotate water and suspended matter to a considerable depth, thoroughly mixing the upper layer of water. In a high wind the upper contour of the wave (its **crest**) is moved ahead faster than its underlying cycle, and it falls over ahead of the wave front. The wave is then said to be **breaking.**

As a wave approaches a beach it breaks in a different manner: the bottom of the rotating cycle strikes the shallowing bottom, the cycle is slowed by the resulting drag, and the moving crest breaks at the beach or the outlying bar (Figure 6.5). These breakers form when water depth is reduced to about 1.3 times the height of the wave. The ordinary maximum height of a wave is about 3 m, but heights of 12 m are often observed. Storm-produced waves in the Pacific reach heights exceeding 13 m. "Tidal waves," better referred to as **sea waves** or (from the Japanese) *tsunami,* are caused by submarine earthquakes or eruptions. They can reach heights of 30 to 70 m, and they move at tremendous speeds (velocities of 800 km/hr, or 435 knots, have been recorded). When it reaches a shore, a *tsunami* can cause overwhelming damage.

The swash and backwash of breakers on a beach can build up a beach by sorting and distributing particles, thus increasing the height of the berm. The pounding and sucking effect of breaking waves is also, however, a powerful erosive agent, and when the waves contain sand or ice the grinding effect is augmented. Floating ice moved by tides and waves can scrape away entire communities of attached organisms—algae, barnacles, mussels, and the like—from the intertidal zone.

Long waves are those which involve the rocking of water around a node that is free from motion; they occur in confined basins or between landmasses, where they can form marine **seiches,** or oscillations (see Chapter 9). Where these long, or standing, waves happen to coincide with the occurrence of tides in a funnel-shaped bay, as in the Bay of Fundy, local tides may reach extreme heights.

Circulation cells

Very gentle winds may not form waves, but instead may set up **convection** or **circulation cells.** A circulation cell is essentially a block of surface water, often exceeding 20 m in length, in which water cycles down and up beneath the surface. Adjacent circulation cells have rising currents coming up together, and sinking currents on their opposite sides. Water cycling in this manner to depths of 10 to 30 or more meters may cool and pick up nutrients and drifting organisms; thus the rising current between cells is often cooler and richer in food than the surrounding water. This may affect the distribution of autotrophic organisms in the surface layers; the Portuguese man-of-war (genus *Physalia*) is reported to be particularly abundant along such nutrient-rich regions of upwelling between circulation cells.

Settling velocity

The rate at which a particle settles toward the bottom of a body of water is referred to as its sinking rate or **settling velocity.** Settling velocities are important to the distribution of suspended particles and small organisms throughout marine waters. Plankton, which drift with the currents described above, also move in accordance with their varied settling velocities, and the entire biota of the deeper waters is dependent, ultimately, upon a rain of nutrient materials from the surface regions. Thus a detailed study of marine ecology requires an understanding of this parameter.

For small particles, which may be thought of for the moment as tiny spheres, the settling velocity can be estimated by a formula used in fluid dynamics and referred to as **Stokes's Law:**

$$v = \frac{2}{9}(d_1 - d_2)gr^2/\mu$$

where v is the settling velocity, d_1 is the density of the liquid, d_2 is the density of the particle, g is the gravitational constant, r is the radius of the sphere, and μ is the viscosity of the liquid. Since viscosity increases sharply with a drop in temperature, we can see that tiny particles will sink very slowly indeed through the cold lower regions of the oceans.

Few particles are actually spherical in shape. To take variation in form into account, Ostwald (1902) added another factor, **form resistance,** to the denominator of the equation. Thus it is possible to calculate the settling velocities of particles, such as shells, exoskeletons, and cases, formed by living organisms. Such particles may have complex arrangements of spines, needles, or "wings," and they often assume elaborate geometrical forms.

Pressure

Pressure increases tremendously at great depths. At sea level, normal air pressure is 1.034 kg/cm² (one atmosphere, or **1 atm**). Under water, pressure increases at a rate of 1 atm for each 10 m of depth; at the bottom of the Puerto

Rico Trench, at a depth of 9219 m, the pressure is 1070 kg/cm², or over 7 tons per square inch.

Water pressure affects different organisms in surprisingly diverse ways. Some of the more delicate plankton organisms, for instance, will burst or be crushed if there is even a slight change in ambient pressure, while a number of small organisms apparently thrive on the deepest ocean floors, protected by their lack of any compressible gas cavity. Blue whales (Scheffer, 1969) are known to dive as deep as 800 m with no apparent difficulty, a feat for which no explanation is yet available.

Osmotic pressure

Osmotic pressure is usually defined as the force that could develop when water diffuses through a semipermeable membrane from a region where it is pure to a region where it is relatively impure; the force will become apparent when the volumes of two bodies of water (such as in cells) changes. Water will flow through such a membrane from the **hypotonic** to the **hypertonic** solution. The forces developed by osmosis may be very great. Some marine organisms are very sensitive to osmotic pressure, and may die if the concentration of solutes in water is changed (as, for example, when seawater is diluted by rainwater). Others have little difficulty adapting from salt to fresh water. Most marine organisms are nearly **isotonic** to seawater—their internal osmotic pressure is nearly equal to that of seawater—and so they face little danger from osmotic pressure. Estuarine and inland aquatic organisms, on the other hand, require mechanisms to prevent severe damage from excessive water uptake, and most such organisms have evolved **osmoregulatory** systems that permit them to excrete excess water.

ECOLOGICAL VARIABLES OF INSHORE WATERS

The ecology of the inshore environment is affected by the interactions of water with the bottom (in shallow water), by rocks, beaches, and shoreline structures, and by the special conditions of semi-enclosed or protected marine waters. In some ways, then, inshore ecology is close to that of estuarine or inland waters. Inshore water is warmed by absorption of heat from sunlight; the bottom, too, is often somewhat warmed by sunlight. Thus, inshore environments are likely to be warmer than those farther offshore. In shallow regions there is also little or no loss of nutrients by sinking, since waves or currents transport organic materials back to the inshore water column. Thus, inshore waters tend to be significantly richer in nutrients than offshore waters. Very close inshore, the physical effects of tides and waves are more violent, as has been noted earlier. Further, as was mentioned in our discussion of estuarine basins, local species must be able to tolerate a constantly changing environment: the most important changes are variations in temperature, in sunlight, and in salinity (resulting from freshwater inflow and rains).

In inshore environments, **benthic** (bottom-dwelling) organisms contribute to the accumulation of metabolic products and selectively remove nutrients and organic compounds from the water; this biological activity can have large effects on the composition of the water. Stream runoff also brings nutrients and organic detritus—as well as soil and fresh water—from the terrestrial ecosystem. The proximity of estuarine spawning grounds augments the variety and abundance of animal species in inshore waters. As will be explained in more detail in later chapters, inshore waters are thus more productive (by as much as 1000 percent) than offshore waters.

It is apparent that typical marine conditions are not characteristic of inshore waters. In semi-enclosed waters this holds true with even greater force. In protected coves, for example, temperatures may be quite high, organic matter may accumulate in high concentrations, and microbial activity may become extreme, depleting the supply of dissolved oxygen and releasing products of decomposition into the water. In arid regions, osmotic pressure may be altered by rapid evaporation. Even more distinctive conditions occur in enclosed waters, such as lagoons or tidal pools, which are filled with seawater only periodically, and sometimes at long intervals. We are dealing here with an ecosystem somewhat beyond the scope of our text; nevertheless, the ecology of different types of tidal or spray pools, with their unique and specialized biota, is a subject of great interest.

TRENDS IN MARINE ECOLOGY

Marine ecology is a field of diverse and complex growth. Researchers are pushing ahead in many areas, including submarine geology, techniques for rapid assay of chemical constituents, and measurement and tracing of radioactive fallout. Such investigations, though hardly ecological in themselves, are often of significance to the living system.

Recent federal approval of the "Sea Grant" proposal has resulted in the establishment of several Sea Grant universities comparable to the long-established Land Grant colleges; this provides financial support for investigation and action in developing uses of the sea. A monthly report of on-going projects, *Sea Grant 70s*, gives an idea of the diversity of current research activities. Problems discussed in recent issues include the utilization of economically valuable by-products from shrimp processing, methods for growing oysters and hardshell clams in enclosed inland seawater systems; the development of cheap—but effective—break water material, the relationship of upwellings to the annual crab catch along the Pacific coast, the effect of toxaphene contamination on estuarine ecology, and a set of field guides for identification of various types of marine life.

Understandably, there has been great pressure to apply the findings of marine researches to practical problems. Since use of the seas and marine resources has long been of tremendous economic and military importance, such interest in applied science is inevitable. Perennial problems include damage to wooden structures (boat bottoms, pilings, traps, etc.) by the shipworm *Teredo navalis* and fouling of ship bottoms and underwater gear by growth of barnacles and other organisms. The switch away from wooden vessels has cut back shipping losses from shipworm damage, but there is still no satisfactory method of controlling shipworm populations. Antifouling measures have had somewhat greater success. Common methods of avoiding fouling damage are use of copper sheathing below waterline and use of antifouling paints. The latter work by slowly releasing heavy metals such as red lead, or other toxic substances, into the water; recent studies of the adverse ecological effects of these substances have led to restrictions on their use. Thus the search for effective but environmentally-safe methods of controlling plant and animal growth is still very active.

Marine ecology will undoubtedly be of practical value to the fisheries industries, which have recently experienced severe reductions in harvest—whales in particular are becoming scarce, anchovy populations have fluctuated severely off the coast of Peru, fish catches have declined along the coast of central California, and competition for diminishing numbers of cod has led to hostilities between British and Icelandic fishing fleets in the North Atlantic. In all these cases some interference with the marine ecosystem seems to be at least a major cause of the declining yield (H.T. Odum, 1969). Such damage is often the result of simple overharvesting, and can easily come about through the use of modern, high-technology equipment. In some cases, however, man's effect is more indirect, being caused, for example, by changes in the fish species' spawning grounds, or by damage to a species which serves as food for the fish. Damage to coastal waters from dumping of industrial and human sewage is often an important factor: Long Island Sound and the waters near New York Harbor, for example, are so heavily polluted that they have been described as "dead areas." Thus studies of inshore spawning grounds and of the food webs supporting commercial marine life are of primary importance. Such studies, together with measures regulating harvesting and other uses of marine resources, may point the way to ecological recovery in such damaged areas.

The first Law of the Sea Conference, held in Venezuela in 1974 under United Nations auspices, was an effort to establish a spirit of international cooperation in maritime affairs, and to delineate areas of jurisdiction over international waters. The conference was concerned not only with restricting military use of the ocean floor and with regulating development of mineral resources, but also with clarifying the responsibilities incurred by nations that engage in large-scale harvesting of marine life.

Applied marine biology is obviously of tremendous scope, ranging from an intense search among marine organisms for pharmaceutically valuable substances to the development of **mariculture**—the use and manipulation of marine waters and biota to "farm" seaweeds, shrimp, lobster, bivalves, and the like. Inevitably, though, concern has come to focus on the effects of polluting the marine environment. Surface slicks and flotsam, for example, have been reported to extend almost continuously across the Atlantic Ocean (Heyerdahl, 1971); offshore drilling for oil, and oil transport in modern supertankers, create serious threats to environmental quality in the form of oil spills. Particular difficulties are beginning to appear in connection with the increasing use of nuclear materials for military and industrial purposes. Nuclear waste disposal is one side of the problem: for some time, long-lived radioactive wastes have been sealed in concrete and stored at special sites along continental shelves; in some cases the concrete has deteriorated, raising the possibility that nuclear wastes might be released into the environment and absorbed into marine food webs. There is also some apprehension about possible accidents at the other end of the process. A leak at a nuclear power plant could lead to release of radioactive contaminants into the hydrologic cycle, with serious consequences for all aquatic life. One of the major functions of marine ecology is to provide a correct assessment of these and innumerable other problems.

SUGGESTED FOR FURTHER READING

Carson, R. L. 1961. *The sea around us,* revised edition. Oxford University Press, New York. First published in 1951, this book, written by the author of *Silent spring,* was a winner of the National Book Award and has long been regarded as a classic for its combination of scientific accuracy and extraordinary literary style. *The sea around us* has probably introduced more readers to oceanography than any other single book, and is strongly recommended to the general reader.

Fairbridge, R. W., editor. 1966. *Encyclopedia of oceanography.* Van Nostrand Reinhold, New York. A single-volume reference book, with articles by more than 130 contributors, covering every major aspect of oceanographic research. Although a great deal of technical information is given, many of the articles are introductory in style, and may be very useful for nonspecialists and undergraduate students.

Idyll, C. P., editor. 1972. *Exploring the ocean world,* second edition. Thomas Y. Crowell Company, New York. A wide-ranging collection of well-written and often entertaining articles, covering the major areas of marine research. Included are the biology, physics, and chemistry of the sea and uses of the sea for the development of food and mineral resources; a chapter on "Marine ecology and pollution" has been added to this edition. Intended for the nonscientist, and generously illustrated, including 32 pages of color plates.

Nybakken, J. W., editor. 1971. *Readings in marine ecology.* Harper and Row, New York. A collection of articles selected and edited by a chief scientist of Stanford University's research vessel *Te Vega.* The articles, from various publications, are well-chosen to reflect current research trends.

PART THREE
Environmental Variables of Natural Waters

7 | Solar Radiation and Natural Waters

In Chapter 1 we stated that four basic principles of ecology would be major themes throughout our study of aquatic ecosystems. These were the **fitness of the environment**; the **adaptation** of each species to the demands of its particular ecological niche; the principle of **limiting factors**; and **traffic in energy and matter.** Of these, the first and the last are directly affected by the intensity, duration, and distribution of solar energy. The reasons for this are evident. The capacity of natural water to support life—its fitness as an environment—is the product of many interacting physical, chemical, and biological processes, all of which are affected by heat; and heat, under natural conditions, is derived mainly from solar radiation. The pathways by which energy and matter are moved through the ecosystem take the form of intricate **food webs.** Such food webs are ultimately dependent upon the energy and materials captured by green plants in photosynthesis and later taken up, in the form of food, by all animal species; the source of photosynthetic energy is, again, solar radiation.

In a single year the amount of radiant energy that reaches the earth from the sun is enormous—about 1.3×10^{21} kcal. Some solar radiation is reflected back from the earth by the atmosphere, clouds, seas, lakes, mountains, and vegetation. In this chapter we shall be concerned with the fraction that does not escape. This fraction of solar energy is used by green plants in photosynthesis, and enters into the meteorological processes—such as wind, rain, and glacier formation by snow accumulation—that create new aquatic basins and channels and modify old ones.

The electromagnetic spectrum emitted by the sun extends from the extremely short gamma rays (with a wavelength of approximately 0.0001 nm) to the

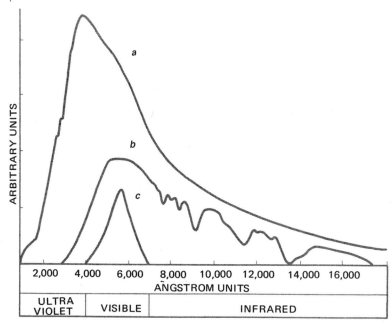

Figure 7.1 Distribution of energy of the solar spectrum above the earth's atmosphere (a), and at the earth's surface (b). The range of the spectrum to which the human eye is sensitive is shown at (c). (After Hutchinson, 1957.)

extremely long hertzian waves (which reach lengths of several km).[1] Much of this total radiation never reaches the surface of the earth. The direct solar radiation that does reach the surface is in the 13,500 to 286 nm wavelength range, and the most intense radiation is limited to the range of the spectrum between 300 and 1300 nm. The peak of radiation distribution is in the blue-green range, at about 550 nm. The wavelengths that are responsible for heating water fall in the range between 0.1 mm and 770 nm—the **infrared** spectrum. A relatively narrow part of the solar spectrum—between 770 and 400 nm—is visible to the human eye and forms what we call "light"; below the visible spectrum, from 400 to about 286 nm, lies the **ultraviolet** spectrum. Figure 7.1 indicates the approximate proportions—in wavelength—of the electromagnetic energy that reaches (a) the outer atmosphere and (b) the surface of the earth.

The earth also emits electromagnetic radiation; because the earth is a cool body, the emissions fall in the infrared (long-wave) range. These radiations are not completely lost back to outer space, because some of the energy (heat) is absorbed by moisture in the earth's atmosphere. At the same time the atmosphere

---

1. In describing electromagnetic wavelengths the **Ångstrom unit** (**Å,** equal to one ten-billionth of a meter) is often used, especially in physics. For work in biology and ecology a larger unit, the **nanometer** (**nm,** equal to one billionth of a meter or 10 Å) is usually more convenient. In older writings, still another unit, the **millimicron** (**mμ**) is often found; it is equal to one nanometer: 1 mμ = 10 Å = 1 nm.

differentially absorbs solar radiation, retaining some 15 percent of incoming energy. The net result is the retention of a large portion of the radiation emitted by the earth and the retention, and inward transmission, of about 85 percent of incoming solar radiation by atmospheric water vapor.

This phenomenon, whereby the earth's atmospheric water vapor transmits most of the radiant energy of sunlight while absorbing the radiant energy of the earth, has been called the "greenhouse effect." The principle of the botanical greenhouse is based on the fact that glass of the sides and roof permits the passage of solar energy, but inhibits the transmission of a great part of the "black body" energy given off from the plants and other structures inside the greenhouse. In other words, the contents of the greenhouse absorb solar energy which passes through the glass; the contents in turn radiate an energy of longer wave length which is not transmitted through the glass. This long-wave heat is therefore retained inside the greenhouse. Actually, greenhouse thermodynamics is more complicated than here indicated, but this simplified statement should point up the analogous roles of greenhouse glass and the layer of atmospheric water vapor in retaining heat.

Of interest to us in terms of light available for photosynthesis and for heat in maintaining the environment is the mean value of solar radiation falling upon the earth's surface. It has been calculated that an average of approximately 0.15 g-cal/cm²/min of radiation is incident upon the earth at sea level. This value is to be compared with the so-called **solar constant** of 1.92 g-cal/cm²/min reckoned as the radiation received on the outer surfaces of the earth's atmosphere. Actually, the light falling upon any area of the surface of the earth may be composed of direct sunlight and scattered light from the sky, or indirect solar radiation. Both of these are subject to considerable variation. The magnitude of direct solar radiation striking a given point depends upon season, geographical location with respect to latitude, time of day, the angular height of the sun and elevation of the point under observation, and the transmission quality of the atmosphere. The importance of latitude in the amount of direct radiation is shown in Figure 7.2. The quantity of indirect solar radiation received at a given

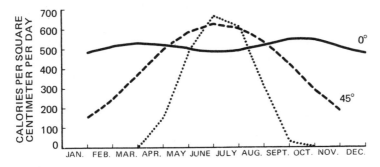

Figure 7.2   Total energy received from direct solar radiation on the fifteenth day of each month at sea level. Solid line indicates energy received at the equator; other curves are for north latitudes 45° and 90°. (Data from Hutchinson, 1957.)

point is quite variable, but generally constitutes about 20 per cent of the total radiation.

Not all of light received at the surface of a body of water enters the water. Some is reflected, the amount being a function of the angle at which the light strikes the surface of the water and the condition of the surface. From an undisturbed surface, the greater the angle of incidence from the perpendicular, the greater the reflection. When the light rays strike the surface at a very low angle, as much as 35 per cent of the light may be reflected. At high angles the reflectivity is on the order of 5 to 10 percent. We are concerned here only with the light that penetrates the water surface.

In Chapter 2 we noted that one of the conspicuous features of pure water is its high transparency. Pure water does not exist in nature, however; yet the penetration of light into water, and the fate of the spectral components within water, are important in ecosystem relationships, and therefore of primary importance to students of aquatic communities. The deviation of transparency of natural waters from pure, laboratory water ranges widely both quantitatively and qualitatively. The differences are determined primarily by dissolved substances, suspended materials, organisms, latitude, season, and the angle and intensity of the entering light. In order to more fully appreciate optical phenomena in, and optical qualities of, natural water, let us first consider some principles derived from laboratory studies of pure water.

Light entering and passing through a given column of pure water undergoes a reduction in total intensity and a change in spectral composition with depth. These processes of reduction and change result from scattering and differential absorption of light by water. The reduction in intensity, or **quenching,** of the light entering the surface is expressed in the following relationships based upon Lambert's Law:

$$I_d = I_0 e^{-kd} \text{ or } \frac{I_d}{I_0} = e^{-kd}$$

in which I_0 is the original intensity of the entering light, I_d the measurement of intensity incident at depth d, and e is the base of natural logarithms (taken as 2.7 in this instance); k becomes a constant for any particular wavelength, known as the **extinction coefficient**, or the proportion of the original light held back at depth d. The proportion of light passing through d is termed the **transmission coefficient.** These equations are based on the assumption of pure water and monochromatic light. Since neither of these actually occurs in nature, factors which further reduce transmission through the distance $I_0 - I_d$ must be applied for natural waters. Such factors would include, in addition to water, suspended materials and dissolved substances.

It has been demonstrated that in pure water at a depth of 70 m the original intensity of blue light is reduced some 70 percent, and the yellow component shows only 6 percent transmission beyond 70 m. The red component is not transmitted beyond 4 m, and orange is quenched at about 17 m. In general, approximately 53 percent of the total incident light is transformed into heat and undergoes extinction in the first meter of water, and approximately 50 percent of the remaining light is quenched with each additional meter of depth (Figure 7.3). The longer wave lengths (red and orange) and the shorter rays (ultraviolet and violet) are reduced more quickly than the middle-range wave lengths of blue, green, and yellow.

LIGHT IN LAKES

In natural waters, the blue segment is transmitted farther than all others. At depths beyond about 100 m blue light is the sole illumination. In highly colored or stained waters, orange and red are transmitted more deeply than other components, but all are rather quickly reduced. In many moderately transparent waters, the greatest transmission is in the yellow.

The most accurate and precise measurements of solar radiation in natural bodies of water are best made with photoelectric apparatus, employing various color filters. A photocell contained in a waterproof compartment is connected with a galvanometer and the cell lowered to desired depths. Direct readings of underwater radiation falling upon the cell are made from the galvanometer at the surface.

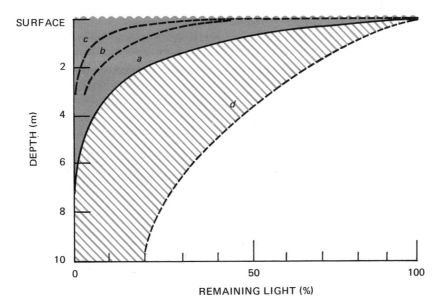

Figure 7.3   Curve of light penetration in natural waters. As depth increases, remaining light decreases toward zero, the asymptote. To see this as the usual mathematical graph, turn page toward the left, with 100% on top. Note that the shape of the curve conforms to the formula for exponential decay: $I_d = I_0 e^{-Jd}$ (see text). In a typical curve (a) light intensity is reduced by 50% in the first meter. In more turbid or colored waters (b and c) the light drops off more quickly. Curve (d) represents abnormally clear waters, as in the Sargasso Sea.

An American, Professor E. A. Birge (Figure 1.2), pioneered in the use of electrical equipment for measuring light in lakes; an instrument developed by Birge and physicists at the University of Wisconsin was first used in 1912. Called by Birge a **pyrlimnometer,** it measured the total radiation that impinged upon the surface of a thermopile, the thermal current produced being read from a galvanometer. In some ways, particularly with respect to uniform sensitivity and the expression of energy units, thermopile equipment remains unsurpassed. But it is costly, and thus the relatively inexpensive photocell has come to be used widely in limnological work. Welch (1948, 1952) explains in great detail this and other methods of light measurement.

About 1905, Birge was joined in Wisconsin by Chancey Juday and together their researches and reports gave impetus to limnology. Some of their data pertaining to four Wisconsin lakes are illustrated in Figure 7.4. The effects of several of the previously discussed factors which relate to transparency of natural waters and comparison with pure water are seen in studying this figure. Total solar radiation is shown as incoming sunlight by the circle with a shaded area at

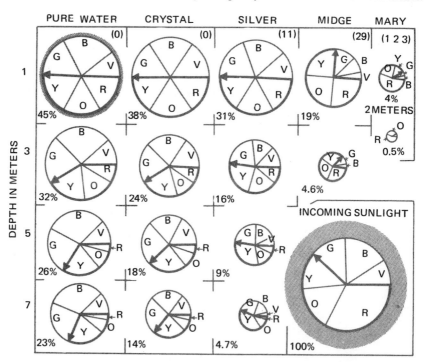

Figure 7.4   Graphic representation of solar radiation at a lake surface (inset, lower right), and transmission in pure water (left-hand column) and in four Wisconsin lakes. Data for incoming sunlight and pure water are approximate. Circle areas and percentages represent total light at indicated depths. The relative proportions of each of the spectral components are shown as circle sectors. Figures in parentheses refer to color as measured on platinum-cobalt scale of U.S. Geological Survey. The shaded area surrounding the graph of incoming sunlight and pure water at 1 m of depth indicate the invisible spectrum; note that it is completely absorbed in the first meter of lake water. (From Mortimer, in Sellery, 1956.)

the bottom right. The invisible part of the spectrum, mostly infrared, is represented by shading and includes about half of the entire area. In this, and in the other circles, the portion representing the visible spectrum is divided into sectors proportional to the quantity of each of the six colors measured by Birge and Juday. Beginning at the heavy, horizontal line in the "three o'clock" position, observe that the segments are labeled red, orange, yellow-green, blue, and violet in order of decreasing wavelength. The enclosure of the red, orange, and yellow segments in a heavy line and the arrow at "nine o'clock" will be referred to later.

The circles representing pure water and the four lake examples at given depths are drawn to scale proportionate to the total light present at the depths indicated and to the circle representing incoming sunlight. The unabsorbed light remaining at each depth is given as a percentage outside the circle. Variations in each of the six components of the spectrum are indicated by the area of the circle segments.

Immediately apparent is the rate at which transmission decreases with depth, and the variations among the lakes. Similarly, great differences in color changes are seen. It is interesting to note that the overall rate of light extinction and the proportion of the color segments for Crystal Lake are not too different from pure water. In both Crystal Lake and pure water, the yellow-orange-red segments decrease, and the green and blue increase with depth. Accompanying this is a swing of the arrow in a counterclockwise direction. The data indicate that red waves are absorbed rapidly in the first 3 or 4 m and that with increasing depth the light becomes greenish-blue. In the series for pure water the proportions represent the absorption of direct sunlight; the lake determinations include mixed light from sun and sky. Since mixed light contains proportionately more blue, this light would be transmitted more effectively than sunlight alone in pure water.

As the diagrams for pure water and Crystal Lake would indicate, these waters have no stain. The figures for the other lakes, Silver, Midge, and Mary, indicate increasing amounts of stain and color and a proportionate decrease in transparency and light transmission. There is also an increase in absorption and reduction of the green, blue, and violet components. In contrast to pure water and Crystal Lake, the arrow swings clockwise in these stained lakes, especially in Midge and Mary. The last two are bog lakes and colored dark brown by dissolved and suspended organic materials from the surrounding bogs. In these lakes, orange and brown are the conspicuous spectral components, but nearly all of the incoming light is absorbed in the upper 2 m.

SECCHI DISK TRANSPARENCY

A much older method of measuring transparency of water, and one that is still of considerable value, is the use of the **Secchi disk.** The method was devised by A. Secchi, an Italian, in 1865. In practice, a white disk, 20 cm in diameter, is lowered into the water by means of a line with measured intervals. The arithmetical mean of the distance at which the disk disappears from view in descent and that at which it reappears in ascent is given as the **Secchi disk transparency.** In using this device, care must be taken to standardize techniques and conditions (see Welch, 1948); surface waves are a common source of error. Reflected light from the bottom will also introduce an error, since the technique involves comparison of the brightness of the disk with the bottom brightness. It has been calculated that the disk disappears at approximately the region of transmission of 5 percent sunlight.

In very turbid lakes the Secchi disk transparency is quite low. In Lake Texoma, on the Texas-Oklahoma border, values on the order of a few centimeters sometimes occur (Sublette, 1955). Similarly, artificial farm ponds of the southeastern United States often give readings of less than a meter, particularly following fertilization, when blooms of plankton appear. Secchi transparency values near 18 m are characteristic of certain lakes in the Convict Creek Basin of California (Reimers, 1955). Generally, values above 30 m are unusual. Crater Lake, Oregon, however, has a transparency near 40 m.

COLOR AND TURBIDITY OF LAKES

Color and turbidity are here treated together because of their somewhat similar roles in giving various hues and other optical qualities to lakes; one distinction will be made, however. Both of these qualities determine light transmission in natural waters and consequently "regulate" biological processes within the bodies of water. To varying degrees, both give some qualitative indication of the productivity of the waters when simply viewed from above. The general chemical nature of lakes may often be deduced from the color and turbidity of the water.

Before proceeding further, we should define our terms in order to avoid confusion.

COLOR

The color that we perceive is made up of unabsorbed light rays remaining from the original entering light, but now passing out of the lake. Completely pure water should absorb all light components and appear nearly black. This is not seen in natural bodies of water, however, for lakes containing little suspended materials usually appear blue. The blue hue is probably the result of scattering of light by water molecules in motion, the effect being proportional to the fourth power of the wavelength of the light components (Hutchinson, 1957). Since the scattering of light of long wavelengths is less than that of short, the blue predominates.

Such hues as we normally see, however, may range from green-blue through blue-green, green, yellow, yellow-brown, to brown. Such vivid, poetic, and not necessarily imaginative impressions as the "land of sky-blue waters", Thoreau's "pure sea-green Walden water," and the "silver" of Seneca Lake suggest the rich variety of color interpretations possible, depending upon one's point of view.

In aquatic ecology the **true color** of natural waters, also referred to as **specific color,** is derived from substances held in the water in solution or in colloidal suspension. **Apparent color,** which may be quite different, is usually the result of interplay of light on suspended particulate materials together with such factors as bottom or sky reflection. In order to determine true color, a sample of the water should be filtered or centrifuged, thereby freeing the water of sources of apparent color.

In Figure 7.4 the numbers in parentheses near the names of the lakes refer to the color of the water in units of **platinum-cobalt scale** of the United States Geological Survey. The basic technique involves comparison of lake waters with a series of dilutions of a solution of potassium chloroplatinate (K_2PtCl_4) and crystalline cobaltous chloride ($CoCl_2 \cdot H_2O$). The units are called platinum-cobalt units, based upon 1 mg Pt/l as a standard, and range from zero in clear waters (Crystal Lake) to over 300 units in very dark waters of bog communities (Mary Lake: 123). The U.S. Geological Survey set of glass disks corresponding to the platinum-cobalt series is now rather widely used in place of the liquid dilutions (see Welch, 1948, for methods).

The major classes of dissolved and particulate matter found in typical lake waters may be stated as follows:

DISSOLVED SUBSTANCES
proteins and related compounds
fats and related compounds
carbohydrates and related compounds
break-down derivatives of all three of the above

PARTICULATE ORGANIC MATTER (**seston**):
living organisms (**plankton**): mostly microscopic forms
    phytoplankton: plant types
    zooplankton: animal types
nonliving particles (**tripton**): dead organisms, detritus, colloidal substances

The phytoplankton and zooplankton components will be described in detail in Chapters 14 and 15, respectively.

The average year-round percentage composition (by weight) of these categories is a "typical" lake is shown in Figure 7.5. For any particular lake the bar lengths would doubtless vary greatly; swamp or bog lakes would show a higher percentage of dissolved organic matter; alkaline lakes would be richer in dissolved salts. The important point here, however, is that all elements of these classes, individually or in concert, contribute to the color and turbidity of natural waters.

For the most part, lake colors are determined by the predominating components of the **seston,** the particulate mass of various living and nonliving substances in the water. Plankton algae are most frequently responsible for certain colors; an abundance of blue-green algae imparts a dark greenish hue; diatoms give a yellowish or yellow-brown color. Zooplankton, particularly certain of the microcrustaceans, may tint the water red. Humus often causes water to be green, or yellow-brown, the darkest brown coming from extractives of peat.

Suspended colloidal inorganic substances may account for certain tints. Calcium carbonate in lakes of limestone regions results in a greenish color. Volcanic lakes may be yellow-green from sulfur, or red from ferric hydroxide.

Generally, a rich, highly productive lake may appear yellow, gray-blue, or brown due to quantities of organic matter. Less productive lakes tend toward blue or green.

We have just seen that the color of lake waters differs widely, depending upon the nature and quantity of dissolved and suspended materials, the quality of the light, and other factors. It is worth noting also that color within a given lake may not be uniform from surface to bottom. Studies in Weber Lake, Wisconsin, revealed that in August the epilimnion was transparent and contained but little organic material; the hypolimnion, on the other hand, was highly colored. At that time, the surface water had a color rating of 12 units, the lower limits of the metalimnion (about 8 m) 81, and at the bottom (nearly 12 m) the color had increased to 93 Pt-Co units. A similar phenomenon has been observed in a New

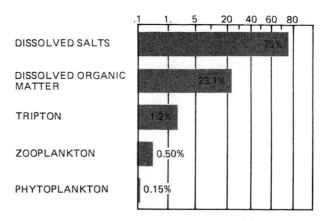

Figure 7.5   Year-round average percentage composition of sestonic components and dissolved organic and inorganic material in a "typical" Colorado lake; logarithmic plot. (Data from Pennak, 1955.)

Jersey stone quarry lake. It is suggested that the increase in color in the depths of such waters is caused by organic substances derived from bottom sediments and also by increased phytoplankton populations supported by such substances.

In a similar vein, the color of lakes may change periodically. Seasonal increases in surface runoff contribute great quantities of inorganic and organic substances which, as we have seen, impart various colors. Summer or early autumn production of phytoplankton "blooms" causes a lake to become a "soupy green," which disappears later in the season. Exposure to light causes the bleaching of certain colors in natural waters. This process also results in variations in the color of a given lake, such variations appearing seasonally with cyclic fluctuations in light intensity and angle of light incidence. The bleaching effect should occur vertically within the lake, depending upon transparency.

TURBIDITY

We have seen that color in natural waters is attributable in great part to suspended materials of varying particle size and composition. **Turbidity**, on the other hand, is the term used to describe the degree of opaqueness produced in water by suspended particulate matter. While the nature of the materials contributing to the turbidity is mainly responsible for the color quality, the concentration of the substances, if sufficiently high, determines the transparency of the water by limiting the light transmission within it.

The kinds of materials creating turbid conditions in a given body of water are as varied as the biotic and abiotic composition of the surrounding terrain, inflowing streams, and the lake itself. Substances such as various grades of humus, silt, organic detritus, colloidal matter, and plants and animals produced outside and brought into the lake, are termed **allochthonous**. Turbidity-creating matter produced within the lake is said to be **autochthonous**. Both contribute to the total quantity and quality of lake turbidity.

Although several methods for measuring turbidity have been devised, two are perhaps most widely used today. One technique involves the U. S. Geological Survey **turbidity rod**, which consists of a marked staff calibrated against known concentrations of a standard material (usually fuller's earth) with a length of platinum wire at one end and an eyepiece at the other. The end bearing the platinum wire is lowered into the water until the wire vanishes from view. The calibration mark at the level of the water surface gives the turbidity in parts per million (ppm). Devices such as the Jackson Turbidimeter and the Hellige Turbidimeter make use of light penetration through a sample of water. In the former apparatus, the disappearance of the flame of a standard candle when viewed vertically through a column of water is the measure (ppm) of turbidity. The Hellige instrument compares a vertical beam of light through a column with the Tyndall effect produced by lateral lighting from the same source, usually an electric bulb. An adjustable aperture regulates the light reaching the eye of the observer and the aperture size is calibrated for turbidity and determined in parts per million. For details of these instruments refer to Welch (1948).

Turbidity is not a uniform parameter even within a specific lake. Seasonal increases in stream discharge, for example, may introduce considerable amounts of silt and other sediments and materials, thereby altering the lake color and turbidity. With decrease in stream discharge, much of the allochthonous matter begins to settle in the lake basin. The rate of settling is not the same for all classes of materials, grading from sand (30 cm in 3 sec in pure, calm water) to colloidal particles (30 cm in 63 yr).

In natural waters the settling process is complicated by certain attributes of water itself and by the dynamics of water movement, some of which have already been considered (see Chapter 2). In accordance with Stokes' equation, the rate of settling (velocity of fall of a spherical body through a liquid) is dependent upon gravity acceleration, the radius of the body, the viscosity of the liquid, and the specific gravity of the body and the liquid. Recall that viscosity is related to temperature and that temperature is not necessarily uniform throughout a lake, this being especially important during summer stratification. Turbulence and other water movements contribute to the rate of settling in lakes, particularly during spring and fall overturns.

Some classes of matter that occur in lakes never settle under normal conditions. These include true colloidal systems of very fine particles of favorable specific gravity, and animal and plant forms capable of modifying their specific gravity and having other mechanisms (flotation devices, locomotion) that prevent settling. The particles in colloidal systems of silt are kept in suspension by the combined effects of turbulence and mutually repelling negative charges on the colloidal particles. It has been observed that the addition of acids to turbid ponds causes flocculation and precipitation of silt particles. These processes are brought about by neutralization of the negatively charged particles by the postively charged hydrogen ions of the acids. The mechanisms and dynamics of settling and the relationships of particle size to distribution by water movements in lakes are not well known—this in spite of the obvious importance of turbidity in such processes as energy circulation, transfer of nutrients, and sedimentation in lakes.

One more important role of turbidity in the aquatic ecosystem arises from the effect, previously mentioned, of suspended materials on transmission of light. This aspect has received considerable attention, especially as it relates to productivity and energy flow within the community. At this point we wish to consider only the effects of turbidity upon light; turbidity as a factor in productivity will be considered later.

The damping effect of suspended particles on the transmission of solar radiation, also termed "light-quenching," can be determined effectively from the use of the formula expressing Lambert-Beer's Law,

$$\frac{I_d}{I_o} = e - kdc$$

in which the concentration (mg/l) of light-quenching material c, and the thickness of the column in which light is quenched d, are related to the ratio of observed light I_z to incident light I_o to determine the partial extinction coefficient k of suspended particles in natural waters. (This k is not to be confused with that representing over-all coefficient in some formulas.) The parameter I_z may be measured with an underwater photometer. The value of c is the dry weight of suspended matter per unit volume of water after centrifuging. The advantage in the use of Lambert-Beer's expression over Lambert's Law lies in the recognition of variable c; in Lambert's Law this factor is included in the extinction coefficient k.

The Lambert-Beer expression has been used to interpret data on light quenching in western Lake Erie and other lakes. In the Lake Erie investigation, I_0/I_z was set at 100 (observed light equals 1 percent of surface light), and d represented the depth at the incidence of 1 percent of surface light. During the study the highest value for c was found to be 35.9 mg/l with a d value of 1.2 m; thus the extinction coefficient k became 0.107. The highest observed value for d was 7.6 and that for c was 5.4; k, at 0.112, was virtually unchanged. A significant datum was obtained in spring when it was found that the average depth associated with 1 percent of the surface light was 3.5 m; this degree of turbidity represented a load of suspended matter near 10 mg/l. The rapid absorption of light and the quenching effect resulting from turbidity factors are vividly demonstrated in these figures.

It might appear that color and turbidity act together to exert relative effects on light penetration. Such may not be the case in all instances, however. Studies in Atwood Lake, a reservoir in Ohio, have indicated that color and turbidity are very probably independent variables exhibiting no interaction with respect to transparency. It was further found that color may be the major factor affecting light penetration, except of course during periods of introduction of large amounts of silt during heavy rainfall.

COLOR AND TURBIDITY OF STREAMS

COLOR

Color in stream waters is determined by the same physical laws—light transmission, differential absorption by substances in the water, reflection, and back-radiation—that operate in other natural waters. However, as a result of several factors streams generally fail to exhibit the great variety of color that may be seen in lakes. The upper reaches of most streams are characterized by clear waters, at least during the nonflood season. In these regions the streams lack a true plankton which, as we have seen, may be responsible for certain apparent colors in lakes. Moreover, relatively "pure" shallow streams appear clear partly because light is absorbed quite rapidly in the first meter or so.

In many areas, particularly in the southeastern United States, small, shallow streams which drain "flatwoods" and swamps are often colored a light to dark amber (about 160 on the U.S. Geological Survey Scale). This color is probably attributable to dissolved plant substances such as tannin; the streams are typically acid. Similarly, certain small creeks in southern Indiana become "inky black" in the autumn due to extractives from large accumulations of leaves in the water. Various extrinsic factors may also account for apparent color in streams. Rich growths of diatoms on rocks on the stream bed may lend to the water a brownish hue; algae may impart a greenish tint, and sulfur bacteria a yellow color. Certainly pollution, from a variety of sources, can contribute to both true and apparent color in streams and should not be disregarded in field investigations.

TURBIDITY

In the lower stream courses (and in spring in upper stream courses) turbidity becomes a dominant and characteristic feature of most running waters. Depending upon the chemical nature of the material in suspension and the particle size, colors may range from near white through red and brown. In streams where turbidity is not excessive, a plankton may develop and lend to the stream a greenish color.

In the monumental report of his investigations of the plankton of the Illinois River from 1894 to 1899, C. A. Kofoid drew a clear distinction between lake and stream turbidity.

> In this matter of silt and turbidity the river as a unit of environment stands in sharp contrast to the lake. Deposition of solids and clear water are normal to the environment of the lake, while solids in suspension and marked turbidity are the rule with river waters. Owing to their varied occurrence these elements, silt and turbidity, also add to the instability of fluviatile, as contrasted with lacustrine, conditions.

As was indicated in Chapter 4, most streams flowing near base level carry considerable loads of silt and other fine particles. Such high turbidity states result in a decrease in phytoplankton due to rapid quenching of light. Throughout most of the lower Mississippi, light at depths of 200 to 400 mm is only one millionth of that entering at the surface. In many of the larger rivers of the interior of the United States, the turbidity frequently exceeds 3000 ppm. Such concentrations of particulate matter often serve to absorb heat, thereby raising the temperature of the water.

COLOR AND TURBIDITY OF ESTUARIES

COLOR

The causes and chemistry of colors in marine waters are not thoroughly known. It is likely, however, that dissolved substances impart certain hues or enter into color shifts. For example, it is known that water-soluble yellow pigments are common in coastal areas and may contribute to the various shades of green in offshore waters. The blue of the sea results, as in inland waters, from the scattering of light by water molecules. Suspended detritus and living organisms give colors ranging from brown through red and green. The Red Sea derives its name from a brownish color due to great numbers of particular algae.

Coastal estuaries generally lack the brilliant colors of the open seas; in some estuaries of moderate tide and current action, blue and green hues are noted during the non-flooding periods. Most estuaries, however, are characterized by dark colors resulting from typically high turbidity. On occasion various planktonic forms become so numerous that they give a reddish or greenish tint to the water. Dissolved materials such as tannic acid delivered by the inflowing stream of the estuary or from local decomposition of organic substances cause estuaries to be light brown (Emery and Stevenson, 1957).

TURBIDITY

The major component of turbidity in estuaries is, of course, silt. The volume of silt transported into estuaries by streams fluctuates seasonally, with the maximum discharge taking place during the wet season. Some materials may be brought in from the sea, but these are usually minimal. In addition to allochthonous particles such as silt, much of the material that contributes to turbidity originates from erosion within the estuary itself; the proportion of autochthonous substances appears to be dependent upon the shape of the basin and prevailing currents.

Throughout the year the amount of materials in suspension in an estuary decreases from the upper reaches to the mouth of the basin; that is, transparency increases downstream. This is due to diminution in velocity and carrying capacity of the inflowing stream current, and to the electrolytic effect of seawater salts. The latter process involves the coagulation of negatively charged particles of colloidal silt by positive ions of certain metals present in seawater.

Even though there is typically a decrease in turbidity seaward along the axis of an estuary, its waters are decidedly more turbid than the sea. Evidence of this is seen in large areas of highly discolored water lying opposite the mouths of estuaries, contrasting vividly with the clearer sea water. Such masses of estuarine waters often extend many miles seaward or, depending upon currents, for great distances along the coast. The major effects of high turbidity in estuaries are, first, the quenching of light penetration, thereby inhibiting photosynthesis and the production of plants, and, second, the building of deep zones of mud, silt, other sediments, and detritus. In many estuaries, notably during periods of considerable stream inflow, light is reduced to 1 percent of surface radiation at depths of less than 3 m. In certain regions of the open sea, in contrast, the yellow-green light components are not diminished to 1 percent until a depth of nearly 100 m is reached. Transparency decreases shoreward such that in coastal waters the 1 percent illumination level is generally reached at a depth of 15 to 30 m.

COLOR AND TURBIDITY OF MARINE WATERS

COLOR

The water of the open sea appears to be a deep blue. At times it may appear intense green, or brown or red-brown in coastal waters. The deep blue is believed to result from light-scattering by water molecules or by minute suspended particles. The brownish and greenish colors are caused by the yellow materials released from living things in coastal regions; the brown is caused by sediment particles. Red or brownish water generally results from blooms of such plankton as blue-green algae, dinoflagellates, and diatoms, and brown-green color over banks results from diatom blooms. The violet and green colors of deep marine waters represents the fraction of the solar spectrum that remains after all other colors have been filtered out, rather than any pigmentation of seawater itself; this can be shown by flash pictures taken with balanced light at great depth: in such pictures there is no tint due to water coloration.

A different phenomenon occurs in some shallow tropical waters where, as in the Bahamas, the color is remarkably bright and varied because of the effects of bottom color and reflectance. The skilled islander can use color to judge water depth and to identify dangerous underwater reefs—the water is deep mauve in the open ocean, light blue over the edges of banks, light green at depths of 8 to 10 m, and varies in other ways according to the nature of the bottom.

TURBIDITY

Moreso than in the case of freshwater, the clarity of seawater is usually expressed in terms of extinction coefficient (page 134). Since the extinction coefficient is the slope of the curve of light absorption, the smaller the coefficient, the less the absorption and the greater the clarity. Typical figures are 0.02 for pure water, 0.1 for seawater, and 0.4 for coastal water. In terms of Secchi disk transparency, these figures mean that a disk 20 cm in diameter would be seen to a depth of 50 m in pure water, to 10 m in marine waters, and to 2.5 m in coastal waters.

Under certain conditions seawater can be extremely clear. The water of the Sargasso Sea is clearer than the "crystal" waters of Crater Lake and Lake Tahoe. The clarity of the Sargasso Sea is a result of its purity. It is assumed that in the absence of surface currents (and consequently of nutrient inflow) all available organic matter and biogenic salts have been depleted by the regional biota, leaving water of high purity and clarity.

APPLICATIONS OF SOLAR RADIATION STUDIES

The penetration of light into water is a key factor in aquatic ecology, since the availability of light makes photosynthesis possible, and photosynthesis is the basis of all food webs. The warming effect of sunlight (which will be explored in the following chapter) is also of great importance.

Certain straightforward uses of the effects of turbidity have long been standard practice in some areas. Reduction of light by increasing turbidity is a common agricultural procedure in Asia, where it is used to reduce the growth of algae in irrigation ditches; in India, nigrosine, the dye used in India ink, is often used for this purpose. In areas where light is so intense as to limit algal growth, turbidity may be similarly increased in order to increase production. In aquaculture projects, light control may augment harvests or control the growth of epiphytes that interfere with the crop; such procedures are being used in the lagoons of the Philippine Islands, where the crop plants in mariculture facilities may be hung some 30 cm below the surface.

Considerable effort is now being made to determine the nature of the dissolved solutes and complexes that occur in natural waters, to determine their sources, and to understand their effects; the consequences of variations in the quality and intensity of light are not yet well understood at the ecological level, and problems in this area constitute a challenging field of research.

SUGGESTED FOR FURTHER READING

Bainbridge, R., G. C. Evans, and O. Rackham, editors. 1966. *Light as an ecological factor*. Blackwell Scientific Publications, Oxford. Twenty-two papers from the 1965 Easter Symposium of the British Ecological Society. Ten of the papers deal with one or another aspect of measuring sunlight or assessing its role in aquatic ecosystems. Most of these deal with marine or estuarine environments, although one, ''The light climate for plants in rivers,'' deals exclusively with the lotic habitat. A very useful set of readings.

Thermal Relations in Natural Waters

8

In Chapter 2, attention was drawn to several unusual, and ecologically important, physical qualities of water. In particular, we noted the high specific heat of water, and its behavior under changing temperatures. From the point of view of ecology, the thermal properties of liquid water are the single most important factors in determining the fitness of water as a natural environment and in regulating the relations among aquatic organisms. The high specific heat of water results in an extremely high capacity for heat absorption, which in turn has important consequences for aquatic species. It should also be recalled from Chapter 2 that pure water has the unique attribute of reaching its maximum density at 3.98°C rather than at its freezing point, 0.0°C. These two characteristics largely determine the physical effects of thermal energy in natural waters, and we will refer to them repeatedly in our discussion of thermal relations in inland and offshore waters.

Chapter 7 indicated that a broad spectrum of solar radiation reaches the surface of the earth. Our attention in that chapter was directed, for the most part, toward the components of sunlight that are important in photosynthesis. Mention was made, however, of **infrared** radiation—that portion of the electromagnetic spectrum with a wavelength greater than 700 nm, and which is sensed by living things as **heat**. We know from Lambert's Law (page 143) that light absorption by water increases exponentially with the length of the light path: in a column of water 1 m in length, 91 percent of light with a wavelength of 820 nm is absorbed and only 9 percent transmitted; at a depth of 2 m the amount absorbed increases to 99 percent. More than 50 percent of all solar radiation is genrally absorbed within this region, the upper one or two meters below the surface of a body of water.

It is this light with which we shall be concerned in this chapter, for it the absorption of this light energy that results in the heating of natural waters. Our primary concern will be with **temperature,** considered as an **intensity factor** of heat energy. Another ecologically important factor of heat energy, the **capacity factor,** has already been considered briefly (page 20) in our discussion of specific heat, and will be referred to again as we discuss the absorption and release of calories from natural waters.

TEMPERATURE AND HEAT IN LAKES

One of the most striking thermal phenomena of lakes, and one of the most important in terms of ecology, is the relationship between water and temperature as observed in seasonal variations. In many lakes, these variations take the form of pronounced changes in the overall thermal structure and dynamics. During winter, the temperature of the water in moderately deep to deep lakes is relatively uniform from surface to bottom; or, if ice forms, this colder layer floats on the underlying waters. In spring, circulation and mixing of water results, typically, in a uniform temperature from surface to bottom. During summer, the vertical distribution of temperature may come to resemble that shown in Figure 8.1. In this case, the lake is essentially stratified. From the surface to about 15 m, temperature changes little with depth. Between 15 and 30 m, the temperature drops rapidly. The region of the lake below approximately 30 m is rather uniformly cool. In the fall, the summer condition is broken up by circulation and mixing similar to that of spring, resulting once more in uniform temperature of the water. How do these conditions come about? The answers are found in light absorption, heat dynamics, density phenomena, and wind action. These are best appreciated by considering in more detail the **annual temperature cycle** of a lake.

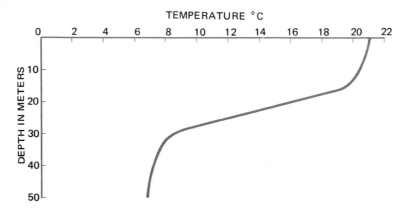

Figure 8.1   Summer temperature conditions in a typical (hypothetical) temperate-region lake.

SEASONS IN LAKES

In winter, as shown for March at the extreme left of Figure 8.2, ice at near 0°C covers the surface of the lake. Recall that ice floats because it has a density less than that of water. Below the ice the temperature to the bottom is relatively uniform and, as suggested by the temperature of 1.5°C at about 15 m, gives evidence of being rather thoroughly mixed—the mixing having occurred in the fall before the ice formed. The density is nearly uniform below the ice. Because of the low angle of the winter sun and the shading of the water by the snow and ice cover, photosynthesis is inhibited. This, together with respiration of organisms and lack of oxygen replenishment from the atmosphere, may result in low oxygen content of the water. In some cases, it has been noted that the slowing of respiration in aquatic organisms (caused by low temperatures) prevents much oxygen depletion. More commonly, however, by late March oxygen is nearly depleted, especially toward the bottom.

In the spring, a higher sun and increased day length brings about the melting of the ice. As the surface waters warm up to 4°C and become more dense, a slight, and temporary, stratification exists which sets up convection currents.

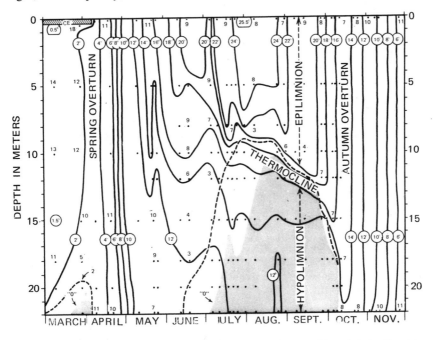

Figure 8.2   The seasonal cycle of temperature and oxygen conditions in Lake Mendota, Wisconsin, during 1906; studies of this lake by E. A. Birge (page 5) became the basis of limnology in America. Values in italics are for dissolved oxygen (ppm); shaded areas represent the seasonal occurrence of near-anaerobic conditions (0.2 ppm O_2 or less). The broken line represents the O_2 value of 2 ppm. (After C. H. Mortimer, 1956)

These currents, aided by wind, serve to mix lake water throughout until it is uniformly at 4°C and at its maximum density. Note from the vertical isotherms in Figure 8.2 that in Lake Mendota the mixing continued into early May even while the water was warming to 10°C. It is apparent that the amount of dissolved oxygen had increased throughout the lake. This vernal process of mixing has been termed the **spring overturn,** or **spring circulation period.**

Now we have seen that heat is absorbed very rapidly in the surface waters; therefore solar radiation alone cannot be responsible for heating the deeper waters. Some other agent must contribute to the circulation of warm waters into the deeper regions. In lakes, heat transfer is accomplished mainly by winds. It is only by this force that lighter, warmer water can be made to circulate with the denser underlying water at 4°C. Winds blowing across the surface of water set up a current resulting from frictional differences between moving air and water. Upon reaching the shore this current moves downward and across the bottom, thereby setting up the spring circulation, and ultimately resulting in a thorough mixing of the lake waters.

As summer approaches, the weather warms, the longer days mean longer periods of insolation, and the brisk spring winds subside. Under these conditions the surface waters warm rapidly, expand, and become lighter than the lower waters. Although the wind may continue to blow, its contribution to mixing of lake waters diminishes, for now the thermal density gradient opposes the energy of the wind. With the progression of the summer season the resistance to mixing between two layers of different density (resulting from increased temperatures) becomes greater than the force of winds—the significant density differences having been built up during periods of summer calm. A condition called **thermal stratification** is now evident and the temperature curve resembles that in Figure 8.1.

Figure 8.2 indicates that Lake Mendota became thermally stratified during June and July. In late July, the uppermost region of warm, homothermal water extended downward some 8 m. This region is termed the **epilimnion.** Below the epilmnion there developed a zone of rapid drop in temperature with depth. This zone, only 2 to 3 m thick, is the **thermocline.** In ecological terminology the thermocline is defined as "the plane of maximum rate of decrease in temperature." The **zone** of rapid drop in temperature, including a gradient on either side of the thermocline, is termed the **metalimnion.** Observe in Figure 8.2 that the temperature in this zone decreases as much as six degrees. The zone below the metalimnion is designated the **hypolimnion.** In many lakes the hypolimnion becomes devoid of dissolved oxygen and high in carbon dioxide during a portion of the summer.

The condition of thermal stratification is very stable. The metalimnion constitutes an effective barrier between the epilimnion and the hypolimnion. Currents stimulated by wind, and convection currents derived from cooling at the surface, move freely within the epilimnion but are limited in depth by the metalimnion. This means that heat and nutrients present in the region of maximum light are prevented from mixing throughout the body of the lake. Similarly, hypolimnetic water substances and many organisms of little or no mobility are restricted to the region. For photosynthetic organisms this state of affairs is particularly critical. Depending upon the depth of the metalimnion, of course, photosynthesis is essentially inhibited in the hypolimnion. This means that oxygen available for animal respiration is reduced, carbon dioxide is increased, and that microscopic algae (phytoplankton) as food for hervivorous animals is scarce. Indeed, it has been found in many lakes, Fayetteville Green Lake, New York, for example that minute animals (zooplankton) eat great quantities of purple bacteria which are abundant in the deep, oxygenless waters.

The impedance to overall lake current flow presented by the metalimnion has been called the **thermal resistance to mixing.** What, we might ask, is the character of this resistance, and how is such a barrier maintained? The answer to the first question is found in the density-temperature relationships mentioned briefly a few paragraphs back. In Figure 8.1 we see that between depths of about 15 m and 30 m the temperature declines from near 20°C to approximately 8°C, or about 12°C in 15 m. Recall that the density of water is dependent upon temperature, and that the difference in density per degree change is fairly great. For example, as water warms from 4° to 5°C the density changes 8×10^{-6}g mass/ml. Therefore within the rapid and wide temperature change in the metalimnion a considerable range of density also exists. Consider also the density difference between the temperature at the epilmnion-metalimnion inter-face and that at the hypolimnion-metalimnion interface in Figure 8.1. This difference represents a strong value of 0.00164 gm mass/ml. Where temperature-depth data are available, a graph of the **relative thermal resistance** can be drawn. This parameter represents the ratio of the density difference between water at the upper and lower faces of each stratum to the difference between water at 5° and 4°C.

We might conclude, then, that resistance to mixing presented by the metalimnion region is primarly the result of temperature decrease with depth and the related density differences within the zone. Now, what perpetuates the metalimnion and its resistance to mixing? We have already learned that the epilimnion is a region of relatively free circulation and considerable turbulence, and further, that the metalimnion acts to limit the downward extent of water motion. As wind-blown surface currents move against the shore of the lake they are deflected downward and encounter the metalimnion. The metalimnion barrier in turn causes the current mass to spread out horizontally at the upper zone of the metalimnion. Each increment of warming of surface waters in the summer adds a proportionate measure of relative thermal resistance between the epilimnetic and metalimnetic waters. Thus the barrier effect is somewhat self-perpetuating in the sense that density differences between the two regions are increased as the continually warming waters circulate toward the cooler metalimnion.

Throughout the great range of lake morphology, climate, and topography, the various attributes of the thermocline are subject to considerable variation.

The thickness of the metalimnion may fluctuate with season, becoming thinner as summer progresses. This is shown rather clearly in Figure 8.2, but in other lakes the effect may be more pronounced, the thickness ranging from about 12 m in early summer to around 5 m in late summer; a very thin metalimnion of some 2 to 5 mm has been reported in Barber's Pond, Rhode Island (Wood, 1957). Across this layer, visible to underwater observers, the temperature decreased almost 8°C. The depth of a metalimnion is determined by a number of influences such as length and time of seasons, and by basin morphology. Within a particular lake the metalimnion usually becomes deeper in late summer (Figure 8.3a). Local variations in metalimnion level in response to wind action and to the temperature of incoming waters may be expected. One additional point should be considered here; smooth temperature curves are not always obtained. Typically a late-developing stratification results in the formation of a thick epilimnion and more than one metalimnion. In Figure 8.3c, note that two such regions are present. More will be said about the metalimnion subsequently with respect to some general aspects of stratification.

With the approach of autumn, the angle of incident light decreases, day length decreases, and cooling begins. This is to say that the lake loses heat faster than it is absorbed. As the cooling extends toward the deeper regions of the lake, the density differences between isothermal strata become less. Thus the thermal resistance to mixing is weakened. Wind-generated currents carrying cooler, oxygenated waters reach ever more deeply into the lake. The metalimnion sinks rapidly. As shown in Figure 8.2, overturn during the period of **autumnal circulation**—or **fall overturn**—began in Lake Mendota in early October and continued into December. Observe in the figure the uniform, and relatively high, oxygen content from surface to bottom accompanied by a temperature difference of less than 1°. The period required for autumnal circulation is dependent upon local climate and the depth and morphology of the lake. The deeper the lake, the greater the time required for cooling to become uniform. Two points are important here. First, cooling (like heating) takes place only at the surface; second, cooling to 4°C is accomplished primarily by convection currents.

An **inverse temperature stratification,** not depicted in our figure, may exist during winter as the water cools below 4°C. Remembering that water reaches its most dense state at 4°C, we can easily appreciate that as surface waters cool beyond that point another "density difference" occurs, in which the lighter, cooler waters float on more dense, "warmer" waters. Although the winter temperature conditions of stratification are reversed with respect to summer, the density relationships are similar—the less dense mass occupies the upper part of the lake. The inverse stratification is most pronounced following the formation of the ice cover. Immediately below the ice (0°C) the temperature of the water rises sharply to near 4°C, or whatever the temperature of the water body. Inverse stratification is usually of rather short duration and, indeed, may not occur every year in a given lake. The result is a period of **winter stagnation**.

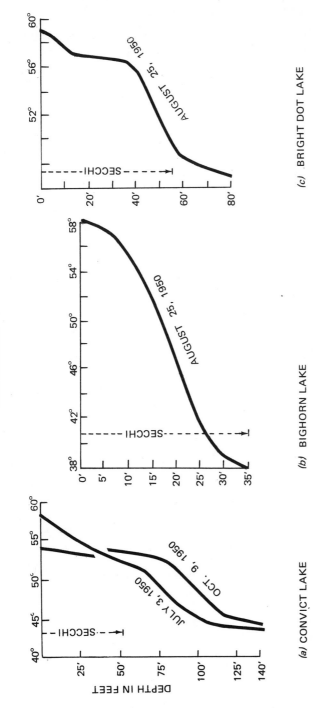

TEMPERATURE °F

(a) CONVICT LAKE

(b) BIGHORN LAKE

(c) BRIGHT DOT LAKE

Figure 8.3 Temperature profiles for three lakes of various depths in the Convict Creek basin of the Sierra Nevada in California. Secchi disk transparencies are also indicated. (After Reimers *et al.*, 1955.)

The sources and processes involved in the heating of lake waters beneath ice have received considerable attention. It appears, however, that heat from direct solar radiation through the ice, and heat derived from bottom muds are the major sources of energy for warming the lake. The heat-distributing mechanisms are apparently density currents of two types: the **temperature-dependent** currents, which are derived from heating of water through the ice in the shallow zones and which flow into the deeper parts of the lake; and **chemical** density currents derived from dissolution of bottom substances in water—more dense by virtue of the dissolved materials, these currents may also transport heat.

THERMAL CLASSIFICATION OF LAKES

As a result of extensive studies of temperature phenomena, heat, and stratification, since the time of Forel, a great store of terminology and schemes of classification of lakes based on thermal characteristics has accumulated. In an attempt to keep terms and classifications to a minimum we have adopted a scheme proposed by Hutchinson (1957). By no means universally applicable, by virtue of the variable nature of lakes, Hutchinson's system is nevertheless useful, and it points to the more modern concepts and approaches to thermal properties and dynamics in lakes. Earlier terms and classifications, certainly not useless, are to be found in older texts and other works on ecology and limnology.

The following types are proposed for lakes occupying basins of sufficient depth to allow for stratification mixing ("mixis"), and the formation of a hypolimnion. The classification takes into account altitude, geographical location with respect to latitude, and depth of the basin:

Amictic lakes Lakes insulated and protected by ice cover from outside influences of weather and other factors; only a few examples, all poorly known lakes of the Antarctic and high altitudes.

Cold monomictic lakes Lakes of the polar regions in which the waters at any depth never exceed a temperature of 4°C; ice-covered and exhibiting an inverse temperature stratification in winter; one mixing ("monomixis") at temperatures not greater than 4°C in summer. Also called **polar lakes. (Subpolar** lakes are similar, but may have midsummer water temperatures exceeding 4°C.)

Dimictic lakes Lakes (also called **temperate lakes**) in which two circulations take place each year, in spring and autumn; thermal stratification is inverse in winter and direct in summer; typical of lakes in the temperate zone and in higher altitudes in subtropical regions. Three orders of dimictic, or temperate, lakes are recognized. First-order lakes have bottom-water temperatures of about 4°C in summer, with little or no circulation. Second-order lakes are stratified and have bottom temperatures well above 4°C in summer. Third-order dimictic lakes are unstratified and their circulation is continuous.

Warm monomictic lakes Lakes of the warmer latitudes in which the temperature of the water never falls below 4°C at any depth; also called

subtropical lakes. One circulation each year in winter, directly stratified during the summer. Lake Providence, Louisiana, stratifies from May through September, and circulates continuously from October to April.

Oligomictic lakes Warm lakes in which the water temperature is considerably higher than 4°C; circulation occurs rarely at irregular periods; also called **tropical lakes,** they are typically found at low elevations in tropical zones.

Polymictic lakes Lakes in which mixing is continuous but occurs only at low temperatures, usually just over 4°C; characteristic of high mountains in the equatorial regions. Stratification does not develop because of heat loss to a relatively uniform environmental temperature.

HOLOMIXIS AND MEROMIXIS

With respect to circulation, or mixing, generally, most lakes are said to be **holomictic** (''wholly mixing''); that is, if circulation takes place it is complete and extends the entire depth of the lake. In these lakes the temperature of the hypolimnion usually decreases uniformly to the bottom. However, some lakes are known in which the summer temperature curve resembles that of a holomictic lake except that temperature increases slightly through a considerable distance. How is it that warm water is lying underneath cold water? What of the density relationships? This bottom water is not less dense than the upper water. It has been found that the bottom stratum usually contains large quantities of dissolved salts, and these increase the density sufficiently to prevent mixing with the upper layers. We have, then, a lake in which circulation is not complete; such lakes are called **meromictic** (''partly circulating''). The bottom, noncirculating layer is called the **monimolimnion**, and the density gradient becomes a **chemocline**. The waters above the chemocline in which thermal stratification such as in holomictic

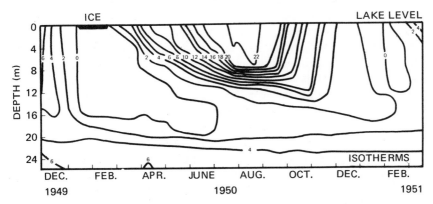

Figure 8.4 Seasonal temperature variations in a meromictic lake (Soap Lake, Washington). The period of complete ice cover is indicated by the shaded bar. (After Anderson, 1958.)

lakes can occur is termed the **mixolimnion**. A lake in western Austria (Längsee) possesses a monimolimnion that has not mixed with the upper waters for over 2000 years. The monimolimnion contains a heavy concentration of salts derived from the sediments by biochemical processes. Heat in the bottom layer is apparently gained from bacterial activity and from insolation. Since there is no circulation by currents, heat is lost mainly through conduction. The annual temperature cycle of a meromictic lake is shown in Figure 8.4.

HEAT BUDGETS AND LAKE STABILITY

Having considered the annual temperature cycle in a typical lake of the temperate zone, as well as some unusual cases of stratification, let us now give attention to some of the broader aspects of heat and heating. As stated earlier, temperature is the expression of the **intensity** phase of heat energy, while heat, as such is the **capacity** factor.

For purposes of comparison of lake qualities and expression of the nature of the lake waters as environment with respect to organisms, the calculation of heat income and heat budgets becomes a useful tool. **Summer heat income** is the quantity of heat delivered to a lake necessary to warm the waters from the homothermal spring condition of 4°C to the summer maximum. Although direct solar radiation contributes essentially all of this heat, the distribution of it is accomplished by wind-driven mixing of waters of differing densities. The work of the wind in summer heat income is done against gravity because of density differences arising from heating of surface layers by direct insolation. The **annual heat budget** takes into account the total quantity of heat taken into the lake to warm the waters from the lowest temperature of winter to the maximum summer temperature. Either parameter can be determined in absolute terms such as calories (**cal**) or gram-calories (**gcal**).[1] Usually, however, we wish to compare heat characteristics of several lakes, in which case it is well to introduce area. The result is then expressed in calories per square centimeter of water surface. The heat content per unit surface area of a lake at any given time can be determined by totaling the product of temperature change times volume for declared depth intervals, and then dividing by the surface area of the lake. The gain in heat content per unit area beyond the homothermous 4°C state is the summer heat income, and is more a function of mean depth than of temperature differences. The annual heat budget may be derived in the same way, except that the lowest winter temperature is substituted for 4°C. Table 8.1 gives morphological and thermal data for five lakes in the Convict Creek Basin of California, and shows the relationship between depth and annual heat budget. Comparisons of annual heat budgets of lakes must consider depth differences. Shallow lakes are

---

1. The gram-calorie (gcal) is defined as the amount of heat needed to raise the temperature of one gram of water from 15° to 16°C.

TABLE 8.1 Five lakes in the Convict Creek basin, California; physical features, water temperature, and heat dynamics[a]

Lake	Elevation	Surface area (ha)	Mean depth (m)	Transparency (m)	TEMPERATURE, °C Bottom to surface	TEMPERATURE, °C Lake mean	HEAT INTAKE, GM CAL/CM² Summer heat income	HEAT INTAKE, GM CAL/CM² Annual heat budget
Convict	2275	68.6	26.4	15.3	6.6-14.7	10.9	18,173	22,984
Mildred	2970	4.3	8.1	12.0	10.0-12.5	11.1	5,822	7,298
Witsanapah	3228	1.8	8.4	9.0	5.5-10.0	8.5	3,827	5,358
Edith	3030	7.3	17.7	13.8	4.1-14.1	8.7	8,601	11,826
Dorothy	3102	43.8	40.8	20.1	4.4-14.7	8.8	17,967	25,402

[a] Data from Reimers et al., 1955.

not able to store all the heat transmitted into them. During the period of maximum temperature, the lakes become vertically homothermous and heat is transferred and lost to bottom sediments.

Within temperate regions the annual heat budgets of typical lakes range from about 30,000 to 40,000 cal/cm². Much of this heat (33 to 75 percent) enters the lake after the spring circulation period, and work energy is involved in the circulation of the heat to lower lake levels. Lake Michigan, with an annual heat budget of 52,400 cal/cm², and Lake Baikal, whose annual budget amounts to 65,500 cal/cm², are the highest known. The summer heat income for Lake Michigan is 40,800 cal/cm², and Lake Baikal receives 42,300 cal/cm² between spring isothermal 4°C and summer maximum. To bring about the thermal conditions shown in Figure 8.2 for Lake Mendota an annual budget of about 23,500 cal/cm² and a summer income of some 18,240 cal/cm² are required.

In view of the importance assigned to the work of wind in distributing heat in a lake, we should question how much work a wind of a given velocity can do and how much work is required to mix a given quantity of heat. As Hutchinson has pointed out, little work has been done on the problem, and our knowledge is therefore meager. However, some data from various sources as compiled by Hutchinson will serve to convey a general idea. With respect to the first question, Langmuir (1938) has stated that winds of velocity 300 to 700 cm/sec developed forces of 0.65 to 6.3 dynes/cm² on lake surfaces. The summer heat income of Green Lake, Wisconsin, is 27,316 cal/cm² and the heating period is 122 days. Calculations show that the mean rate of work directed toward heating the waters is 0.02 dynes/cm². These data indicate that the work of the wind in summer heat distribution is very small in comparison to the force of wind on the lake surface.

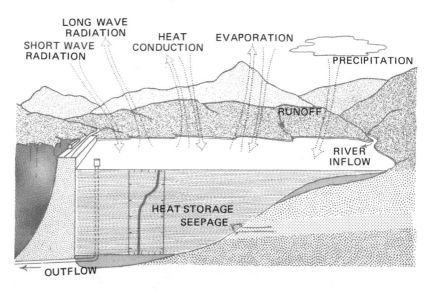

Figure 8.5   Major processes of heat transfer in a lake. Inset shows a temperature profile. (After Saur and Anderson, 1956.)

The loss of heat during the period of cooling from the summer maximum temperature to the winter minimum must be of the order of the annual heat budget. This loss from the lake is mainly through radiation to the surroundings. Cooling by evaporation also expends heat. The waters of effluent streams remove considerable quantities of heat from lakes, especially from impoundments where warm surface waters are removed while cooler waters remain behind. Heat exchange processes in a lake and impoundment are shown in Figure 8.5.

From the foregoing considerations we sense that a certain quantity of energy is expended in developing summer stratification in holomictic lakes. We have also made reference to the stability of the stratification and to the resistance offered by the lake to mixing. This, in essence, is one definition proposed for the term **stability.** More properly, **stability of stratification** is that energy of resistance which the lake offers to oppose an upset of density stratification. As warm, less dense water comes to overlay cool, more dense water during stratification the center of gravity is shifted downward. Stability then becomes a measure of the amount of work required to raise the center of gravity or to displace it to its original position. Stability decreases in autumn as the thermocline sinks below the center of gravity and the waters above and below this center attain similar densities. The greatest stability is probably reached just prior to maximum heat content in summer.

While we have been considering temperature and heat relationships in relatively large bodies of standing waters, we should not lose sight of the fact that pools and small ponds also exhibit responses to seasonal and daily temperature influences. Such responses often limit activity of some organisms and should certainly be studied in limnological investigations of small situations. "Microthermoclines" and odd temperature stratifications can be observed in certain small bodies of water under given conditions. A reversed stratification has been shown to exist at times in small Burt Pond, Michigan. The data for this pond are given in Table 8.2 and indicate, among other things, how rapidly the temperature

TABLE 8.2   Summer pondwater temperature at varying depths below algal mat, Burt Pond, Michigan[a]

	Temperature, °C
Mat surface	21-22
2.5 cm beneath mat surface	23-24
5.0 cm below algal mat	25
7 to 8 cm below algal mat	21-23
15 cm below algal mat	20

[a] Data from Young and Zimmerman, 1956; all temperatures were recorded at 1:30 PM.

of a small body of water responds to atmospheric conditions. It is suggested that the reversed stratification results from heating of the surface by direct radiation in the morning, followed by cloudiness and winds which cool the surface in the afternoon.

THERMAL PROPERTIES OF STREAMS

The basic thermal characteristics and processes in the water in streams are not different from those of lake waters. Penetration of light, absorption, specific heat, and other factors attributable to the nature of the water molecule operate in determining the fitness of the stream as an environment just as in lakes. As an environment, however, a stream presents a considerably different set of temperature conditions. These conditions derive from variations in velocity, volume, depth, substrate, cover, water source, and a number of additional features operating seasonally, daily, and even longitudinally along the stream course at a given time. Recall that we have termed a stream community an ''open system'' by virtue of the considerable land-water interrelationships; some of these interrelationships have a strong influence on temperature.

The major factor in the warming of stream waters is direct solar radiation. A series of temperature curves recorded over several successive days of cloudless skies followed by an overcast day will show rather regular fluctuations from late afternoon maxima to early morning minima during the cloudless period, but the cloudy-day curve will be flattened. Additional evidence for the importance of direct insolation is found in the frequent occasions when the temperature of the water exceeds that of the air. This usually occurs on very clear days of intense sunlight. The limit of nocturnal cooling is, of course, regulated in small, shallow streams by atmospheric temperatures. In larger, deeper streams it depends upon the rate of heat loss before warming begins again. Additional heat may be gained from bottom sediments in streams, but little data are available to show the extent of such heating.

On the other hand, the temperature of stream water is a measure of the actions and interactions of a wide variety of factors. Consider a stream rising in a rocky, wooded highland and flowing down over the coastal plain to the sea. In the upper reaches the waters are cooled by the substrate, by the shading provided by vegetation, and possibly by the entrance of spring-fed tributaries. Turbidity may be quite low. As the stream approaches the lowlands it becomes wider and deeper, and more water is exposed to direct sunlight. This, and increased silt content (which absorbs considerable heat), result in the development of a segment quite different with respect to temperature from the upper reaches.

Depending upon size and origin of streams, their diurnal and seasonal temperatures follow atmospheric temperatures more closely than do those of lakes. Diurnal temperature variations at a given point in a stream may be related to two major sets of factors: (1) conditions at that point, and (2) conditions upstream from the point. Under (1) we should consider velocity and discharge, season and hour, and the daily range of fluctuations of air temperatures at the point. Factors pertaining to (2) include the nature of upstream environment, substrate (and impoundments, if any), atmospheric conditions and temperatures upstream, and distance and time of flow from critical upstream situations.

With respect to size and water temperature fluctuations, we can conclude that the smaller the stream, the greater the temperature variations and the more rapid the response to environmental fluctuations. Furthermore, these fluctuations exist throughout the year, but are minimized beneath ice cover. The range of daily variation of water temperature is maximal when there is the greatest differential between mean diurnal air temperature and mean water temperature. With increased volume and turbidity the range of fluctuation diminishes. Because of the wide range of temperatures exhibited by small, cold streams, a single temperature reading may not at all approximate the average daily figure.

The source of the stream's waters and the nature of the drainage pattern in some instances determine the thermal properties of the stream. Many spring-fed streams and those originating from surface discharge of subsurface aquifers are essentially thermostatic. Lander Springbrook, New Mexico (see Chapter 15), is a small **rheocrene** (''flowing'') spring (Noel, 1954). Seventy meters below the source the annual variation in water temperature is 4°C; air temperatures fluctuate from about 14.3° to 33°C. The annual variation of water temperature of Silver Springs, Florida, a large spring run (daily discharge about 600 million gallons mainly from aquifer openings) varies no more than 1° throughout the year (Ferguson, 1947). Diurnal fluctuation in Silver Springs is also of the order of 1°. These spring-fed streams do not receive significant quantities of surface water which might otherwise bear on their thermostatic qualities. Surface-fed streams typically show wide seasonal fluctuations corresponding with atmospheric conditions. Figure 11.3*f* (page 248) illustrates the nature of seasonal variations and correlation with air temperatures in a stream typical of much of the nonmountainous regions of the United States.

Because of turbulence and the shallow nature of most streams, thermal stratification is not generally an attribute of streams. When stream waters stratify, the process usually takes place in pools along the stream course. In some instances the stratification is a result of inflow of cool spring water at the lower levels of the pool. In Jack's Defeat Creek, Indiana, a pool was found to be stratified, the temperature decreasing 7.5°C from surface to bottom at about 62.5 cm. Other qualities, including free carbon dioxide, dissolved oxygen, and alkalinity were also stratified. Inflow of spring water possibly accounts for the local stratification in this case. If warmed upstream waters are gently introduced into a cool pool of water, the inflowing, less dense waters are apt to flow over the more dense pool waters and on downstream without considerable mixing.

One word on the effects of human endeavors, such as agriculture and industry, on stream temperature and ecology needs to be interjected here. We know that water taken from a stream and used in irrigation is warmed in the process and, upon being dumped back into the stream, increases the downstream temperature for some distance. Similarly, water used in cooling various industrial installations warms the stream. In addition to the simple warming of the stream which, as such, may exceed the temperature tolerance of some of the organisms, the increased heating also decreases the oxygen retention capacity of the water, thereby affecting certain organisms and stream metabolism. Reservoirs, whether for water supply or hydroelectric power, exert considerable influence on the quality of stream water below the impoundment. As spring temperatures of streams approach the favorable point for spawning of warmwater fishes, sudden discharge from the lower levels of an upstream impoundment cools the stream and inhibits reproduction of the fishes. Conversely, the influx of warm, surface waters from an impoundment into a cool trout stream has deleterious effects on the fishes and the stream community. Much precise work needs to be done on these aspects of stream relationships. The results of such research could go far in enlightening the agencies concerned with dam-building projects. More will be said concerning this problem of **thermal pollution** in the closing pages of this chapter.

HEAT AND TEMPERATURE IN ESTUARINE WATERS

Before considering some of the more general thermal aspects of the estuary as a body of water, we should give attention to the effect of increased concentration of salts upon certain basic properties of water. Recall that pure water has a specific heat of 1. As the concentration of dissolved salts increases, the specific heat decreases. For seawater at 17.5°C and salinity of 35‰ (parts per thousand) the specific heat is 0.932. This suggests that less heat is required to warm a given volume of salt water than to warm the same amount of fresh water. Whereas for pure water density is a function of temperature alone, being maximum at 4°C, the presence of salts depresses the temperature of maximum density in sea water. At a salinity of 24.70‰ maximum density occurs at the freezing point, —1.33°C. At salinities higher than 24.70‰ the temperature of maximum density is below the freezing point. Unlike pure water, sea water at high salinities undergoes regular increase in density as it cools to freezing. Generally, the freezing point of sea water varies inversely as the salinity. At a salinity of 10‰, commonly experienced in estuaries, the freezing point is near —0.5°C; at 35‰ salinity, sea water freezes at about —2.0°C. The complexities of estuarine thermal dynamics are easily appreciated when we remember that within a few miles the salinity in an estuary may grade from near 0‰ to 25‰ or higher.

The heat content of estuarine waters is derived mainly from solar radiation. These waters are directly heated *in situ* as they occupy the estuary basin. Heat is also received indirectly from inflowing stream water and from tidal flow from the sea. The temperature of the estuary is, therefore, primarily a function of the temperatures of entering streams and the sea together with tidal stages. If the estuary empties into a relatively deep sea, the seaward temperature is apt to be more stable than in the upper zone where the stream temperature may fluctuate widely. On the other hand, estuaries of spring-fed streams flowing into shallow sea zones may exhibit more stable features in the headward regions.

Since most estuaries are shallow, we should expect considerable diurnal and seasonal fluctuations in surface temperatures. In the East Bay, an arm of the Galveston Bay system of Texas, during July, surface water warms from an early morning temperature of about 27.0°C to 33.0°C in late afternoon; atmospheric temperatures during the same period range from near 28.0° to 34.0°C (Reid, 1955). Seasonal temperature fluctuations depend upon latitude and a number of local factors such as water source, basin morphometry, winds, and tides. In Apalachicola Bay, Florida, the annual temperature range is of the order of 25°C. Temperatures in the Sheepscot estuary of Maine range from about freezing to near 25°C in the upper reaches of the estuary, while in the lower region the range is about 15°C. These temperatures are determined primarily by atmospheric and climatological conditions. A number of instances have been noted, however, in which sudden changes in the pattern of oceanic currents result in the influx of cold masses of water into an estuary, thereby drastically reducing water temperature.

Another factor that affects temperatures in estuaries is the flooding of marshes and mud flats during high tides. In warm, humid regions the surface of an exposed marsh may, during the summer, become quite warm. As the rising tide covers the marsh with estuarine waters, heat is imparted to the waters. In other regions the exposed muds may become cooled through evaporation and thus lower the temperature of incoming water. Studies in the Elkhorn Slough estuary of California have shown that mud gains heat much less rapidly than the overlying water, even in a depth of only 15 cm.

Temperature distribution in estuaries is largely a function of depth together with the relative effects of stream inflow and tidal exchange. In a shallow, mixing estuary the waters tend toward vertical homothermy. Longitudinal temperature patterns vary seasonally. Depending upon the morphology (surface-volume ratio) of the stream basin and the volume and rate of discharge, the upper waters of the estuary may be cooler in winter and warmer in summer than the lower estuary waters. From the mouth of the Patuxent River of Chesapeake Bay to a point about 74 km upstream, the average surface temperature from March to July grades from about 21° to 23°C; bottom-water temperatures similarly increase on the order of approximately 3°C. From September to January, both surface and bottom temperatures increase slightly over 1° from the bay upstream (Newcombe *et al.,* 1939).

Under suitable conditions of climate and depth, estuaries may exhibit a vertical temperature gradient. In summer the surface waters are usually warmer than the underlying layers. With the cooling weather of autumn, the upper waters cool more rapidly than the deeper masses, resulting in overturn and mixing. In midwinter the temperature of the surface waters may hover near that of maximum density, thereby giving rise to convection currents which continue to circulate the waters. Increasing insolation and winds in spring warm the surface, and a temperature gradient is established. Density stratification derived from cold or warm fresh water flowing over the more dense salt water also contributes to vertical temperature differences in estuaries. Such a stratification may be recognized by salinity measurements from the surface downward.

HEAT AND TEMPERATURE IN MARINE WATERS

The most important source of heat in marine waters is radiation from the sun and sky, although heat also enters the seas by convection from the atmosphere and by the condensation of atmospheric water vapor. The radiant energy available at sea level varies—throughout the year, at different latitudes (Table 8.3), and in different climatic regions. The average amount of radiant energy available at a given latitude during each month was calculated long ago by Kimball (1928) for the portion of the globe between latitudes 60° north and 60° south. The greatest available radiation (in gcal/cm²/min) is approximately 450, in January, at 30° south latitude, roughly the latitude of Capetown, Buenos Aires, and Easter Island; at the latitude of New York City (42° north) the maximum, 317, occurs in May.

Thus great expanses of the ocean are exposed to the tropical or summer sun. Much of the heat from this insolation is absorbed by the upper layers of seawater and carried long distances by prevailing currents, thus forming a tremendous mobile heat reservoir system. The warming effect of this system is often a very important influence upon climate: the weather along the eastern coast of North America and in coastal Northern Europe is strongly modified by the Gulf Stream, and the Kuroshio Current, flowing northeast from Japan, has a similar moderating effect upon the climate of the American northwest and Alaska. Where the seawater is warmer than the surrounding air, water vaporizes to form persistent fogs. In the Grand Banks, a region of shoals southeast of Newfoundland in the western North Atlantic, warm vapor accompanying the Gulf Stream collides with the cold air typical of that latitude; the region is usually fogbound. In the north

TABLE 8.3 Mean surface temperature at various latitudes in the North Atlantic Ocean[a]

North latitude	Mean temperature, °C
65°	5.6
55°	8.6
45°	13.2
35°	20.4
25°	24.1
15°	25.8
5°	26.6

[a] Data from Sverdrup *et al.*, 1942.

Pacific in the region of the Aleutian Islands, west of Alaska, similar prevailing fogs can be observed.

Disregarding for the moment the complexities of deep ocean currents (Chapter 6), we can identify three basic temperature layers in the open seas. Near the surface is a shallow region of heat exchange and relatively thorough mixing. This layer is usually relatively warm, and, like many lake waters, it is homogeneous in late winter. (It should be noted, however, that surface waters are extremely variable in temperature, ranging from $-1.9°$ C in the Arctic Sea to higher than $30°$ C in some tropical regions.) Under the extremely calm conditions found in the Sargasso Sea this upper layer extends to a maximum depth of 500 m, but it is usually much shallower (Warren, 1966). Below the upper layer occurs the **main thermocline,** a layer physically similar to a lacustrine thermocline. In this layer the temperature drops rapidly, from about $17°$ C to $5°$ C; the main thermocline varies in thickness from 500 m to a maximum of 1000 m. Below the main thermocline and continuing down to the ocean floor the water is essentially homogeneous and cold, the temperature in this third layer decreasing very gradually with increasing depth.

These three layers account for the basic temperature structure of marine waters, but two other important features also deserve mention. The depth of the main thermocline varies considerably with latitude, rising toward the surface in the tropics and the middle latitudes. At approximately latitude $50°$ the thermocline rises to the surface, and poleward of that point there is virtually no depth–temperature gradient. Further variation is introduced as a result of seasonal warming, and a **seasonal thermocline** is observed in the upper layers in many regions. At depths greater than 200 m seasonal changes in insolation and other weather factors produce virtually no change in temperature (Warren, 1966).

More detailed studies indicate that water temperatures are not really smoothly graded throughout the lowermost layer. Rather, marine waters of different temperatures circulate at different depths, somewhat resembling immense rivers, which flow in different directions at various levels. Antarctic water, for example, flows northward at considerable depth through the Pacific. It wells up off the coast of Peru, bringing cool, nutrient-enriched water to the surface and supporting an extensive fisheries industry. Comparable subsurface flows have been traced in all the major oceans, and in the early 1960s high-velocity equatorial countercurrents—somewhat like aquatic "jet streams''—were shown to exist in the Atlantic and Pacific Oceans.

THERMAL ENERGY IN AQUATIC ECOSYSTEMS

The importance of thermal energy, as a regulating and limiting factor in aquatic (and other) ecosystems, is well known. All known species on this planet are restricted to a relatively narrow range of temperature, from about $-200°$ C up to the boiling point of water, $100°$ C. Some can survive for a time at very low temperatures, usually in an inactive state; and a few species of blue-green algae and bacteria, adapted to the extreme environment of mineral hot springs, can endure temperatures exceeding 90° C and are able to reproduce at temperatures only slightly lower. For the most part, however, any given aquatic species is **stenothermal**—adapted to only a very narrow range of temperatures. Diurnal or seasonal variations in temperature have profound effects on the metabolism and activities of aquatic species, with the most severe effects of temperature change being observed at the upper limits of the range of tolerance. In other words, most species can survive a drop in temperature better than they can an equivalent increase.

Even in everyday usage we tend to make a fundamental distinction between "cold-blooded" (**poikilothermic**) and "warm-blooded" (**homeothermic**) animal species. The activities of most poikilotherms—invertebrates, fishes, amphibians, and reptiles—are slowed down by a drop in temperature; many aquatics, such as fish, reach a state of virtual suspended animation during winter. Many cold-adapted species, however, are active at low temperatures, and in some, growth and reproductive rates increase with cold. In autotrophs especially, the rate of growth is strongly affected by temperature. Many plant species exhibit their maximum growth at low temperatures, because the cold inhibits respiration and yet provides the optimum condition for photosynthesis; conversely, a rise in temperature may slow the rate of photosynthesis but cause an increase in respiration, resulting in a net drop in productivity. This observation is of tremendous importance in applied aquatic ecology. Another extremely important principle is that for many aquatic species the **variation** in temperature may be fully as important as the average temperature of the water. The temperate-zone species characteristic of North American waters are, for the most part, adapted to an environment whose temperature varies seasonally by 10° C or more. If some external agency should dampen this fluctuation—as by impounding a mass of water or by altering the structure of the watershed—many species will be unable to reproduce or, in many cases, to survive.

RESEARCH TRENDS IN THERMAL RELATIONS

The most conspicuous trend in this area of research, especially since the late 1960s, has been the application of theory to practical environmental problems. **Thermal pollution**—the introduction of heated effluent into a natural body of water—was mentioned earlier (page 164), and this has been a continuing, and growing, focus of ecological concern. Some researchers prefer to refer to the process as **thermal enrichment**, certainly a more neutral and possibly a more accurate term. Others contend that such usage simply avoids facing up to unpleasant realities, and insist on the term "pollution." Such disputes are typical of any area of study that is also of great current interest. For our purposes, the term "thermal pollution" will be used to refer to any heating of natural waters that tends to degrade the environment and/or to alter the natural biota.

The question arises: why is thermal pollution a matter of such great concern? The single greatest contributor to thermal pollution is the electric power industry. The production of electricity from fossil fuels—coal and oil—uses tremendous amounts of natural water to generate steam and to cool the generating mechanisms. It has been estimated that the production of one kilowatt hour of electricity entails the release of well over a million gcal of waste heat into the cooling water. As nuclear-powered electric plants come into use the burden on water resources increases—a typical nuclear installation, producing about 3 billion watts of electricity per year, may be expected to release heat at a rate well above 50×10^9 gcal per hour (Odum, 1971). Several estimates suggest that by the early 1980s one-sixth to one-fourth of all runoff water in the country will be needed simply to cool all the power stations existing in the United States at that time (Singer, 1968). Applying the same principle on a smaller scale, one limnologist (Cairns, 1968) has calculated that the average household "consumes" 541 gallons of water per month simply by the normal use of electricity for lighting, television, heating, kitchen appliances, and the like. Most of this water is not in fact consumed, of course; it is actually heated in the generating process, and then returned to the lake, stream, or estuary from which it came. Earlier in the century, when electric power was much less widely used, it was possible to assume that the natural water system was able to absorb and disperse this waste heat, but this no longer holds true. Clearly, an increase in the production of electric power by present methods will have grave effects on all aquatic ecosystems. The effects will be of several types:

Damage in the power plant Minute planktonic organisms, small crustaceans, and the eggs and larvae of fish and shellfish are drawn into the cooling system via the water intake, and are there killed or damaged. The result may be a serious disruption of the natural ecosystem.

Chemical pollution Heavy metals and industrial chemicals are often discharged in the effluent water. This issue will be considered in more detail in Chapters 10 and 11; it is mentioned here only to suggest the complexity of the problem.

Weakening of the aquatic biota Another point is closely related to this last one. When the temperature of water increases, many species become noticeably more vulnerable to damage from toxic wastes—both those introduced with the heated effluent and others already present in the water—as well as to parasites and disease.

Direct damage from heat Heat may directly affect an aquatic species in many ways. The most dramatic effect, of course, is thermal death, as seen in large fish kills in many American rivers, but a great number of internal disturbances are also possible. Changes in respiration, metabolism, growth, behavior, reproductive capacity, and other functional aberrations can often be traced to a change in water temperature.

Interference with development Many species are stenothermal at some critical stage of spawning or development. For example, the eggs of the brook trout *Salvelinus fontinalis* will develop only between 0° and 12° C; at higher temperatures successful reproduction is not possible.

Population shifts As the stenothermal species disappear from heated waters, heat-tolerant species increase in number and displace the original species in the ecosystem. In thermally polluted lakes and ponds normal, and varied, algal flora may be replaced by masses of blue-green algae, which are of little value in the aquatic food webs and may even contribute to further pollution. In a polluted Pennsylvania stream, pollution had a similar effect upon the distribution of fish species. Only three species could be observed in the polluted stream (goldfish, killifish, and a species of shiner), although 20 other species occurred in other, unpolluted parts of the same watershed.

Oxygen depletion A rise in water temperature increases the metabolic rate in most species. At the same time, however, oxygen is less soluble in the warmer water, and increased bacterial activity further reduces the oxygen supply. The water may rapidly become unfit for all but a few largely anaerobic animal species.

Clearly the effect of thermal pollution is complex, and the problems it raises are potentially very serious. To the seven major effects already mentioned, we might add an eighth. Lacustrine ecology, as we have seen (page 152), is strongly affected by the spring and fall overturns and by the thermal stratification of lake waters. These events determine the distribution of dissolved and suspended nutrients, and they are inevitably central to the ecology of any lake. If a power plant were to alter the thermal relations in a lake or pond by adding an upper layer of warm water, there is a strong possibility that normal circulation patterns would be disrupted. The semiannual overturn could then take place out of season or could even fail to occur. The ecological consequences of such a change would be drastic, unpredictable, and almost certainly deleterious.

These considerations have given a strong impetus to recent research in thermal relations. There has been a tendency for research on pollution to focus on only a single aspect of the problem at one time—chemical, physical, thermal and so forth. Relatively little direct attention was given to biological effects. More recently, scientists have recognized the need for a more integrated approach, joining direct biological assessment of an ecosystem to a survey of its physical and chemical parameters. The general conclusion that emerges from such studies is that thermal pollution generally leads to the disappearance of many species, and thus to a simplified ecosystem dominated by a very few species. The simplified environment is ecologically unstable, relatively unproductive, and in most cases inadequate or unreliable for use as a resource for food, recreation, or even waste disposal.

The problem of thermal pollution clearly calls for protection of the environment. Water used in the power-generating process could be returned to the watershed without having been cooled at all; usually it is cooled (and sometimes recycled) in pools, reservoirs, or cooling towers. From the point of view of protecting the aquatic environment, cooling towers are preferable, but they are also quite expensive. Economic decisions obviously have to be made at this point. In England, every generating station is required to use a cooling tower (Cairns, 1968); no such uniform policy has been adopted in the United States. In this country the remedy of choice has been, rather, the formulation of thermal pollution standards to limit the effect of heated effluents on the aquatic environment. New York State, for example, has developed a complicated set of standards that are appropriately adjusted, it is hoped, to the ecological needs of lakes, streams, estuaries, and coastal waters at different seasons. It is impossible to predict how effective such standards will prove to be. In any case, a thorough physicochemical and biological monitoring program is needed, not only to ensure that the regulations are being observed, but also to determine if the standards are adequate to the needs of the ecosystem.

In our discussion of thermal pollution we have been dealing—in some detail—with one aspect of the principle of limiting factors. A more general statement of the principle is given in the **law of tolerance** proposed by the zoologist V. E. Shelford in 1913. Briefly stated, the "law" points out that for a species to be successful, its environment must be "fit" in terms of a number of different, but interrelated, conditions. Light, motion, chemistry, heat, and other factors must be present in certain kinds and amounts, forming a complete complex of suitable conditions. Each species is able to tolerate only a limited range of variation with respect to any one of these factors: thus any single factor may be limiting for a particular species.

If, for example, the water in a hypothetical pond were deficient in oxygen, fish would be unable to survive there, even though all other conditions were optimum. In reality, of course, the other conditions would not be optimum, because a change in one factor is almost always related to changes in others. It is convenient to consider the different parameters of an ecosystem one at a time, but in practice it is necessary to consider the entire complex of environmental variables. In this chapter we have applied Shelford's law of tolerance to a single factor, temperature. In the following chapters we shall examine the reactions of various aquatic species to variations in other physical, chemical, and biological parameters.

SUGGESTED FOR FURTHER READING

Clark, J. R. 1969. Thermal pollution and aquatic life. *Scientific American* 220:14. A survey, for the reader with some background in biological science, of the major causes and effects of thermal pollution.

Drost-Hansen, W. September and December, 1969. Allowable thermal pollution limits—a physico-chemical approach. *Chesapeake Science,* Volume 10, Chesapeake Biological Laboratory, Solomons, Maryland. A two-part article by a specialist in chemical physics who has carried out research on the physical chemistry of biological systems. For advanced undergraduates, this article is an excellent summary and presentation of the major issues in thermal pollution.

9 | Natural Waters in Motion

Although poets may sometimes refer to quiet waters—"the waters that glass the clouds," or the "still waters" of the Psalms—it is a basic feature of aquatic environments that they are rarely if ever completely calm. In the smallest pool or in the most sluggish stream, the water is in motion, to some extent and in some way characteristic of the particular body of water. This motion is caused by temperature and density differences, by gravity, and by wind; these factors may act together or singly, and in a particular situation any given factor may predominate in setting the water into motion. Every process that takes place in natural waters—physical, chemical, or biological—is affected, directly or indirectly, by the movement of water.

In this chapter we will be examining aquatic environments with regard to the most important ways in which they are affected by the movement of water. As we shall see, currents, waves, and other major forms of motion are fundamental properties of an aquatic environment. As much as the morphology of a basin or channel (or its temperature, turbidity, or the like), the motion of natural water determines its fitness as an environment for a limited range of species. In this chapter we shall limit our discussion largely to the physical aspects of motion; in later chapters we shall examine more closely the effects of motion on the biota of different aquatic environments.

MOVEMENT OF WATER IN LAKES

As a general rule, the water of a lake basin is partially or wholly in motion. It is this movement, derived from either internal or external forces, or both, that is responsible for the circulation of heat, dissolved substances, and some organisms in lakes. Turbulent flow is characteristic of the movements of lake water; this results from the action of molecular systems imparting irregular direction and velocity to water mass. Turbulent movement is then incorporated into the more conspicuous forms of motion in lakes. The larger movements of water are called **current systems**, and are frequently defined as two types: **nonperiodic,** or arhythmic; and **periodic,** or rhythmic.

NONPERIODIC CURRENT SYSTEMS

Nonperiodic systems are those often termed simply "currents." The description implies a unidirectional flow of water. This movement may be caused by differential heat distribution within the lake, the passage of stream waters through the lake, and by winds. Broadly speaking, nonperiodic systems are produced and maintained by external forces.

Wind is doubtless the major external force acting to set lake waters in motion. Sustained winds blowing across the surface serve to pile up water at the downwind end of the basin. The result is lowering of the water mass in another part of the lake. As the winds subside, the piled-up water begins to flow along a gradient, or down the slope, so to speak. Eventually, the overall lake level reaches equilibrium and the current system comes to rest. We have seen in the previous chapter that the action of wind is also important in the development of summer stratification of lakes.

A phenomenon familiar to all who have done much boating on lakes is that of **wind streaks**, linear accumulations of foam, floating debris, or oil. The physical chemist Irving Langmuir (1881–1957) demonstrated that surface flow resulting from wind drift is not an even and uniform movement of water containing variable velocities, but rather possesses a definite pattern (Langmuir, 1938). Water flow in wind drift is in the form of helices lying parallel to one another oriented in the direction of the wind with the direction of the helices alternating clockwise and counterclockwise, and the "streaked" appearance is caused by the accumulation of materials in zones of convergence between clockwise and counterclockwise helices. The width of the zone of convergence varies seasonally, being wider in the late autumn than in spring. It appears that the downward vectors of the helices serve to transfer momentum from the surface into the deeper regions of the lake.

In large lakes the effects from stream inflow, wind, and earth's rotational forces often combine to produce a pattern of large current swirls. The surface currents of Lake Constance, in Europe, have been studied rather intensively, especially with relation to the path of the Rhine River through the lake. The general circulation pattern is shown in Figure 9.1. As a result of geostrophic forces and the shape of the basin, the Rhine waters flow across the upper end of the lake and follow the northern shore. The river current, together with wind patterns, apparently produce the large swirl between Langenargen and Rorschach.

The influence of the Rhine on the currents in Lake Constance serve to point up another kind of nonperiodic current in lakes; such may be termed **density currents.** These currents result from the passage of river water of a given density through a lake of differing density. As suggested previously, the route of the stream water is determined mainly by geostrophic forces, the shape of the lake basin, and local meteorological agents. The depth at which the river water flows is a function of density differences. If the density of the river water is greater than any of the lake waters, the stream flows over the bottom of the lake. Similarly, if the lake is generally more dense than the stream, the latter flows at the lake surface. During the summer months the water of the influent is usually cooler than the lake surface, and therefore more dense. Thus the river water flows downward until it meets a region of greater density, whereupon the river flow becomes horizontal. At the mouth of the Rhone River in Lake Geneva, a well-studied subaqueous "waterfall" is formed in this manner. The cold glacial waters of the Rhone, being denser than the lake water, descend to the floor of the lake; in so doing, they have, over a period of years, excavated a channel in the lake bottom (Hutchinson, 1957). In thermally stratified lakes, the river flow is usually above the hypolimnion; thus the lower lake levels receive little of the nutrients carried by the inflowing stream.

In Norris Lake, a Tennessee Valley Authority impoundment, there are indications that density currents move through the lake as the water is drawn down for hydroelectric power. These currents may be poor in dissolved oxygen content due to organic decomposition, causing movement of fishes to avoid the oxygen minima (see Chapter 16). Density currents may be important in ponds and small lakes, and probably relate to distribution of materials beneath ice cover in winter. In Lake Mead, formed by Hoover Dam on the Colorado River between

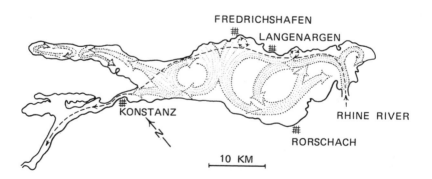

Figure 9.1 Schematic pattern of surface currents in Lake Constance, Switzerland/ Germany. The route of the Rhine river water through the lake (as determined by conductivity measurements) is indicated by a dashed line. (After Hutchinson, 1957.)

Nevada and Arizona, a density current resulting from increased silt load forms during the summer. This **turbidity current** may flow into the lake below the level of the river current (Pennak, 1955). (The effect of turbidity currents on the ocean floor was mentioned on page 113).

One more aspect of nonperiodic movement of water should be considered, that of **turbulence** and **eddy effects**. Although the mechanics of these phenomena are best known from laboratory studies, the applicability of the principles to limnology is vast and deserving of more intensive study in the future. To anyone who has watched a stream plunging through riffles or along irregular shores, the intense turbulence and the resultant eddy systems are obvious. Within lakes, however, such turbulence is reduced and is certainly less conspicuous; nevertheless, eddy effects are observed in lakes, where they are caused by currents such as those previously described and to be discussed later. The great importance of eddy effects is in heat transfer within the water mass (**eddy conductivity**), in the diffusion of dissolved materials (**eddy diffusivity**), and momentum transfer (**eddy viscosity**).

In Chapter 4, we distinguished **laminar** from **turbulent** flow, as these occur in streams. It is agreed that simple laminar flow, wherein sheets of liquid move uniformly without local variations in velocity (but with velocity differences from "sheet" to "sheet"), does not normally occur in nature. The more characteristic flow of natural waters is one of turbulence, in which irregular motion of subsidiary masses is contained within some relatively simple current. The nature of the turbulence is an expression of various factors including flow velocity, velocity gradients encountered, and the nature of the boundaries of the flowing systems. Figure 9.2 can be interpreted in terms of a density current passing through a lake. In this example, the lake proper functions in somewhat the same manner as a stream bank, in the sense that it offers resistance to moving masses of water. The velocity and magnitude of the eddy systems operating at the upper and lower zones of the density current for example, are functions of velocity factors (of the lake proper and of the current), and of boundary

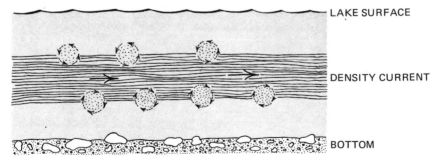

Figure 9.2   Highly schematic representation of eddy systems set up along an interface of lake water and density current, as the current flows through a lake below the surface. See text for discussion.

characteristics (including temperature and chemical composition). It appears, then, that through eddy effects, mass and molecules can be exchanged from one layer to another.

One effect of eddies is that by which masses leaving one layer carry with them momentum from that layer and acquire the momentum of the new layer before returning to the original one. This is a simplified definition of **eddy viscosity.** In practice, a **coefficient of eddy viscosity** may be calculated from the formula

$$r_s = -A \frac{d\bar{v}}{dn}$$

in which r_s is the **Reynolds stress,** or shearing stress per unit surface area, $\frac{d\bar{v}}{dn}$ is the shear of the velocities as measured, and A (from ``Austausch'') is the coefficient of eddy viscosity, that is, the relationship between mass and transverse distance through which a liquid flows per unit time in a path corresponding to that of the main direction of flow. The transfer of motion through eddy viscosity increases with turbulence, and thus eddy viscosity is considerably greater than molecular viscosity, by a factor as great as one million. In natural waters, eddy viscosity is a primary force in reducing the settling rate of sediments and in maintaining the position of microscopic organisms.

In the same sense that eddy viscosity describes the transfer of mass and momentum from one water stratum to another, **eddy conductivity** describes the exchange of heat across surfaces. This parameter takes into account the specific heat of the fluid and the heat gradients encountered between masses. This heat transfer is then proportional to the mass transfer described in the formula for eddy viscosity. Important here is the idea that conduction may be lateral as well as vertical. Eddy conductivity is fundamental in the heating processes of lakes, described in the preceding chapter.

The transfer of dissolved substances, so very important in lake metabolism and in the maintenance of living organisms, is another function of mass exchange, termed **eddy diffusivity.** An important aspect of eddy effects is the simultaneous transfer of more than one substance, or the **principle of common transport.** This principle holds that a motion that transfers one environmental factor may simultaneously carry another, different, factor. For example, while dissolved salts are being transported downward, dissolved gases, or heat, may be carried upward in the same eddy system. Some idea of the rate at which substances may be distributed through this process is gained from the observation that in 1961, summertime eddy diffusivity in the western part of Lake Erie was approximately 25 cm²/sec, or probably some 104 times that in water in a quiet bottle. For the aquatic ecologist eddy effects are important because of the ways in which they contribute to lake dynamics, heat and nutrient distribution, and the like. A fuller discussion of the physics and mathematics of eddy relationships may be found in Hutchinson (1957) or in several more recent works on hydrology.

A mathematical constant, the **Reynolds number** (R_e), has been calculated to determine exactly how fast water must flow before it becomes turbulent. Laminar flow will give way to turbulent flow under conditions of the equation

$$R_e = \frac{d\bar{v}l}{\mu}$$

in which d is the density of the liquid, $\bar{v}$ is the velocity of the liquid, l is the depth of the channel, and μ is viscosity. In shallow channels, much wider than they are deep, turbulence appears when the value for R_e approaches 310.

PERIODIC CURRENT SYSTEMS

To this category belong water movements exhibiting some form of rhythm or periodicity. Two major systems are to be considered; these are: **traveling,** or **surface waves** (also called **progressive waves**) and **standing waves**, or **seiches**. As has been shown for nonperiodic currents in the preceding section, external forces, mainly wind and other meteorological factors (atmospheric pressure), are responsible for periodic currents.

Surface Waves

Lake surfaces are never completely smooth; slight irregularities and turbulences are present. Friction occurs between these surfaces, and wind blowing across the lake results in movement of the water. This motion imparts the first phase of wave formation. Subsequently, a crest is built, and increased wind velocity tends to build up the wave; eventually, an eddy is produced by wind on the trailing slope of the wave, giving a certain thrust to its back, thereby accelerating the motion of particles in their orbits within the wave.

There is no essential horizontal movement to waves in open water. A particle of water moves in a vertical circular pattern, or orbit, returning to its original position as the wave rises and falls. The diameter of the orbits and, correspondingly, the velocities, decrease rapidly with depth. It has been calculated that the diameter of orbits at a depth equal to one-half the wave length is on the order of 1/500 that of surface orbits, or quite imperceptible. Not until the wind develops to such a force as to produce a tumbling crest, or whitecap, is there any forward motion of the water. A similar effect is exerted by the bottom in shallow zones where friction directed upon a wave causes the mass to pitch forward, producing **surf.** These breaking waves are called **waves of translation;** open-water swells are termed **waves of oscillation.**

Both velocity of wind and the time that it has been exerting force upon a wave determine the size of waves. In order to build waves of any great amplitude a considerable amount of time is required to transfer the wind energy into the kinetic and potential energy of a wave. Similarly, space (**fetch**) is required, for a wave must have area in which to move before the wind in order to gain amplitude. One relationship between wave height (h_w) and fetch, during strong winds, may be expressed as

$$h_w = 0.105 \sqrt{x}$$

where x is the fetch, or distance (in centimeters) along which the wind has blown unobstructed before reaching the wave.

The situation as we have described it thus far is highly simplified. In reality, waves traveling from various parts of a body of water cross and recross many times, depending on their velocities and points of origin. The direction of waves can also be changed as they approach a shallowing shoreline, such as a beach; such a change in direction is referred to as **refraction.** Refraction occurs when the frontal part of a wave is slowed by friction against the shallows, while the later part of the wave continues to move at its original speed. In this way, waves that approach the shore at an angle may actually arrive as **parallel waves,** moving perpendicularly to the shoreline. Similarly, waves approaching a point of land will often sweep around it and arrive at the shore on either side of the point in the form of parallel waves (Figure 9.3).

Although waves and wave action are of particular interest to physical limnologists, our interest in terms of the overall ecology of lakes is mainly in the action of waves in circulation of vital materials in lake waters, and the stirring effects of breaking waves in shallow-water zones. From a broader point of view,

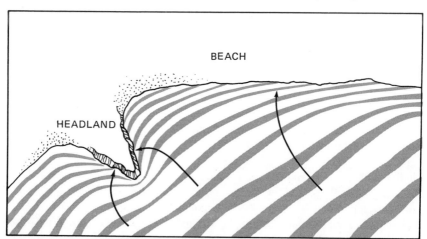

BEACH

HEADLAND

Figure 9.3   Waves approaching land are refracted by the contour of the shoreline. When the shoreline is straight, refraction bends the wave front parallel to the beach. As a wave approaches a point of land (left) the wave front is bent by the shallow bottom, concentrating the force of the wave upon the headland.

wave action is most instrumental in sedimentation and erosional processes which serve to modify the morphology of the lake basin and littoral areas; this topic was discussed in Chapter 3.

Seiches

A **seiche** (pronounced ''sāsh'') is a form of periodic current system, described as a standing wave, in which some stratum of the water in a basin oscillates about one or more nodes. This rocking motion can be seen in a bowl of thin soup passed by a shaky waiter, the seiche being set up by movement of the basin. By blowing upon the soup near one side of the bowl one can create a seiche due to pressure changes, although the basin is fixed. Seiches in lakes have been recognized for many years, and since Forel's pioneering work, beginning about 1870, much attention has been given to this intriguing phenomenon. The name, attributed to de Duillier, writing in 1730, is derived from the Latin *siccus* through the French *sèche,* meaning ''dry,'' and refers to the exposure of low shorelines left dry as the lake water recedes during an oscillation.

We have seen previously that sustained blowing of wind across the surface of a lake will cause water to pile up in the downwind region of the basin. If the wind suddenly subsides, a current will naturally flow toward the area of lowered surface. The water mass does not come immediately to equilibrium. As a result of the energy of motion imparted by the gradient current, an oscillation occurs about a stationary node. This oscillation, a **surface seiche,** continues until damped by contact with the basin proper or by meteorological forces.

The origin of surface seiches may be found in a number of natural mechanisms. From the considerable study of lake oscillations, however, it appears that surface depressions resulting from sustained winds, and from local sudden changes in atmospheric pressure, are the primary agents. There are cases in which local rain showers set up a seiche due to the pressure of the impact of the falling rain. Sudden inflow from a contributing stream has been suggested as the origin of a seiche in Loch Earn, Scotland. Earthquakes have been named also as possible causes of seiches.

Since these standing waves operate according to the laws of oscillating systems, the formation of harmonics is a general characteristic. In addition to the simple, uninodal seiche, binodal and trinodal seiches have been reported commonly. In fact there is at least one record, in Loch Earn (Mortimer, 1953), of a seiche exhibiting a nodality of the sixteenth. Normally, the periods and locations of the nodes are determined by morphological features of the basin, *i.e.,* depth, diameter, and form. The amplitude of the seiche depends upon the source and the intensity of energy giving rise to the oscillation, and to the form of the basin.

Although seiches occur in all enclosed or partially enclosed bodies of water, in small basins the effect is detectable only by sensitive recording devices. In larger lakes the oscillations may be quite conspicuous. Because of the narrow form of Lake Geneva near the city of Geneva, water level fluctuations due to seiches often leave the shoreline dry. Here, in 1841, a seiche was recorded with an amplitude of 1.87 m. In the same lake in 1891, Forel observed a seiche which lasted for 7 days, 17 hr. This standing wave underwent 150 oscillations, the amplitude dropping from a maximum of 20 to 7 cm; the period of oscillations was of the order of 73 min. The Great Lakes exhibit oscillations; the periods of longitudinal seiches in Lake Erie are near 790 min, and for Lake Huron 289 min. Transverse seiches are also known and serve to complicate measurement of lake activities.

Amplitude and period of seiches are determined in practice by the use of a float contained in a protective housing. The float is connected to a counterbalanced pointer over an appropriate scale. Termed a **limnometer**, the device may employ a tracing stylus to give a graphic recording or limnograph. A limnograph record is shown in Figure 9.4.

The periods of seiches can be calculated from various, and often complicated, formulas. For a uninodal seiche in a rectangular basin of regular bottom the period *t* may be calculated using the formula

$$T = 2L/\sqrt{gh}$$

in which L is the length of the basin, h is the water depth, and g is gravity acceleration. Bearing in mind that a seiche is water in motion, it becomes apparent that turbulence will enter into the system as soon as oscillation begins. The effect of turbulence will be to damp out the oscillation as well as to increase the period.

Thus far, our discussion has dealt with the **surface seiche.** Since about the turn of the century, however, it has been known that internal, periodic current systems could be present in lakes under certain conditions. Within a stratified

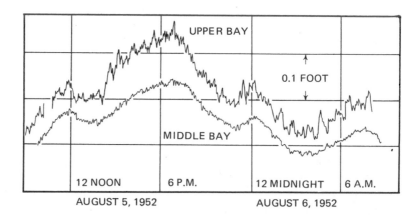

Figure 9.4   A sample of a seiche record for two stations in South Bay, Manitoulin Island, Ontario, at the north end of Lake Huron. The stations, utilizing water-level recorders, are separated by about 11 km. Note that several oscillations are apparent in the record, one at 9 min, one at about 50 min in the early portion of the record (but diminishing), and one long oscillation at about 6 hr. This seiche persisted for only a few days. (After Bryson and Stearns, 1959.)

lake of two or more layers of differing densities the strata may oscillate with respect to each other without being apparent at the lake surface. Recent compilations of data relating to such **internal seiches** indicate that both the ranges and the periods of such seiches are significantly greater than those of surface seiches. During summer stratification the thermocline may, as a matter of character, oscillate between the lighter layer of the epilimnion and the denser hypolimnion. An oscillation of this nature can be clearly and relatively easily detected through temperature recordings at various depths.

Data derived from laboratory and field studies of displacement of lake layers of equal temperature (isotherms) have led to the general acceptance of two major forms of internal seiches. One type, the **thermocline seiche,** results from the movement of water masses between the epilimnion and the most dense zone of the metalimnion. Another system, the **hypolimnion seiche,** moves along the boundary between the metalimnion and the hypolimnion. The time of one complete oscillation (period) is usually greater in the case of the hypolimnion seiche than for the thermocline seiche, both, however, being more or less dependent on the length of the lake. The differences in periods are due to unequal densities of the two strata. In Lake Geneva (about 37 km in length), the period of an internal seiche is about four days.

The mechanisms responsible for setting up interval waves are the same as for the surface seiche. Wind-driven waters of the lake surface pile up at the windward end of the basin, causing the deeper waters at the leeward end to compensate by being pushed up. If the lake is stratified, however, the resultant effects of the displacement are different. Because of greater density difference between air and water than between cool and warm water, slight depression of the water surface by air in motion results in a proportionately greater displacement of the lower water stratum toward the surface in the windward end of the lake. As the wind diminishes, the isothermal strata oscillate about a node near the center of the lake. The metalimnion serves as a sort of "cushion layer" on which the epilimnion moves over the hypolimnion. Figure 9.5 shows the oscillations of

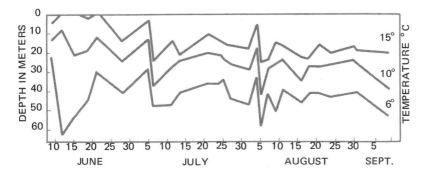

Figure 9.5   Oscillations of three isotherms at a point in the southern end of Cayuga Lake, New York, summer, 1951. Note the development of stratification and the sinking of the isotherms between late June and September. (After Henson, 1959.)

three isotherms in Cayuga Lake, New York, during summer stratification (Henson, 1959).

The questions raised by internal seiches are complicated, both in terms of motion and dynamics and in terms of definition. Not all internal oscillations in lakes are "true" seiches, that is, motions resulting from free oscillations. **Forced oscillations,** that is, those that are sustained directly by wind action, sometimes are present and are mixed with true seiches. In very shallow lakes forced oscillations may damp the effects of seiches, making it very difficult to distinguish between the two forms of motion.

Internal seiches are of considerable importance in the total economy of the lake—probably more so than surface seiches. In Lake Victoria an internal seiche with a period of about 30 days brings about the displacement of water into remote parts of the lake basin. The resultant turbulence mixes lake water over the bottom materials and transports dissolved and suspended materials into the shallow zones. In addition to this more or less horizontal transport, internal seiches serve to distribute heat and nutrients vertically within the lake. One particularly interesting problem is created by the action of internal seiches; this is the determination of the position of the thermocline through temperature measurements. Figure 9.6 shows the variations in position of the thermocline when in oscillation. It has become increasingly apparent that the thermocline is not a stable stratum, but rather is subject to much shifting and modification during the season.

True, or lunar tides, have been reported for large lakes, although the amplitude of lake tides would naturally be small. In Lake Superior the range is

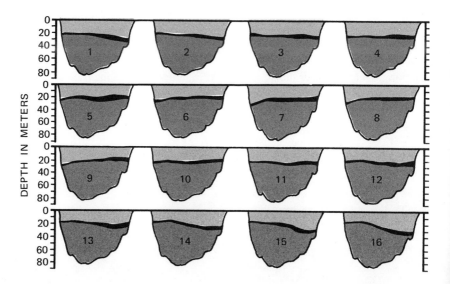

Figure 9.6   An internal seiche. The hourly variations in the position of the thermocline in Loch Earn, Scotland, 9 August 1911. The thermocline (11° to 9° C) is seen to shift greatly. (After Mortimer, 1953.)

about 3 cm. The difficulty in observing and measuring is complicated by the action of seiches. A seiche in phase with lunar time (12.5 hr) would magnify the tide range out of phase; a seiche out of phase with lunar time would inhibit tidal action.

STREAM CURRENT SYSTEMS

In Chapter 4 we discussed some of the major aspects of the dynamics of flowing waters, since these were of considerable importance in the formation, maintenance, and alteration of stream channels. It might be useful for the student to review briefly the section in that chapter dealing with stream dynamics (pages 73–77) before continuing with the following material, for the same processes covered earlier will form part of the basis of the discussion of streams in this chapter.

We have seen that wind is the major force in creating current systems in lakes. In streams the important factor in producing water movement is gravity. So long as a gradient on the earth's surface exists, water flows in response to gravity, to "seek its lowest level" following the route of least resistance. Because the route is seldom straight, the gradient rarely uniform, and the sides and bottom of the channel not often smooth for any distance, varying velocities and turbulences result. Bearing in mind the importance of these factors and activities in channel structure, let us now consider them with reference to the stream as a living-space.

A conspicuous characteristic of stream flow is transport of dissolved substances, suspended materials, and living plants and animals. This ability to move matter also contributes greatly to the unstable nature of bottom types and shore zones, mainly through shifting of bottom materials and through sedimentation. Although these processes and features could result from nonturbulent flow, they are usually of greater magnitude in turbulent waters.

Turbulence in streams is not uniform throughout a cross-sectional area of the stream. Maximum turbulence occurs at some region in the cross section and diminishes from that zone toward the surface or the bottom. Zones of maximum turbulence are usually located on either side of a similarly restricted **thread of maximum velocity.** The number of high-velocity threads and turbulent zones apparently increases with channel width, depth, and irregularity of the bottom. The result is a decrease in stream stability, making for a more rigorous environment for plants and animals. Eventually, however, uniform discharge may contribute toward a balance between the bottom materials and turbulence and the stream bed becomes a more optimal environment for plant and animal species.

Maximum velocity threads and associated turbulent zones do not maintain a constant position in the stream cross section, except perhaps in straight streams occupying channels of uniform shape. Most streams curve and bend throughout their courses, and the velocity threads follow courses that curve even more than do the streams. As a stream flows around a curve, the maximum velocity thread is found on the outside of the curve; if the stream bends in the opposite direction with the next curve, the thread moves across the channel between bends. The actions of the maximum velocity thread, coupled with those of the accompanying turbulent zones, are major processes in the erosion of the outside stream bank (the **cut-away slope**) and deposition of sediments on the inside of the curve (the **depositing slope**).

The presence of one or more maximum velocity threads gives rise to varying horizontal and vertical velocities within the stream proper. Such are in evidence as we view the body of a stream and observe the morphology of the stream bed. We are impressed by the presence of a large sand bar separated by a deep channel from, perhaps, a gravel bed; farther downstream the bottom may be strewn with large rocks or be quite muddy. These features attest to the ever-changing nature of the stream resulting from shifting of the velocity threads and the turbulent zones, and further, to the carrying capacity of the current itself as a function of velocity. A relationship between velocity and the composition of the stream bed is shown in Table 9.1. Obviously, the relationships are not nearly so precise as indicated here. The velocity varies with depth and with obstructions such as boulders. In other words, an organism attached to the bottom or to the downstream side of a large stone may not actually be inhabiting a current velocity indicated by the general nature of the stream bed. Studies of fluvial ecology, moreover, are usually made under conditions of relative calm, whereas the features of the stream bed often reflect the extreme effects of spring thaw, heavy runoff, and flood.

Swirls and eddies of widely ranging proportions constitute another form of current systems in streams. These systems serve to mix stream waters and to effect deposition of organisms, organic debris, and sediments. Many forms of algae are adapted to eddy dispersal of vegetative structures—they take advantage

TABLE 9.1 The nature of a river bed as a function of current velocity[a]

Velocity (m/sec)	Nature of bed	Habitat description
>1.21	rock	torrential
>0.91	heavy shingle	torrential
>0.60	light shingle	non-silted
>0.30	gravel	partly silted
>0.20	sand	partly silted
>0.12	silt	silted
<0.12	mud	pondlike

[a] Data from Tansley, 1939.

Figure 9.7   In watersheds in rocky regions a brief rainfall may greatly increase the discharge of streams. The photographs show Stony Brook, near Princeton, New Jersey, at low-water stage (above) and following a short period of rainfall (below). In the lower photograph the leaves caught on the bush indicate that the water had previously reached an even higher level than that shown. (Courtesy of W. J. Woods.)

of shallow, well-lighted swirls for rapid reproduction and subsequent distribution as the eddy contributes to the main current.

Density currents due to differences in temperature, chemical composition, or to silt load occur in streams, especially large streams with smooth bottoms. Spring-fed tributaries are often cooler than the main stream, and as the colder waters meet with the main stream a stratification results. In the lower reaches of a stream, or (more properly) in the estuary, salinity currents move with the tides underneath the fresh water. Near the mouth of the Escambia River, in Pensacola Bay, Florida, salinity measurements made in the autumn indicated a surface salinity of 4.5 parts per thousand (‰), while at the bottom (about 4.5 m) the salinity was 24.4‰. This sharp salinity stratification was also reflected in the distribution of freshwater and marine fish.

Tides exert influence on river flow, often well beyond the brackish-water zone. This effect serves to increase turbulence in the lower stream course, and also to bring about slight periodic flooding of the valley. This latter process increases the exchange between stream and land. The Paumunkey River flows on the Piedmont plateau and coastal plain of Virginia, and salinity effects are felt some 12.8 km from the coast. Tidal influences, however reach about 64 km upstream.

From the consideration of fluvial processes in Chapter 4, and from the brief review and synthesis given here, it is immediately apparent that current systems in the various stream regions as well as in the stream as a whole are extremely complex. Actually, these physical aspects of stream dynamics have received little extensive study in the field under natural conditions. A greater amount of study has been carried on in hydraulics laboratories, and considerable theoretical work has been accomplished. For further information, textbooks in hydraulics, and professional publications such as those of the United States Geological Survey, should be consulted.

CURRENTS IN ESTUARIES

The three most important factors operating to produce currents in estuaries are oceanic tides, stream flow, and wind. The interactions of these forces, particularly the somewhat antagonistic processes of oscillating tides (their vertical ranges and lengthwise flow) and unidirectional stream flow (its velocity and volume), serve to make the estuary a restless and complex system of water movements. Additionally, the morphology of the basin of the estuary and the channel of the stream modify and determine the stream and tidal dynamics. It should be borne in mind that these forces are not regular and constant. Stream flow varies seasonally with rainfall (Figure 9.7), while tide height and movement are correlated with lunar effects and wind. Winds may serve to increase either tidal movement or stream flow, depending upon the direction and intensity of the moving air mass.

Within most estuaries there exists a relatively regular and uniform rate of water transport. This is to say that the volume of estuarine water discharged to the sea through the mouth of the estuary essentially compensates for the amount of fresh water introduced by the stream into the upper region of the estuary. In an estuary where the incoming fresh water is mixed with the estuarine water, *i.e.*, where no stratification of the two exists, considerable time may be required for the discharge of a particular mass of the fresh water. This time interval is called the **flushing time** (see also Chapter 5). It is an average figure used to describe the period during which a quantity of fresh water derived from stream or seepage remains in the estuary. A simplified formula for determining flushing time (t_F) is

$$t_F = \left(\frac{S_s - S_f}{S_s} \ V_f \right) \frac{1}{v_D}$$

in which S_s and S_f represent the salinity of seawater at the entrance to the estuary and the salinity of the estuarine water, respectively, V_f is the volume of water in the estuary, v_D is the daily volume movement into and out of the estuary. Flushing time in Great South Bay, Long Island, through Fire Island inlet is of the order of 48 days, or approximately 96 tidal cycles (WHOI, 1951). Obviously a parcel of fresh water entering the estuary near the mouth will move out faster than one in the upper reaches. The main point is that, on the average, the downstream movement of dissolved substances and suspended materials, including living plankton, is a slow process. In many instances the time is sufficient for biological events, important in maintaining the integrity of the estuarine community, to occur.

In a **mixing estuary** the stream flow is held back and often reversed by a strong flood tide. Because of the forces of the two opposing current systems, the reversed flow up the stream is normally slow, ceasing completely as the peak of the flood is reached. Upon the ebb, the outward tidal flow is strengthened by stream flow, the velocity decreasing after the mid-time of the ebb and the movement coming to rest at dead low water (**slack tide**). As we have seen for streams, the cross-sectional velocity of an estuary is not constant; a thread of maximum velocity exists. In estuaries, this thread is usually near the surface in the middle of the water course, the precise position depending, however, upon the nature of the shores and bottom.

In many estuaries the typical current pattern is one wherein the lighter, fresh water flows seaward over the upstream movement of denser saline waters (denser by virtue of having a greater concentration of dissolved salts). Under these conditions, a **vertical salinity gradient** exists, and the estuary is said to show **stratification.** Studies on Chesapeake Bay (Pritchard, 1952) have indicated that there is a net horizontal seaward flow in the upper layer, and in the lower layer a net horizontal headward flow. The boundary between the two is not level, but slopes toward the right (looking downstream). In the Tamar Estuary of England, it has been found that at some distance upstream both currents may be seen at the surface (Milne, 1938). During times of considerable freshwater flow, the more saline waters of the incoming tide move to one side of the basin.

Where stratification and net seaward–headward movement occurs, it is possible for certain nutrients and organisms to be transported upstream in the bottom currents, while other materials may be carried to the sea in the upper currents of less saline water. It has been suggested that oyster larvae are moved upstream by lying on the bottom during an ebb tide and then rising into each successive flood tide to be more or less passively transported for the duration of the tide. Figure 9.8 depicts the flow pattern in a typical stratified estuary. Further relationships between currents and salinity distribution will be considered in Chapter 11.

In the preceding paragraphs we have seen some of the effects of oceanic tides on the dynamics of the estuary of a stream. In a great number of instances tidal effects are not restricted to the brackish zone, but occur throughout a considerable length of the stream–estuary–sea continuum. Indeed, in the low-lying course of the Amazon, tides serve to bring about measurable variations in stream depth some 960 km above the saltwater zone. The result is the entrance of tides into the stream before the preceding tide has ebbed. In the Amazon there are said to be eight tides along the stream course at one time.

The morphology of the stream channel and estuary basin regulates, to a great extent, the velocity and magnitude of tidal currents. In a wide, unrestricted estuary, currents are typically slow, without significant turbulence. In a narrow, deep basin, rapid, turbulent currents generally occur. If the mouth of the estuary is restricted by depositional features or land closures, the incoming tide may be held back until it suddenly breaks forth into the basin as a tidal wave, or **bore.** In the Severn Estuary, England (Figure 9.9), the tidal range may approach 15 m. A bore of 3 m (sometimes reaching 7.5 m) has been reported in the Tsientang River of China. Such currents exert profound effects on the nature of the substrate, turbidity, and biota of the estuary.

Seiches similar to those described in lakes may occur in coastal bays and estuaries. The period of these long stationary current systems is determined by the depth and horizontal parameters of the basin. Oscillations are apparently

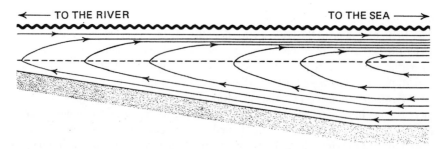

Figure 9.8 Schematic representation of current flow pattern along the central axis of an estuary. (After Pritchard, 1952.)

common in many coastal waters, the effects often obscured, however, by tides. A trinodal oscillation has been recognized in San Francisco Bay, and the great water-level fluctuations of the Bay of Fundy result from the synchronization of the tide and seiche, the latter having an amplitude of some 15 m. The causes of seiches in bays and estuaries are not completely known. It is probable that forces such as variations in atmospheric pressure and wind can supply sufficient energy for the development of standing waves in coastal bodies. As in lakes, friction operates to damp out the oscillation.

The importance of waves and wave action in the estuary is primarily a function of the morphology of the estuary basin. Wherever the mouth of the estuary is restricted and surface area of the body of water slight, there is little opportunity for the development of waves of any appreciable magnitude. Consequently, the role of wave action in watermixing and erosion is negligible. On the other hand, broad estuaries having wide mouths are subject to receiving the full effect of oceanic waves in addition to local disturbances. Generally, however, the effect of moderate waves is to erode the tidal marshes, resulting in the introduction of nutrient materials into the estuary and in modification of the shoreline.

Winds are also influential agents in producing certain currents in estuaries. We have already considered the possible role of wind in the development of seiches. Wind contributes greatly to the unusually high waters moved into coastal features during hurricanes and other storms. Depending upon duration and intensity, wind storms may temporarily disrupt the normal circulation patterns in estuaries. Similarly, changes in wind direction may influence the rate of tidal ebb and flood.

Figure 9.9   Severn Bore, at Gloucester, England. This is a rear view of the wave of translation brought up by high spring tides at the estuary mouth. The bore is usually reduced by the time it reaches Gloucester. (Courtesy of British Information Service.)

MOTION IN MARINE WATERS

As we saw in Chapter 6, marine waters differ from other aquatic environments in one special way; because of the great size of the oceans and seas, their major parameters are not morphological, but rather physical and chemical. For this reason, the most important aspects of movement in marine waters were discussed in some detail in that chapter (see pages 118 to 122). In this section we will recapitulate and modify slightly our earlier discussion.

OCEANIC CURRENTS

The **Coriolis force**—an effect of the earth's rotation (page 118)—is the major factor that sets marine waters in motion and determines the direction of oceanic currents. This basic movement is modified in many complex ways by such other factors as coastline configuration, submarine topography, convection cycling, and the confluence of large masses of water. The basic worldwide pattern of flow—at a rate varying from 0.5 to 6 knots—is westward, poleward, and back toward the equator, describing great oceanic **gyres** such as the Alaska Current in the northern Pacific or those encircling the Sargasso Sea in the subtropical North Atlantic.

The patterns of the major currents are steady and predictable. Their precise paths may vary, however, both seasonally and in response to long-term changes in the environment. For example, the greatest of the permanent ocean currents, the Gulf Stream, is observed, at high latitudes, to wander toward and away from the North American coast, sometimes carrying tropical species, such as the brown alga *Sargassum* or the jellyfish *Physalia,* to the shores of Long Island and New England.

Oceanic currents have also contributed to the distribution of plants and animals along the routes of the major currents. Naturalists have long noticed that organisms from the intertidal zones of widely separated continents are often surprisingly closely related; the explanation seems to lie in a number of species having been transported from one region to another by the persistent oceanic currents. Similarly, the fauna of the Galápagos Islands, some 1000 km west of Ecuador, was observed by Charles Darwin to bear a close resemblance to that of mainland South America. This observation later became a major support for Darwin's theory of evolution through natural selection.

TIDES AND WAVES

The dynamics of tidal and wave movements have been examined earlier; for details, the student is referred to Chapter 6 and to the earlier sections of this chapter. We shall, however, make several additional observations regarding the ocean tides. Tides are caused by the gravitational pull of the moon and sun upon the open seas—the **hydrosphere**—as the earth rotates upon its axis. In terms of the planet as a whole, the effect of the tides is seemingly miniscule. The Pacific Ocean, for example, has a width of 14,483,000 m; normal tides have the effect of raising and lowering a small portion of that area by a maximum of 3 m! Nevertheless, this slight alteration of sea level may be of crucial importance to coastal species.

The cycle of tides is much more complex than most of us realize. The basic pattern is for the tide to rise and fall twice a day (a **semidiurnal** cycle) in response to the earth's rotation. The moon and (to a lesser extent) the sun exert a gravitational pull that results in this movement of the water. Particularly high tides are observed approximately twice each month—at times of the new moon and the full moon—when the moon and sun are more directly aligned with the earth and the waters of the hydrosphere respond to their cumulative gravitational pull. This cycle—from the maximum (**spring**) tide to the minimum (**neap**) tide—takes about 14 days.

A third factor might also be mentioned. The orbit of the earth around the sun, and that of the moon around the earth, do not usually lie in the same plane. At certain times (roughly the times of the spring and the fall equinoxes, in March and September) the planes of the orbits do coincide briefly. The result is a particularly high **equinoxial spring tide,** which recurs in a semiannual cycle.

These phenomena suggest some of the complexity of tidal patterns. Other factors account for such further local peculiarities as **diurnal tides** (occurring once daily) and **mixed tides** (varying between once and twice daily, and being of unequal and varying magnitudes). Ecologically, the many permutations of tide patterns make the intertidal environment very demanding, and they make the survival and success of intertidal species particularly difficult. Not only must an intertidal organism be able to withstand the periodic submergence and exposure of the daily cycle, but it must be able to tolerate the extreme and prolonged variations that occur as a result of the long-term fluctuations in tide patterns.

SUGGESTED FOR FURTHER READING

Batchelor, G. K. 1967. *Introduction to fluid dynamics.* University of Cambridge Press, Cambridge, England. Introduction to the theory and application of fluid dynamics. A very complete text by a mathematician and theoretical physicist.

Defant, A. 1958. *Ebb and flow: the tides of earth, air, and water.* University of Michigan Press, Ann Arbor. An introduction, written for the non-hydrographer, to tides, seiches, and other rhythmic water movements. The final two chapters briefly explore the effect of lunar and solar gravitation on movement of earth's landmasses and atmosphere.

10 Gases in Natural Waters

arlier, in Chapter 2, we laid strong emphasis upon the extraordinary capacity of water to hold substances in solution and to enter into chemical reactions. These abilities, as we explained, are largely the result of the weak hydrogen bonding that links the molecules of liquid water; the molecules may be separated or united with the expenditure of a relatively small amount of energy. Thus all natural waters (including what we think of as "pure" rainwater) contain some materials in solution—dissolved gases and inorganic and organic solids. It has often been pointed out that pure water does not occur in nature, nor would we want it to; the substances held in solution are basic to the nature of life in aquatic communities, and the properties of water as a solvent and a reagent contribute, more than any others, to maintaining that life. The "impurities" in natural waters impart to lakes, streams, and marine waters the physical, chemical, and biological structure that maintains the continuity of the ecosystem.

These impurities are derived from many different sources, they are present in various states and quantities, they undergo different transformations, and they contribute in many different ways to the overall metabolism of the aquatic community. Certain gases, occurring in proper proportion, are essential to respiration and photosynthesis in aquatic situations; other gases are lethal to life. Certain dissolved mineral salts serve as nutrients for free-floating plants; other salts may limit life through osmotic effects. Dissolved organic materials are also present in natural waters, but their reactions are not well known. The nature, quantity, and transformation of substances dissolved in natural bodies of water are generally indicative of the origin of the basin or channel, the climatic regime of the area, and the composition of the substrate in the drainage system.

In this and the following chapter we consider certain properties and reactions of the more conspicuous substances that contribute to the chemical nature of natural waters as an environment. Although our approach will take the form of naming the materials and considering them in turn it is most important to keep in mind that any lake, stream, or other body of water is a dynamic system of chemical interdependencies and interrelationships in association with physical and biological features.

DISSOLVED GASES IN LAKES

A great variety of gases are found dissolved in natural waters. Some (such as hydrogen, nitrogen, NH_4, and H_2S) occur in association with dissolved solids or biological activity; they will be discussed in the following chapter, which is concerned with dissolved solids. The most widespread gases to occur in water under natural conditions are **oxygen** and **carbon dioxide.** In much limnological work, measurement of the amounts of these two gases present is the first step in any detailed study of the ecology of a lake.

OXYGEN

Of all the chemical substances in natural waters, oxygen is one of the most significant. It is significant both as a regulator of metabolic processes—both of community and of organisms—and as an indicator of lake conditions. Hutchinson (1957) has succinctly and aptly indicated the importance of oxygen:

> A skillful limnologist can probably learn more about the nature of a lake from a series of oxygen determinations than from any other kind of chemical data. If these oxygen determinations are accompanied by observations on Secchi disk transparency, lake color, and some morphometric data, a very great deal is known about the lake.

The oxygen available for metabolic relationships in natural waters is the oxygen held in simple solution. This is not, as some beginning students might think, the O in H_2O. The volume of oxygen dissolved in water at any given time is dependent upon the **temperature** of the water, the **partial pressure** of the gas in the atmosphere in contact with the water, the concentration of dissolved salts in the water (its **salinity**), and **biological activity**.

The solubility of oxygen in water is increased by lowering the temperature. For example, the solubility increases by more than 40 percent as fresh water cools from 25°C to freezing. This can be illustrated by a simple exercise with the **nomogram** in Figure 10.1. Place a ruler or other straightedge across the figure to connect a chosen temperature on the uppermost scale with a given percent saturation on the middle, inclined scale, say 10°C and 100 percent saturation. Now read the lowermost scale of oxygen concentration in ml/liter at the point where the ruler intersects the scale. Interpreted in one fashion we see that 7.9

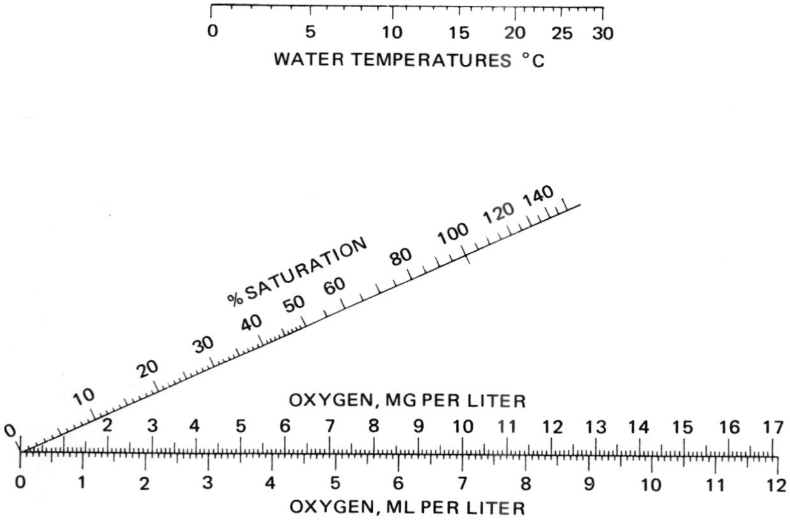

Correction Factors for Oxygen Saturation at Various Altitudes

Altitude		Pressure	
Feet	Meters	mm.	Factor
0	0	760	1.00
330	100	750	1.01
655	200	741	1.03
980	300	732	1.04
1310	400	723	1.05
1640	500	714	1.06
1970	600	705	1.08
2300	700	696	1.09
2630	800	687	1.11
2950	900	679	1.12
3280	1000	671	1.13
3610	1100	663	1.15
3940	1200	655	1.16
4270	1300	647	1.17
4600	1400	639	1.19
4930	1500	631	1.20
5250	1600	623	1.22
5580	1700	615	1.24
5910	1800	608	1.25
6240	1900	601	1.26
6560	2000	594	1.28
6900	2100	587	1.30
7220	2200	580	1.31
7550	2300	573	1.33
7880	2400	566	1.34
8200	2500	560	1.36

WATER TEMPERATURES °C

% SATURATION

OXYGEN, MG PER LITER

OXYGEN, ML PER LITER

Figure 10.1 A nomogram for determining oxygen saturation values at various temperatures and altitudes. A nomogram is designed to be used by connecting the observed temperature and the dissolved oxygen concentration with a straightedge or a dark-colored thread. The point at which the straightedge intercepts the inclined scale indicates the percent saturation. The factors in the altitude/pressure scale are applied to correct for those factors (After Rawson, 1944.)

ml/liter of oxygen constitute 100 percent saturation of 10°C. Now connect 5° C and 100 percent saturation and note that under these conditions a greater concentration of oxygen is dissolved at 100 percent saturation—9.0 ml/liter to be exact. In other words, solubility increases with a drop in temperature. The nomogram can also be used to obtain the value for the **weight** of a dissolved gas, in mg/liter, by reading the upper side of the oxygen concentration scale rather than the lower side; in that case the oxygen values at 10° and 5° C are, respectively, 11.3 and 12.9 mg/liter. It should be noted that this lower pair of scales can be used as a table for converting mg/liter or parts per million (1 mg/liter = 1 ppm) to ml/liter. The original purpose of the nomogram is to determine percent saturation of a given oxygen concentration at various altitudes and temperatures; altitude must be considered in order to take pressure into account.

At a given temperature the concentration of a saturated solution of a slightly soluble gas such as oxygen, which does not unite chemically with the solvent, is very nearly directly proportional to the partial pressure of that gas (Henry's Law). The solubility of oxygen also relates to Dalton's "law of partial pressures" which states that the total pressure of a mixture of gases is equal to the sum of the pressures exerted by each of the component gases. In other words, the solubility of each gas is independent of other gases in the mixture. Under similar conditions of pressure and temperature the solubility of oxygen in water is over twice that of nitrogen and about one third that of carbon dioxide.

The student should note that "100 percent saturation," as used here, does not refer to all the gas that can be held in solution; rather it refers to the amount that occurs **at equilibrium** under a stated set of conditions. Thus, on a sunny afternoon in water containing a large amount of plankton, the observed "saturation" value may range upward to well above 150 percent.

The effect of the third factor, the concentration of dissolved salts, upon the solubility of oxygen is that of decreasing the oxygen concentration as salinity increases. At 0°C fresh water at saturation contains slightly over 2 mg/liter more oxygen than does average sea water (35‰ salinity); at 15°C the difference is about 1.5 mg/liter.

Having given attention to the major factors responsible in regulating the solution of oxygen in waters, let us now consider the sources of this gas. Obviously the atmosphere in contact with the lake surface is a potentially inexhaustible source of oxygen. The volume percent of oxygen in the atmosphere is calculated to be 20.99, or approximately 210 cc of oxygen per liter of air. This is some 25 times the concentration of oxygen in the same volume of fresh water.

The rate at which atmospheric oxygen passes across the air–water interface and becomes dissolved in the water is dependent upon a number of factors. For one thing, increased wave action or other disturbances at the lake surface results in greater passage of the gas into solution. Second, the greater the difference in partial pressure between air and water, the greater the rate of solution. Third, the less the moisture content of the gas, the more rapid the solution of that gas. Bear in mind, however, that in all of these processes there may be a "two-way" movement, that gas can also be lost from the lake to the atmosphere. The **direction** of movement, as well as the **rate**, is determined by the foregoing factors.

Oxygen in natural waters may also be derived from photosynthetic activity. In shallow, nonstratified ponds lacking significant wave action, oxygen may be present mainly as a byproduct of carbohydrate synthesis by rooted plants and by phytoplankton, according to the formula

$$6CO_2 + 12H_2O \rightarrow C_6H_{12}O_6 + 6O_2 + 6H_2O$$

or, more simply,

$$6CO_2 + 6H_2O \rightarrow C_6H_{12}O_6 + 6O_2$$

Photosynthesis is also a source of dissolved oxygen in large deep lakes, but photosynthesis in such a lake is restricted to a relatively small zone, which is delimited by the vertical range of transmission of light effective in photosynthesis. This lighted region is called the **euphotic zone**. It extends horizontally

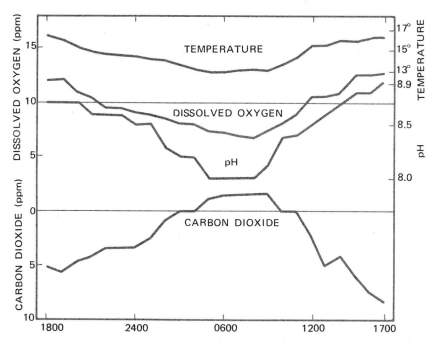

Figure 10.2   Diel fluctuations in temperature, dissolved oxygen concentration, pH, and carbon dioxide content in the summer surface waters of Buckeye Lake, Ohio. Negative carbon dioxide values represent the amount of the gas needed to make the water neutral to phenolphthalein (see p. 206). (After Tressler et al., 1940.)

from shore to shore and vertically from the surface to a level beyond which photosynthesis-effective light fails to penetrate. As we have seen, turbidity, color, and the absorptive effect of water itself serve to quench light; thus they essentially determine the euphotic zone. In the shallow shore region (the **littoral**), submerged rooted plants along with phytoplankton contribute oxygen to the lake. In the euphotic zone of the open lake (the **limnetic**), phytoplankton contributes the autochthonous oxygen. In most temperate lakes, during summer thermal stratification the euphotic zone corresponds closely with the epilimnion. This accounts mainly for the low-oxygen conditions in the hypolimnion mentioned on page 151. Within the near-surface region of maximum photosynthesis and oxygen gain from the atmosphere, the water often becomes supersaturated with oxygen at the height of diurnal concentration cycles. Maximum oxygen production usually occurs in the afternoon on clear days, the minimum immediately after dawn. Cyclic daily fluctuations in oxygen concentration, called the **oxygen pulse**, have been observed when conditions are suitable. Figure 10.2 illustrates hourly changes in dissolved oxygen concentration along with fluctuations of other physical and chemical features of a lake.

The lower limit of the euphotic zone is marked by a level at which organic respiration and decomposition consume oxygen at a rate equal to that at which oxygen is produced over a 24-hr period. This is the **compensation level**; it will enter significantly into our later discussion of productivity. Below the compensation level in our typical lake lies the **aphotic zone**, a zone in which there is insufficient light to maintain oxygen production at compensation. This is the region of low oxygen content, often depleted. It frequently corresponds with the hypolimnion.

A vertical oxygen distribution pattern such as we have been considering is characteristic of moderately productive lakes of small size. These lakes, termed **eutrophic** (''well nourished''), typically exhibit a hypolimnetic loss of oxygen, the curve of the oxygen distribution dropping sharply through the metalimnion and described as **clinograde.** Figure 10.3(a) represents a clinograde oxygen distribution in a eutrophic lake. It seems generally agreed that temperature and oxidation of organic materials are the major factors that cause oxygen depletion in the hypolimnion. The rate and magnitude of this oxidative breakdown are dependent upon the volume of organic substance supplied to the process. This substance for the most part is produced in the upper lighted zone, which, in relation to the organic synthesis, is called the **trophogenic region.** From here the materials settle into the zone of decomposition, or **tropholytic zone.** It is now apparent that correlations exist among epilimnetic circulation, light transmission in the euphotic zone, and the production of organic substance in the corresponding trophogenic region.

At somewhat the other extreme of oxygen distribution and biological processes we find relatively deep, unproductive, or **oligotrophic** (''poorly nourished'') lakes. Such lakes are usually rather clear for considerable depths and turbidity is low. Since light transmission is high, the euphotic zone is deep and photosynthesis may occur, although reduced, at some distance below the upper strata. As a result of low productivity per unit volume there is little oxidation in the hypolimnion. The oxygen curve may show a decided increase through the metalimnion and a high oxygen content in the hypolimnion. An **orthograde** (vertical) oxygen curve in an oligotrophic lake is shown in Figure 10.3(b). Between the two extremes in oxygen distribution, production, and uptake described here exists a vast range of intermediate and intergrading patterns, many of them exhibiting various complications in oxygen states and metabolism.

In certain clear lakes during summer stratification very high oxygen concentrations may occur in the metalimnion (Figure 10.4). This great concen-

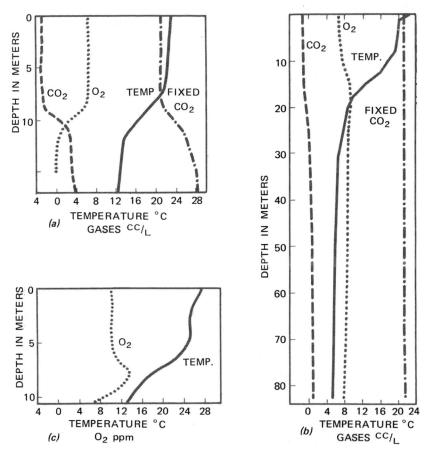

Figure 10.3 Various typical oxygen curves in lakes. (a) Clinograde curve of vertical oxygen distribution in Conesus Lake, New York, August, 1910; (b) Orthograde curve for Skaneateles Lake, New York, August 1910; (c) Orthograde curve showing decrease in dissolved oxygen near lake bottom in a New Jersey quarry (no date). (After Birge and Juday, 1914.)

tration of the gas is apparently derived from levels in and below the epilimnion in which the transparency of water permits a high rate of photosynthesis. The **metalimnetic oxygen maximum** may be followed by a corresponding marked reduction of oxygen in the metalimnion. The mechanism, or mechanisms, responsible for this **oxygen minimum** cannot be described with certainty. It has been suggested that rapid oxidation of slowly settling materials from the epilimnion might account for the low metalimnetic oxygen concentration. Such a process would necessitate that the position of the metalimnion be below the trophogenic zone or at least that photosynthesis proceed at a reduced rate.

The annual cycle of oxygen distribution in a dimictic lake generally follows the pattern shown in Figure 8.2 (page 151); note that underneath the ice in March the concentration of dissolved oxygen is 14 ppm. This saturation developed during autumn circulation, when falling temperatures increased the solubility of the gas. In winter some of the oxygen is consumed by organism respiration and by organic decomposition. In deep lakes oxygen consumption is relatively small, but in shallow bodies of water the oxygen depletion may kill a great number of animals. Vernal circulation redistributes the oxygen as shown for late April and early May. With the development of summer stagnation the hypolimnion becomes quite poor, and often completely lacking, in oxygen.

In meromictic lakes the oxygen distribution reflects the essential two-region structure of the water mass, with the denser, more saline waters lying at greater depths. Depending upon the circulation pattern above the chemocline, the oxygen content may take the form of a clinograde curve. The waters of the monimolimnion are usually anaerobic. Lakes of the tropical and subtropical regions exhibit widely diverse oxygen relationships. Polymictic lakes, by virtue of continuous circulation, exhibit a generally high, uniformly distributed, oxygen content throughout the year. Oligomictic waters, in which circulation seldom occurs, are typically characterized by a condition of hypolimnetic oxygen

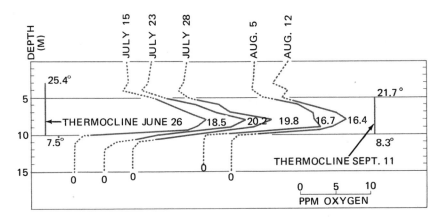

Figure 10.4   Metalimnetic (thermoclinal) oxygen maxima in Myers Lake, Indiana, during the summer of 1952. (After Frey, 1955.)

depletion. In shallow lakes of such colder regions as Alaska, oxygen at near saturation appears to be uniformly distributed by circulation from surface to bottom.

The transport of dissolved oxygen throughout a lake is accomplished primarily by currents set up within the lake and by eddy conduction. Lake currents, as we have previously seen, may be set into operation by wind action on the surface and by density differences between layers in the lake. Circulation in the epilimnion during summer distributes oxygen within that zone. Eddy systems thrown up along the interfaces between thermal layers or along stream density currents moving through the lake serve to move substances across boundaries, usually at right angles to the current. During seasonal overturn periods the sinking of upper, dense water masses also delivers oxygen to the underlying regions, thereby accounting for the homogeneous oxygen content during those times of uniform density. Simple diffusion of oxygen molecules plays a minor role, indeed, in oxygenation of natural waters. The process is very slow. As a matter of fact, it has been shown that if diffusion from the surface were the only process operating, it would require over 600 years to raise the oxygen concentration at a depth of 10 m by only 0.4 mg/liter.

With respect to the utilization and ultimate fate of dissolved oxygen in lakes, many aspects have already been considered. In summary, we might say that the major processes acting to consume oxygen are animal and plant respiration, and organic decomposition. During long periods of cloudiness or where waters may otherwise be shaded, for example underneath broad floating mats of vegetation, organismal respiration and decomposition may rapidly take up oxygen. Where decay bacteria occur in great quantity, as in lakes or streams highly polluted with organic sewage, the water often becomes completely anaerobic.

Although we have been concerned primarily in this section with the maintenance of oxygen in lake waters, the source of the gas, and its distribution and fate, we have had to venture into the subject of productivity. This is a natural consequence because there is an obvious direct relationship between the dissolved oxygen content of natural waters and the amount and rate of energy fixation. Recall that the form of the curve of vertical oxygen distribution indicates a great deal about photosynthesis and decomposition at various levels within a single lake, as well as providing a point of comparison of several lakes. A fuller account of these important aspects of natural waters will be given in the later sections of the text.

REDOX POTENTIAL

It has been known in chemistry for many years that oxygen plays an important role in chemical changes in various kinds of solutions. Only within the past 40 or so years, however, has investigation of certain oxygen-related phenomena been transferred from laboratory beaker and battery jar to natural waters of lakes and seas. One aspect of these laboratory and field investigations has been concerned with oxidation and reduction processes. This line of research has already been fruitful in contributing to our knowledge of aquatic chemistry and its influence on the activities and distribution of certain organisms.

The term **oxidation** is applied to the process in which oxygen is added to a substance, or in which hydrogen is lost from a compound, or in which an element loses electrons. Conversely, the loss of oxygen, the addition of hydrogen, or the gain of electrons is termed **reduction**. At the elemental level an electron transfer in which **ferrous** iron is oxidized to **ferric** iron may be stated

$$Fe^{++} \leftrightharpoons Fe^{+++} + e$$

Under certain conditions, the oxidized iron may undergo a change back to the reduced state by a shift in electrons; this is suggested in the reversibility of the process shown above.

The extent to which a substance can undergo oxidation–reduction processes is dependent upon the concentration of other oxidizing–reducing systems and their products in the solution. Within a given solution, the proportion of oxidized to reduce components of a particular system in relation to other systems constitutes the **oxidation–reduction potential**, or **redox potential**. In a solution, for example lake water, a system with a given redox potential undergoes reduction and oxidizes a system of lower redox value. It may also cause a reduction to take place in a system of higher redox potential.

In practice, the redox potential is determined by immersing a nonreactive electrode (bright platinum is often used) in a solution. The second electrode, necessary to complete the circuit, is usually the hydrogen electrode. The presence of the electrode sets up an electron flow, the direction of which depends upon the proportion of oxidized to reduced material. Where an excess of the reduced state (Fe^{++}) exists, electrons flow to the electrode and oxidation in the system results. In a solution containing a surplus of the oxidized state (Fe^{+++}), flow from the electrode contributes electrons and reduction occurs. The relative excess or deficit of electrons can be measured in volts or millivolts (1 mv = 0.001v) on a potentiometer, as an electromotive force about the electrode. This measure gives the intensity of the force (E_h), either positive or negative, and is the redox potential. A positive E_h reading results from a state tending toward oxidation; a negative E_h indicates a system causing reduction. In addition to **intensity**, the oxidation–reduction dynamics also has the attribute of **capacity**. The capacity of an oxidation–reduction system refers to the ability of a system to undergo a certain amount of oxidation–reduction transformation without an intensity change.

Oxygen in natural waters produces a redox potential which is influenced considerably by temperature and the hydrogen ion concentration (pH). In order to allow for pH effects, the potential is measured at the prevailing pH and then referred to pH 7, this measurement being called the E_7. Correlations between ranges of E_h at pH 7 and oxidation–reduction reactions in certain systems have been established as follows:

$$NO_3^- \text{ to } NO_2^- \quad : 0.45 \text{ to } 0.40 \text{ v } (+450 \text{ to } +400 \text{ mv})$$
$$NO_2^- \text{ to } NH_3 \quad : 0.40 \text{ to } 0.35 \text{ v } (+400 \text{ to } +350 \text{ mv})$$
$$Fe^{+++} \text{ to } Fe^{++} \quad : 0.30 \text{ to } 0.20 \text{ v } (+300 \text{ to } +200 \text{ mv})$$
$$SO_4^= \text{ to } S^= \quad : 0.10 \text{ to } 0.06 \text{ v } (+100 \text{ to } +60 \text{ mv})$$

At a temperature of 25°C and pH 7, well-aerated lake waters exhibit a redox potential of about 0.5 v(+500mv). This potential remains relatively steady as long as the oxygen content is above approximately 1 mg/liter. In other words, the redox potential, as such, is affected but little by the oxygen concentration, except under the near-anoxic conditions.

Generally speaking, the curve of vertical distribution of the E_h in lakes follows that of dissolved oxygen and may be a "mirror image" of that for ferrous iron. In stratified lakes exhibiting a somewhat orthograde curve of oxygen, the redox potential gives essentially the same pattern. In small lakes in which the oxygen in the hypolimnion is quite low, therefore giving a clinograde oxygen curve, the E_h usually, but not always, appears in a clinograde distribution. Gradations between these forms are common.

Oxidation-reduction activities at the water-mud interface of the lake bottom bear markedly upon the lake chemistry, particularly in the deep hypolimnion, and upon the type of organisms present in the shallow part of the sediments. The E_h of mud exposed to oxygenated water varies near 0.5 v (+ 500 mv). This potential may extend for several millimeters through an oxidized zone of brownish mud at the surface of the bottom sediments, the **oxidized microzone**. Accompanying the decrease in hypolimnetic oxygen with summer stagnation is a diminution in the depth of the oxidized microzone. As the E_h of the interface approaches 0.2 v (+200 mv), the oxidized microzone may disappear (Figure 10.5). Below the microzone, the sediments of deep-water lakes are usually highly reducing in nature, the value of E_h approaches zero. Still deeper, in the anoxic region, E_h values are negative, and values in the -100 to -200 mv range are not uncommon. An oxidized microzone may be persistent in lakes of size and depth sufficient to exhibit an orthograde oxygen distribution. The processes responsible for maintaining the integrity of the microzone are not completely understood. It was early suggested that molecular diffusion of oxygen through the mud depended upon the reducing state of the sediment. Turbulence may also be important.

We are agreed that low oxidation–reduction potentials suggest the presence of reducing substances which would, in all probability, utilize such free oxygen as might be brought into the solution. For organisms such as anaerobic bacteria, or other organisms not requiring free oxygen for respiration, a low E_h would pose no problem. Anaerobic bacteria exist where the E_h lies below -400 mv.

Oxygen-dependent plants and animals would be excluded from such a zone, although they may be found inhabiting places where the E_h is as low as -200 mv. Zonation of certain insect larvae in lakes has been correlated with E_h. The midge *Calopsectra (Tanytarsus)* sp. dominated the insect fauna in bottom muds over which the E_h of the water was $+400$ mv or above. Another midge, *Tendipes (Chironomus)* sp., characterized muds in waters in which the E_h was below $+300$ mv.

CARBON DIOXIDE

The very great importance of carbon dioxide as a contributor to the fitness of natural waters as environment derives from essentially three factors. In the first place it serves in a more or less purely chemical sense to "buffer" the environment against rapid shifts in acidity and alkalinity. In this sense carbon dioxide ameliorates the chemical environment through the ability of the gas to combine with water to form an acid, and to react to give a neutral salt, or a base. We shall give attention to these reactions later. A second contribution of importance by carbon dioxide pertains to regulating biological processes in aquatic communities. Seed germination of some plants, as well as plant growth, is determined by the concentration of carbon dioxide. Various animal processes such as respiration and oxygen transport in blood are related to carbon dioxide. A

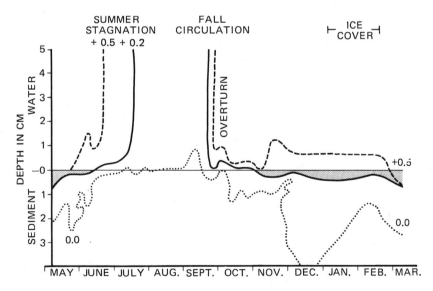

Figure 10.5   Seasonal variation in redox potential at the mud/water interface in Esthwaite Water, England; depth is 14 m. The heavy line represents the isovolt line $E_h = +200$ mv. The portion of the sediment indicated by light shading is (oxidized) brown mud, and the remainder is (reduced) black mud. (After Mortimer, 1942.)

third and most important contribution by carbon dioxide lies in the fact that it contains carbon. Carbon is one of the most versatile of all elements, because its nuclear structure allows it to combine with many other elements to form an enormous variety of compounds, some of them exceedingly complex. The ability to form many compounds is due largely to the asymmetrical nature of the bonds and to the fact that carbon can form chains of atoms almost without limit. This latter characteristic distinguishes carbon from other elements and from inorganic substances, although many inorganic compounds, of course, also contain carbon. Carbon dioxide and water supply the carbon, hydrogen, and oxygen which are major components of protoplasm.

The numerous and varied activities of carbon dioxide in the aquatic ecosystem are made possible primarily by the very high solubility of the gas in natural waters. The solubility of carbon dioxide varies considerably in nature, because carbon dioxide normally exists in a delicate equilibrium with several other compounds, forming a complex buffer system. Under laboratory conditions, however, the solubility of the gas varies inversely with temperature. It can safely be said that at normal ''natural'' temperatures, carbon dioxide is much more soluble than oxygen in water. At 20°C and atmospheric pressure of 760 mm Hg, water in equilibrium with atmospheric carbon dioxide contains about 0.88 vol of the gas; only 0.031 vol of oxygen are contained in water under similar conditions. Although air contains some 700 times more oxygen than carbon dioxide (by volume), the proportion in water at equilibrium is more nearly equal, about 4 ml carbon dioxide per liter to 6 ml oxygen per liter. If we consider the amount of carbon dioxide ''locked up'' in various combined forms, the total amount of the compound in water, particularly the oceans, is much greater than in air.

Carbon dioxide in natural waters is derived from a number of sources. Bacterial decomposition of organic matter in the tropholytic zones, and respiration by animals and plants, contribute to the store of carbon dioxide. In the case of plants the greater net contribution is at night, when photosynthesis is not occurring. Ground waters flowing or seeping into lakes and streams may carry carbon dioxide, the amount being determined by the extent of decomposition in the topsoil and, as we shall presently see, by the chemical nature of the underlying rocks. Within the body of water, certain chemical reactions between acids and various compounds of carbonates release carbon dioxide. Finally, the atmosphere directly furnishes some carbon dioxide to natural waters, and rain, as it falls through the atmosphere, dissolves some of the gas and delivers it to lakes and other waters.

The importance of rain in supplying carbon dioxide to inland waters lies in reactions wherein carbon dioxide is maintained and transported in forms other than as a gas. As rain percolates through soil containing carbon dioxide of decomposition, some of the gas becomes dissolved in rain water. The reaction between CO_2 and H_2O results in the formation of carbonic acid (H_2CO_3). If this weak acid encounters carbonate-holding rocks, limestone ($CaCO_3$) for example, the latter dissolves as calcium bicarbonate, or $Ca(HCO_3)_2$. The solution of Ca $(HCO_3)_2$ remains stable only in the presence of a certain amount of **free,** or **equilibrium, carbon dioxide.** Free carbon dioxide represents the CO_2 in H_2CO_3 plus that in simple solution. The formation of H_2CO_3 and its dissociations are shown in the reactions

$$CO_2 + H_2O \leftrightharpoons H_2CO_3 \leftrightharpoons H^+ + HCO_3^- \leftrightharpoons 2H^+ + CO_3^=$$

Note in the above reactions that carbon dioxide is contained in two states not previously encountered, i.e., as bicarbonate and carbonate radicals, HCO_3^- and $CO_3^=$ respectively. This carbon dioxide is called **combined carbon dioxide.** Solution of $CaCO_3$ is dependent upon the addition of CO_2 in an amount greater than that of free carbon dioxide; this additional CO_2 is known as **aggressive carbon dioxide.** The major points to be gained from these considerations include (1) the relations between rain and soil in supplying compounds containing carbon dioxide to natural waters, (2) the chemical reactions by which those compounds are formed, and (3) the occurrence of carbon dioxide in its three forms: free (CO_2 in solution plus that in H_2CO_3), half bound (HCO_3), and bound ($CO_3^=$). The direction of reactions involving the forms of carbon dioxide and the very occurrence of these in natural waters are reciprocally related to acid-base relationships in the medium. When carbon dioxide disolves, the reaction and end products depend upon the nature of the solvent, particularly with relation to the hydrogen-ion concentration. Under acid conditions, the combination is as shown above; H_2CO_3 is formed, followed by dissociation to $H + HCO_3^-$. Under highly basic conditions the reaction is

$$\text{Base } OH + H_2CO_3 = H_2O + base^+ + HCO_3^-$$

Chemical Buffering

The maintenance of near-neutral conditions in mineralized waters is due to **buffering** by chemical systems such as the carbon dioxide–bicarbonate–carbonate complex. Other systems may involve magnesium, sodium, or potassium. In other words, as acid conditions arise, the reaction between acid and base from the bound carbonate, for example, brings about an increase in neutral bicarbonate. Continued acidification releases carbon dioxide and carbonic acid from the bicarbonate accompanied by loss of carbon dioxide from the system. This reaction is the basis for certain techniques used in aquatic ecology to determine the so-called **alkalinity** of natural waters. In these tests a quantity of strong acid is added to water in the presence of a proper indicator. The amount of acid necessary to convert any carbonate or bicarbonate present to free CO_2 is a measure of the HCO_3^- and $CO_3^=$ in solution. In this sense, alkalinity refers to the quality and quantity of compounds which bring about a shift in the pH of a solution toward the alkaline side of the pH range. Although not always the case, alkalinity usually reflects the activity of calcium carbonate. Therefore, three forms of alkalinity may be recognized: bicarbonate (originally determined by the use of methyl orange as indicator, and thus referred to as **M.O. alkalinity**), normal carbonate (**phenolphthalein alkalinity**), and hydroxide. See Welch (1948) or American Public Health Association (1971) for descriptions of methods and interpretations of results. At the other end, increased alkalinity of the solution brings about a reaction between the base and carbonic acid to hold the departure from neutrality to a smaller value than would otherwise be the case if the buffer system were not present.

Definition of pH

The pH is the logarithm of the reciprocal of the hydrogen-ion (or more properly, the hydronium-ion) of a compound. The pH may be expressed mathematically as follows:

$$pH = \log \frac{1}{(H^+)}$$

where (H^+) is the amount of hydrogen ions in a solution in moles per liter. In a liter of pure water there is 0.0000001 of hydrogen mole/ions (and a corresponding quantity of (OH^-)). The pH of pure (neutral) water is, therefore,

$$pH = \log \frac{1}{0.0000001} \text{ , or 7}$$

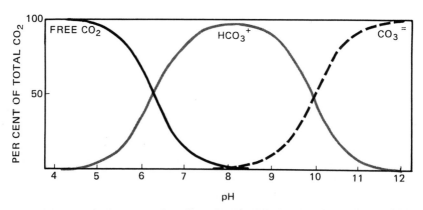

Figure 10.6   Hydrogen ion concentration (pH) as related to the occurrence of three forms of carbon dioxide in water. Heavy curve = free carbon dioxide; light curve = bicarbonate; broken line = carbonate radical. (After Emerson and Green, 1938.)

Increase in the concentration of H^+ ions results in a lower pH value, or, conversely, reduction in the H^+ concentration brings about a higher value. From this relationship, a pH scale has been devised. This scale is usually limited to a range from pH 0, corresponding to a solution with $(H^+) = 1$, through pH 7 (or neutrality) to pH 14, corresponding to a solution with $(H^+) = 10^{-14}$. From pH 0 to pH 7 solutions are **acid.** From pH 7 to pH 14 reactions are **alkaline.** With respect to the buffering system, we note that only free carbon dioxide is of any import in natural water systems below pH 5, that bicarbonate dominates the range from pH 7 to pH 9, and that the carbonate radical is most important in the range above pH 9.5 or 10. These relationships are shown in Figure 10.6.

The pH range of lakes having some degree of flow through the basin is generally from about 6 to 9. In limestone regions the dissolved carbonates may extend the pH range considerably beyond 9. In basins lacking outlets, evaporation may concentrate alkaline substances, resulting in pH readings of over 12. At the other extreme, accumulation of acids such as sulfuric acid in volcanic lakes and mine drainage gives a pH as low as 1.7.

Occurrence of Carbon Dioxide in Lakes

Having given attention to some of the major aspects of the sources of carbon dioxide in natural waters, as well as to the complex and often critically balanced reactions of the forms of the compound, let us now turn our attention to general considerations of the compound in lakes. From what we have seen of the close relationship between the chemical nature of the drainage basin substrate and the chemistry of waters of the basin, we should expect wide regional variation in carbon dioxide content of lakes. Newly formed lakes in regions of weakly soluble rocks typically contain little carbon dioxide in any form. In these situations the low quantity of carbonates in the substrate results in little of the bound or half-bound forms of carbon dioxide. Similarly, the paucity of soluble minerals as nutrients in biological processes inhibits the development of large biotic populations that would contribute carbon dioxide through respiration, and, of course, decomposition is reduced. Lakes of this type are usually slightly acid, the pH ranging near 6. Lakes of higher acidity (pH 4 to 6) are common in regions of lowlands and bogs. In these waters the free carbon dioxide content is usually quite high, ranging to nearly 200 ppm. As would be expected in view of the low pH, the concentration of bound carbon dioxide as carbonate is low, usually less than 9 or 10 ppm. Lakes having these characteristics are termed **soft-water lakes.** A large number of lakes fall into a class in which the pH is circumneutral, and which may be called **medium-water lakes.** Free gaseous carbon dioxide in medium-water lakes varies widely, frequently showing supersaturation relative to the partial pressure of the gas in the atmosphere. These lakes may contain bound carbon dioxide up to 30 or 35 ppm. In regions where the substrate contains easily dissolved minerals, **hard-water lakes** occur. These lakes are characterized by negative values for free carbon dioxide due to withdrawal of bicarbonates at a greater rate than carbonates are precipitated, and by pH values ranging from about 8.5 upwards. Bound carbon dioxide content amounts to over 35 or 40 ppm, often reaching 200 ppm or more. Calcium or magnesium carbonate is often precipitated as **marl**, which was mentioned earlier (page 60) and will be discussed further in Chapter 11. In some highly saline lakes, carbonate concentration may be 8500 ppm or above.

Seasonal Cycle of Carbon Dioxide and pH

During winter in the colder regions ice cover forms over the lake, thereby inhibiting exchange of materials across the air-water interface. Thus the lake becomes essentially "sealed in." Oxidation of organic substances, particularly in the depths, consumes oxygen and increases the carbon dioxide content. This often results in the building up of a gradient of metabolic substances from surface to bottom, especially if there is some transmission of light through the ice permitting photosynthesis in the upper, unfrozen waters. Under such conditions the waters immediately below the ice may contain a considerable quantity of oxygen, sometimes reaching near saturation, and little or no free carbon dioxide. This zone may show slightly alkaline conditions, with low bicarbonate concentration and absence of carbonates. Remember that we should expect to find carbonate only in the absence of free carbon dioxide. In the deep waters, bicarbonates and free carbon dioxide are increased, while oxygen may be nearly or completely depleted; the pH may drop, as during summer stratification.

In the warmer regions of the temperate zones, nonfreezing lakes typically contain quantities of carbon dioxide more or less uniformly distributed throughout the waters during winter. This is, of course, due to circulation during the cold season. At this time the phenophthalein alkalinity of surface waters is usually nil, and carbonates are absent.

In holomictic lakes the period of vernal mixing brings about the relatively even distribution of all dissolved materials. During this overturn the pH of the water is uniform from surface to bottom. Free carbon dioxide and other gases, such as methane and hydrogen sulfide, derived from winter decomposition processes, are lost to the atmosphere. Following the loss of the free form, bound carbon dioxide reappears as carbonate in medium and hard-water lakes as summer stratification begins.

We have already seen that in productive holomictic lakes exhibiting two circulation periods the oxygen concentration diminishes rapidly with depth during summer, giving a clinograde curve. In these lakes the carbon dioxide and carbonate concentration show a general inverse relationship to the oxygen; that is, the concentration of carbon dioxide and bicarbonate increases slightly with depth (Figure 10.3). On the other hand, an orthograde oxygen distribution is usually accompanied by only slight increase, if any, in carbon dioxide. During the summer the epilimnion of medium and hard-water lakes is typically devoid of free carbon dioxide, and contains measurable quantities of carbonates. It has been shown, for example, that in Douglas Lake, Michigan, in July, free carbon dioxide may be absent to a depth of about 14 m, and in this same distance carbonates range from about 10 to 8 ppm. The pH in the upper waters is slightly over 8. At a depth of 20 m, the free carbon dioxide content reaches 11 ppm at a pH of 7.1. As a point of interest, records over a 30-yr period show that the surface alkalinity due to half-bound carbon dioxide varied only from 110 to 128 ppm.

Autumnal circulation brings about uniform physical and chemical conditions throughout the lake. Vertical gradients in density, temperature, dissolved gases, and solids, which became established during summer stagnation, are broken down. The hypolimnetic gases of decomposition are thrown off at the lake surface. The lake's oxygen content is circulated, and with lowering temperatures of autumn an increased supply is dissolved in the water.

OTHER LAKE GASES

Methane is an organic gas (CH_4) widely called "marsh gas," common in many alkaline lakes, ponds, and swamps, particularly during summer stratification. It is produced by bacterial decomposition of organic substances in the tropholytic zone, primarily in the lake bottom. The process involves a multistage breakdown of complex organic material to organic compounds of simple molecular structure and then further decomposition, releasing methane and carbon dioxide. It occurs only under anaerobic conditions. In carbohydrates, for example, cellulose is attacked by bacterial enzymes and hydrolyzed to a simple sugar. Anaerobic decomposition of the sugars may result in the formation of hydrogen and methane. The time of highest production appears to be during summer stagnation when the bottom muds have become significantly reducing, that is, when the redox potential is low. The process can proceed at relatively low temperatures, about 5°C, because of the tolerance of at least one of the methane-producing bacteria.

In shallow bodies of water such as ponds and swamps, and in the shore zones of lakes, bubbles containing methane and other gases are often seen rising to the surface and erupting. The formation of bubbles is apparently due to insufficient water in the shallow situations to dissolve the gas as it is formed in bottom mud and debris. It follows then that the occurrence of bubbles at the surface over deep water would be unlikely. Under winter conditions methane bubbles may be present in the ice cover. Analysis of the gases in bubbles in a Russian lake (Beloye) revealed a methane proportion of from 74 to about 84 percent, and 5 to 18 percent hydrogen. As the bubbles rise to the surface, the hydrogen is lost and the methane volume diminishes to near 24 percent. In the ascent, nitrogen and a small quantity of oxygen are gained. These transfers of gases into and out of the bubble operate, of course, under the law of partial pressures. Although the full story of methane and its role is not known, it does appear that some of the gas is oxidized by organisms in oxygenated zones of lakes as bubbles rise. Here we see yet another phase in lake metabolism serving to decrease the oxygen content of deep waters.

Additional gases occur in natural waters and should be mentioned at this point, even though certain of them occur in small quantities. These will be considered in Chapter 11 in relation to dissolved solids. As we have already seen, hydrogen is formed as a decomposition product in the anaerobic zones of bodies of water. It acts, in part, to form methane and also occurs free in bubbles. Free ammonia may, under certain conditions, be present in small quantities in lakes and streams, and elemental nitrogen, derived mainly from the atmosphere, is highly important in lake metabolism; these substances will be taken up in connection with nitrates. Hydrogen sulfide, a decomposition product, is frequently present in the hypolimnion of certain lakes during summer; it will be considered in the following chapter along with sulfur.

DISSOLVED GASES IN STREAMS

The behavior and basic relations of gases in streams follow the same fundamental physical and biochemical laws as operate in lentic situations. However, temporal and spatial relationships of dissolved substances generally are variously modified and complicated by the special features that characterize streams. In other words, the presence of a current with its inherent turbulence effects, the considerable exchange between stream and surrounding terrain, and the variations in water volume and chemistry associated with climate and drainage basin morphology all serve to make a momentary or long-term picture of stream conditions quite different from a lake. During our time another factor, an "unnatural" one, has become important in relationships in streams. This is pollution, and although the act of ravaging our waters by dumping wastes into them is not to be condoned, we have learned much about biochemical reactions from pollution studies. This topic will be studied more fully in a later chapter.

OXYGEN IN STREAMS

There are three primary sources of oxygen in stream water, the contributions of each being far from equal and indeed varying greatly with time of day, season, current velocity and stream morphology, temperature, and biological characteristics.

Ground Water and Surface Runoff

For most streams these sources are relatively insignificant in supplying oxygen. Water issuing from springs, subterranean channels, or seepages is typically low in dissolved oxygen, often to the point of being anaerobic. Not only do these ground waters fail to provide oxygen to the spring run, but also the run itself may dilute the oxygen content of a parent stream at the point of junction of the two streams. If, however, subsurface waters flow over broken rocks shortly before reaching the surface, the waters may be near saturation. Similarly, if surface runoff is rapid and vigorous, the water may be high in oxygen content; but sluggish surface drainage seldom contributes great oxygen stores to running waters.

Photosynthesis

In less turbid streams vegetation contributes oxygen to the waters during the day. Rich growths of algae on submerged objects such as rocks or logs, and algae floating free in the water, together with higher plants growing beneath the surface, produce high amounts of oxygen, particularly on cloudless days. During the night and on cloudy days this production may be somewhat balanced by respiratory consumption of the oxygen by plants and animals. In the shallow headwater reaches of a stream a net production of oxygen by photosynthesis contributes to the downstream content. In the lower, more turbid regions of most streams, however, local photosynthesis probably contributes little to the downstream oxygen content. Since turbidity is to a great extent a function of stream discharge and capacity, we can again recognize the importance of stream channel and watershed features in the chemistry and biology of streams.

Physical Aeration

The introduction of a large amount of organic substance, such as sewage or debris from swamp or marsh, flooding into a stream may bring about a depression of the dissolved oxygen content below the saturation value. The difference between the actual oxygen content and the amount that could be present at saturation is called the **saturation deficit.** This deficit is incurred, of course, through the uptake of oxygen by aerobic decomposition of the organic materials in the stream. Yet downstream, barring no further immediate depressions, the stream shows evidence of regaining its earlier level of oxygen concentration. The oxygen serving to offset the oxidative loss is absorbed from the atmosphere through reaeration of the stream waters. Reaeration is, therefore, a process by which streams secure oxygen directly from the atmosphere, the gas then entering into the biochemical oxidation reactions in the stream. Within the stream, distribution of the oxygen derived from the atmosphere is accomplished by turbulent transport. The rate at which reoxygenation of a given parcel of water takes place depends upon a number of factors, including temperature, degree of turbulence, depth of the parcel, magnitude of the saturation deficit and, naturally, the momentary oxygen demand by decomposition processes. The importance of temperature lies in its inverse relationship to oxygen solubility, and to its influence on metabolic demands of organisms. Turbulence is a highly variable factor in the oxygenation–deoxygenation relationship, varying from negligible in quiet pools to highly important in riffles and rapids.

How effective is reaeration as a method of oxygenating flowing waters? Figure 10.7 gives data for a 27-mile segment of Holston River below Cherokee Dam, one of the Tennessee Valley Authority impoundments, during a time of uniform flow. Note particularly the relationships between discharge and saturation deficit, and the dissolved oxygen content. The increase in oxygen concentration is greater at the lower discharge than at the higher. The total mass picked up, however, is greater in the higher discharges.

Under natural conditions, the waters of streams typically contain a relatively high concentration of oxygen tending toward saturation. However, a number of factors operate to varying extents to reduce the oxygen content and to contribute to the loss of the gas from streams. The most important are listed below.

Turbulence We have just witnessed the fact that turbulence plays an important part in the aeration of streams. It follows, therefore, that physical aeration is reduced with decrease in turbulent flow. In stream areas below riffles

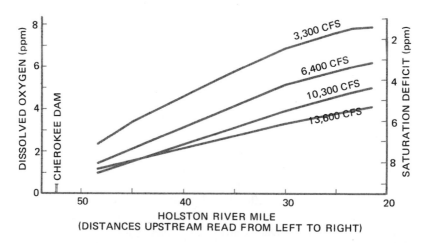

Figure 10.7   Rate and extent of reaeration as a function of river discharge. Data were collected over a 27-mile stretch of the Lower Holston River, below Cherokee Reservoir, Tennessee, during a time of uniform flow. The vertical axes indicate dissolved oxygen and oxygen saturation deficit, in parts per million; the horizontal scale indicates distance downstream from the dam. Discharge rates are given in cubic feet/second. See discussion in text. (After Churchill, 1958.)

and falls, the water is often saturated with oxygen throughout the day and night. In quiet reaches and in streams of low velocity, the waters may be below saturation during the night.

Respiration of organisms Respiratory activities of plants and animals, and oxidation of organic matter utilize dissolved oxygen of streams. The effects of these processes are more conspicuous at night, being masked during the day by photosynthesis.

Photosynthesis In the more stable zones of streams, submerged plants contribute in a major way to the oxygen content of the water. Consequently, fluctuations in photosynthetic activity will be reflected in the amount of dissolved oxygen present.

Temperature In streams, as in other waters, solubility of oxygen varies inversely with temperature. Thus, raising of water temperature could result in loss of oxygen from streams.

Atmospheric pressure Since the solubility of oxygen bears a direct relationship with atmospheric pressure, reduction of pressure would bring about a decrease in the amount of dissolved oxygen.

Inorganic reactions Certain inorganic activities, such as the oxidation of iron, may contribute to the loss of oxygen from polluted streams.

Inflow of tributaries The introduction of tributary waters of low oxygen content serves to dilute the concentration of oxygen in the receiving streams. This effect is especially noticeable with the entrance of spring or some seepage waters.

The annual cycle of oxygen of streams is closely correlated with temperature conditions. Studies of large rivers and small streams in warm southern

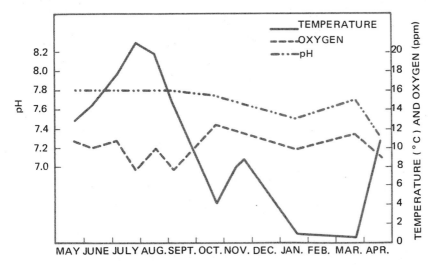

Figure 10.8   Seasonal variation in temperature, dissolved oxygen, and pH in Mad River, Ontario, during 12 months in 1930-1931. (After Ricker, 1934.)

regions of moderate temperature regimes and in northern climates of broad temperature fluctuations have shown that the oxygen content of flowing waters is generally highest in winter and lowest in late summer (see Figure 11.3(a), p. 248). This primary temperature-oxygen relationship may, however, be tempered by a number of factors acting throughout the year. In small, slow streams of northern latitudes, decreased day length and ice and snow cover may serve to inhibit photosynthesis, thereby bringing about a short-term depression of the winter peak of oxygen concentration (Figure 10.8). The vernal decline of oxygen content in slow, clear streams may be attributed to the action of spring floods in removing vegetation. Further, decrease in oxygen content toward late summer, in streams generally, may be due to one or more of several factors. Water temperatures reaching their maxima in late summer hold less of the gas in solution; decreased discharge results in diminished physical mixing and reoxygenation; greater decomposition of summer-produced organic material utilizes some of the available oxygen. On the other hand, a rapid and abundant growth ("bloom") of phytoplankton often increases the oxygen content during the summer. The effect of such a bloom on the oxygen curve for the St. Johns River, Florida, is shown in Figure 10.9; this phenomenon is most likely in streams in which the discharge is sufficiently low to permit the building up of a considerable phytoplankton mass.

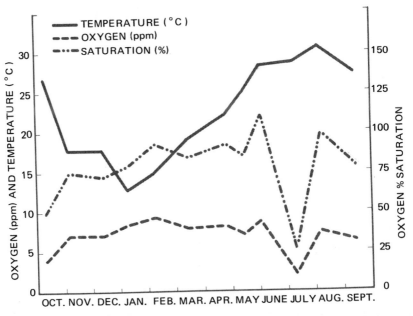

Figure 10.9   Seasonal variation in temperature, dissolved oxygen, and percent oxygen saturation in St. Johns River, Florida, during 12 months in 1939-1940. (Data from Pierce, 1947.)

A daily oxygen rhythm, the diurnal pulse, is largely a reflection of temperature fluctuations and photosynthesis–respiration relationships. In mountain and
hill country the water is typically at between 95 and 105 percent saturation at all
times, so long as there is not a dense growth of rooted aquatic plants. Turbulence
is of great importance and easily maintains an average of 100 percent saturation.
In a slow, shallow stream containing a fair abundance of rooted plant growth,
daytime photosynthetic production of oxygen exceeds turbulent diffusion into the
air and the respiratory consumption of the gas. The result is frequent supersaturation with oxygen during the day, followed by a drop to 100 percent saturation at
night. In winter in the temperate zone the time of maximum daily oxygen
production corresponds closely with the time of highest temperature. In summer,
oxygen production lags behind rising temperature, often by two or three hours,
thus resulting in high saturation values even after sundown. Depending upon
oxygen demand, the minimum level usually occurs prior to the early-morning
temperature low. In streams of this nature the oxygen concentration may be
acutely affected by sunlight intensity. In Figure 10.10 the depression in the
diurnal curve between about 1300 and 1500 hr is doubtless due to cloudiness
shown for the same periods on the light-intensity curve. Such data as these
indicate the importance of insolation and photosynthesis in maintaining the
oxygen content of slow streams. The considerable diurnal oxygen range in such
streams, as for example that in Figure 10.10, also points up the importance of
more than a single oxygen determination in a stream study. Obviously an average
of several measurements would be much more meaningful than one, and a diel
series would most satisfactorily describe the oxygen conditions in this stream
type. We have already learned that oxygenation through turbulent mixing is a

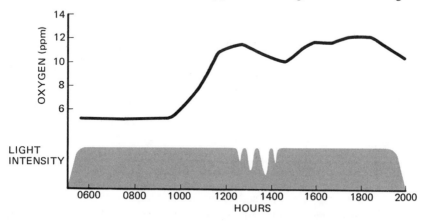

Figure 10.10   Oxygen production and light intensity, Honey Creek, Oklahoma, in summer
(daylight hours only). In such a slow, shallow stream, with rooted plant growth, isolation
plays a major role in determining dissolved oxygen content. Note particularly the sharp
effect of intermittent cloudiness, which reduced light intensity between noon and 1400
hours. (After Hornuff, 1957.)

prominent feature of fast waters and even highly turbid streams of low discharge. The diurnal oxygen pulse of such streams as these is usually less pronounced, since mechanical aeration and oxygen consumption are essentially uniform within a given stream segment.

The distribution of dissolved oxygen throughout the linear extent of a stream is subject to many variables, so much so that generalizations relating to the feature are not readily apparent. As we have seen, the upper reaches of streams are usually well oxygenated, the concentration being a function of turbulence and water temperature. Farther along the gradient, the stream becomes slower and rich growths of submerged plants may develop. In this region the oxygen content is influenced less by physical aeration and more by organic oxygen production and respiration. The maximum–minimum relationships here may be the reverse of those found upstream, in that maximum saturation values will be found in the afternoon and the lowest values in the early morning. In the lowlands, increased turbidity and organic decomposition produce a generally low oxygen content. In clear, slow streams supporting an abundant flora, oxygen may be added to the waters along the stream course, thereby resulting in a net increase in downstream segments. The introduction of organic pollutants and inflow from marshes and swamps may serve to decrease the oxygen content at and below the point of inflow of these materials. Thus the amount of dissolved oxygen distributed through the course of a natural stream is strongly determined by channel and flow characteristics—considered in Chapter 4—and by biochemical interactions and processes.

Although our knowledge of reservoir limnology is meager, we know that the effects of the release of impounded waters into streams below the impoundment are varied and important, and worthy of considerably more research in the future. For example, we know that release of water from a reservoir through deep-water outfalls during summer stratification introduces poorly oxygenated water and organic materials from the hypolimnion into the stream below the impoundment. Under extreme conditions this actual dilution of the stream and introduction of oxidizable matter could have deleterious effects on the biota. On the other hand, draw-down from the epilimnion through shallow outfalls may serve to increase the dissolved oxygen content of the downstream waters.

CARBON DIOXIDE AND pH IN STREAMS

In streams the occurrence and abundance of components of the bicarbonate buffer system and the pH condition are determined primarily by current, biological processes, and the chemical nature of the substrate. The role of current is that of ameliorating the chemical climate of the stream. This is accomplished through mixing and moving concentrations of substances, but usually within a relatively restricted segment of the stream. Over a considerable distance, and depending upon the volume of introduced materials, the chemical composition of stream waters is subject to considerable change. The biological processes acting to influence the nature of the water are those previously considered—photosynthesis and respiration. In general, the behavior of carbon dioxide in streams is similar to that of oxygen, but with inverse properties. In stream areas of abundant plant growth the concentration of carbon dioxide is minimum during most of the day and maximum in the early morning hours. In sluggish streams a phytoplankton bloom may indeed exhaust the free carbon dioxide supply and obtain further carbon dioxide from bicarbonates. In the lower, sluggish stream course under conditions of high turbidity due to organic suspensoids, high carbon dioxide content occurs in the presence of depressed oxygen concentration, owing to bacterial action.

The chemical composition of mineral-bearing rocks in the stream valley and channel, and also the drainage nature of the valley, may act in a major way to determine the water composition. Under certain conditions these factors may somewhat offset the influences of biological processes. For example, in limestone regions a high concentration of carbon dioxide is not apt to develop in the stream because, as we have already seen, any excess of gas would enter into combination with lime in the substrate to give a carbonate. The direct strong influence of the substrate rests primarily upon the solubility of the buffer substances in rocks. Streams issuing from or flowing over relatively insoluble igneous formations of high silica content are typically soft, because the bicarbonate content is insufficient to buffer pH changes due to accumulation of carbon dioxide. As a result of a shift of the buffer system toward carbonic acid, the pH of such streams is below neutrality. However, atmosphere–water equilibrium essentially adjusts the carbon dioxide content in such a way that the pH value does not go far toward the acid side, usually coming to balance at about pH 6. If, on the other hand, the stream encounters water from bogs, which commonly occur in areas of siliceous formations, the pH may be further lowered.

In many regions, the Appalachian slopes of North America for example, the fate of soft-water streams is to flow eventually over sedimentary formations in the lower reaches. These formations are normally rich in soluble carbonates, and the addition of carbonate ions serves to shift the pH value toward the basic range of the scale, usually to pH 8 or higher. The increase in pH value is accompanied by a marked rise in total alkalinity and a decrease in carbon dioxide.

Highly acid streams occur primarily on low marshy or swampy terrain, on poorly drained sandy "flatwoods," or under special conditions. These steams are usually stained brownish and support a relatively meager biota. The nature of the acids contained in such streams is not fully known, but frequently considered to be humic, yellow organic, or perhaps sulfuric. In the southern United States, the pH value in sluggish acid streams may range near pH 4, with a concurrent high concentration of free carbon dioxide (25 ppm or more) and low bicarbonate content. Oxygen in these streams is often low, being on the order of 20 to 30 percent saturation at 17° to 18°C.

Hard-water streams occur commonly as spring runs or surface drainage streams in regions of soluble basic geologic formations. Calcium and magnesium carbonate, often occurring together, are prominent sources of ions which, in solution, contribute to the hardness of streams; however, sulfates, chlorides, and other compounds may also contribute to total hardness. The combination of high carbonate concentration and low carbon dioxide content, characteristic of these streams, apparently constitutes a more favorable environment for plants and animals than do acid conditions; generally speaking, hard-water streams support a varied and abundant association of organisms. Most unpolluted major streams exhibit a pH value on the alkaline side of neutrality, and, indeed, tend toward uniformity of composition generally.

Although the relationships of carbon dioxide, bicarbonates, and carbonates will be considered a bit further with respect to dissolved solids in Chapter 11, we might briefly summarize their activities in relation to pH in lakes and streams: first, the pH value varies inversely with the dissolved carbon dioxide concentration, and directly with the bicarbonate concentration; second, the critical value relating to the presence or absence of free carbon dioxide is pH 8, the free gas being absent above that value; finally, the absence of free carbon dioxide does not limit photosynthesis of certain algae and higher plants, some being adapted for utilization of carbon dioxide from carbonates, usually resulting in very high pH values. Certain of these relationships as observed in a New Jersey stream are shown in Figure 11.3 (b and c), page 248.

Methane, hydrogen sulfide, and other gases derived from breakdown of organic substance may be present in high concentrations under proper conditions in very slow streams and in stagnant regions of stream pools. In moving water, however, turbulence due to current tends to diffuse and eliminate the gases.

TABLE 10.1 Saturation coefficient values for oxygen and carbon dioxide in water at varying levels of temperature and salinity[a]

Salinity (‰)	Temperature (°C)					
	0°		12°		24°	
	O_2[b]	CO_2	O_2	CO_2	O_2	CO_2
0 (freshwater)	49.24	1717	36.75	1118	29.38	782
28.9 (brackish water)	40.1	1489	30.6	980	24.8	695
36.1 (seawater)	38.0	1438	29.1	947	23.6	677

[a] Data from Sverdrup et al., 1942.
[b] All concentrations expressed in mg / liter at sea level atmospheric pressure.

DISSOLVED GASES IN ESTUARIES

The considerable differences between the chemistry of fresh water and that of seawater bring about complex relations of dissolved substances in general within an estuary. The occurrence of dissolved materials relates to the disproportion of dissolved materials found in the waters of the inflowing stream at the upper end of the estuary and in the seawater at the mouth of the estuary. Between these two relatively uniform states there exists a considerable gradient in processes and conditions. Dissolved gases are distributed in an estuary in accordance with turbulence and current factors, biological activities, and salinity and temperature effects. We have previously considered most of these factors and their influence on dissolved gases with respect to lakes and streams; in general, the interrelationships are similar in estuaries. One new factor is quite influential in estuaries; this is salinity, the total amount of dissolved inorganic solids.

OXYGEN

In salt water the solubility of oxygen decreases as water temperature and salinity increase. If we visualize cool, fresh (low-salinity) water entering the uppermost reaches of an estuary and grading toward somewhat warmer and more saline seawater at the lower extreme, we can begin to appreciate the adjustments taking place throughout the length of the estuary. Table 10.1 gives saturation (absorption) coefficient values of oxygen in water at various temperatures and salinities. The values show not only that temperature is the most important factor in determining oxygen solubility, but also that salinity manifests considerable influence. We see then that less oxygen can be dissolved in seawater than in fresh water. At saturation at 15°C, a liter of seawater (at 36‰ salinity) contains 5.8 ml oxygen per liter of water; a liter of fresh water at the same temperature holds 10.3 ml.[1] The importance of these relationships lies not only in the aforementioned possibility of a considerable linear oxygen gradient in a mixing estuary. There is also the possibility of fluctuations associated with stream flood seasons and the inflow of large quantities of fresh water, or with dry seasons when tidal flow of sea water dominates.

In nonmixing estuaries salinity stratification during summer often results in conspicuous differences in dissolved oxygen content in deep and in surface waters. In parts of Chesapeake Bay during summer the oxygen concentration may range from 90 to 100 percent saturation at the surface, while bottom waters

1. These values, correlated with Table 10.1, were revised some time ago (see Richards and Corwin (1956) for a review of the subject). However, Steen (1958) has found good agreement with the earlier data used above. Since these earlier data have been widely employed in many computations, we have continued their use.

show from 40 to 50 percent saturation. The surface–bottom differential decreases upstream as the depth decreases. In these shallow zones there is probably more mixing than in the bay proper. The range of oxygen concentration in the bay reflects the activity in the lighted trophogenic zone and the relatively slow replacement of seawater in the lower level. It may well be that even with more rapid replacement the water from the sea might be poor in dissolved oxygen.

Along with vertical differences, oxygen characteristically varies diurnally and seasonally within estuaries. The ranges of such variations differ, depending upon the nature of the freshwater source, the morphology of the basin, and effects of tides. In deep, turbid estuaries lacking the contribution from an abundant bottom flora, diurnal oxygen pulses are apt to be relatively slight. In Chesapeake Bay the range for surface waters was found to be from about 85 percent of saturation in early morning to about 115 percent in late afternoon. Shallow, clear estuaries may contain bottom growths of algae on which oxygen bubbles may be seen. In waters such as these the minimum–maximum diurnal range may exceed 200 per cent. The effects of incoming seawater during a tidal cycle may serve to mask the local conditions. Seasonally, the oxygen dynamics in an estuary may be influenced by variations in river discharge, tide, day length, and biological effects. Surface waters in one part of Chesapeake Bay were found to have 143 percent saturation with oxygen in April and about 42 percent in August. Bottom waters in the same area reached 133 percent saturation in October and 24 percent in June.

THE CARBON DIOXIDE SYSTEM AND pH

The solubility of carbon dioxide in estuarine waters is determined primarily by the amount of seawater mixing with fresh water, and secondarily by temperature. The importance of seawater as a solubility factor rests, of course, upon salinity. In Table 10.1 we see that at any temperature the absorption coefficient value of carbon dioxide decreases as the salinity increases. The high solubility of carbon dioxide is due to its chemical reaction with water. Although some of the carbon dioxide in seawater is in the form of the free gas and as carbonic acid, much more is present as bicarbonate and carbonates. This condition results from the fact that seawater contains alkaline radicals in excess of the equivalent acid radicals which, as we have seen, shift the carbon dioxide system toward the carbonate formation, thereby reducing the free carbon dioxide concentration. The presence of the excess bases in seawater—boric acid and its borates, carbonic acid and the carbonates—serves to buffer the water against great changes in pH that might develop from the addition of acids or bases. The pH of seawater at the surface is very stable, usually ranging between pH 8.1 and 8.3.

Since carbon dioxide uptake is greater in the presence of excess base, and since river waters usually contain a lower concentration of excess bases than seawater, we should expect to find a lower content of free carbon dioxide in the mouth of the estuary than in the upper reaches. Similarly, because river waters are seldom buffered, the free carbon dioxide concentration and pH should be more variable in that part of the estuary dominated by stream influences. Streams transporting large quantities of humic material in colloidal suspension are frequently slightly acid. Upon meeting seawater the colloidal particles are coagulated and the pH shifts toward the alkaline side of neutrality. East Bay, Texas, receives considerable runoff from organically rich salt marshes, and it was found that during summer the pH ranged from 6.9 throughout much of the bay to 7.8 near the mouth where it discharges into the Gulf of Mexico. Gulf waters during the same period gave a pH value of 8.0.

In view of what we have learned of oxygen–carbon dioxide–pH relationships generally, we should expect the vertical, diurnal, and seasonal distributions of carbon dioxide and pH to operate according to certain principles. Although few actual studies have been made of carbon dioxide in estuaries, the pattern of distribution is expected to be the reverse of that of oxygen. By the same rule, pH values should vary inversely with the free carbon dioxide content and directly with the dissolved oxygen concentration.

DISSOLVED GASES IN OCEANS

All the atmospheric gases occur in seawater—nitrogen, oxygen, carbon dioxide (usually in the form of carbonates or bicarbonates), and others, including traces of ammonia, argon, helium, hydrogen, and neon. Hydrogen sulfide and methane occur locally as an effect of bacterial decomposition. Observation of dissolved gases has a particular utility in physical oceanographic studies. Particular combinations of gases, in predictable concentrations, seem to be permanent characteristics of certain masses of water; movements of such masses (such as the Gulf Stream) can be traced, in part, by chemical analysis of dissolved gases.

OXYGEN

The amount of dissolved oxygen in seawater varies between 0 and 12.6 mg/liter. Along a north–south transect in the eastern Atlantic Ocean, for example, the amount of dissolved oxygen in the tropics, at 15° north latitude, was found to be low near the surface (oxygen values of 1.4 to 2.9 mg/liter). Higher values (4.3 to 5.7) occurred at 1000 m and a value of 7.2 was found at a depth greater than 2000 m. At 40° north latitude oxygen values were 8.6 to 10 at the surface and 7.2 at 1000 m or deeper; at 40° south latitude, oxygen values in surface water ranged from 7.2 to 8.6 (Wattenberg, 1933). The characteristically low concentration of dissolved oxygen in tropical waters is now recognized as a permanent characteristic, the **oxygen minimum layer**. This layer is a tongue of water at intermediate depth, 600 to 800 m thick, projecting from Central America into the Pacific, in which the concentration of dissolved oxygen is about 0.25 mg/liter. The oxygen minimum layer was at first believed to support only anaerobic bacteria, but it is now thought that the low oxygen concentration is due to extremely efficient respiratory withdrawal of oxygen by marine zooplankton.

In bays where bacterial activity is high and circulation low, surface oxygen concentrations become very low, especially in hot regions. In areas of heavy phytoplankton growth, in proximity to masses of algae, or (often) in lagoons, high levels of photosynthetic activity result in a local excess of dissolved oxygen. The redox potential is fairly constant in aerated seawater (in the $+500$ mv range), but near and in reducing sediments, values drop to the -100 mv range.

CARBON DIOXIDE AND pH

Carbon dioxide values are fairly constant in aerated surface waters, usually ranging between 24.7 and 28.5 mg/liter. Because of the very uniform pH of seawater (8.1 to 8.3), the amount of free carbon dioxide is virtually zero, and most CO_2 occurs as HCO_3 and CO_3. Off southern California, at a depth down to 200 m, CO_3 values decrease to zero while those for HCO_3 increase steadily; below 200 m the amount of H_2CO_3 increases with depth. The $CO_2 \rightleftharpoons HCO_3^= \rightleftharpoons CO_3^=$ equilibrium is very effective in seawater. Carbon dioxide is accumulated and held partly in the form of soluble carbonates, while much is converted to insoluble $CaCO_3$ and $MgCO_3$ by corals, coralline algae, and coccolithophores; ultimately, a great deal of dissolved carbon dioxide is sedimented into limestone.

Seawater is strongly buffered, which accounts in large part for its constant pH. The buffer system primarily involves salts of carbonic and boric acids. Thus, alkalinity (as measure of buffering effect) is quite high, and in open water it bears a close relationship to salinity.

$$\frac{\text{alkalinity}}{\text{chlorosity}} = 0.120$$

In brackish water, this relationship does not hold. It also does not hold in estuaries fed by rivers high in bound carbonates.

SUGGESTED FOR FURTHER READING

Golterman, H. L., ed. 1969. *Methods for chemical analysis of fresh waters.* Blackwell Scientific Publications, Oxford, England.

Rainwater, F. H., and L. L. Thatcher. 1960. *Methods for collection and analysis of water samples.* U.S. Geological Survey, Washington, D.C.

Strickland, J. D. H., and T. R. Parsons. 1968. *A practical handbook of seawater analysis*. Fisheries Research Board of Canada, Ottawa.

Three up-to-date and widely used reference books, describing most of the currently used methods of water analysis; useful supplements to Welch (1948), the leading source on limnological methods. (The student should note that each of these books covers analysis of solids as well as gases in natural waters, and thus may be used in conjunction with either Chapter 10 or Chapter 11 of the present text.)

11| Dissolved Solids in Natural Waters

Of all the elements that occur in nature, more than half have been detected, in some form, in fresh or marine waters. Specialists agree that virtually all of the others could probably be found as well, although some might occur only in trace amounts (Chapter 2). In Chapter 10 we considered the more important gases that are found dissolved in, or combined with, inland or marine waters. Of the solid materials (the subject of this chapter), the more conspicuous include carbonates, chlorides, sulfates, phosphates, and often nitrates. These anions occur in combination with such metallic cations as calcium, sodium, potassium, magnesium, and iron, to form ionizable salts. As a result of the high availability and solubility of carbon dioxide, carbonates are usually the most abundant salts in freshwater environments. Both the quantitative and the qualitative aspects of the chemical composition of inland waters are influenced to a high degree by the geochemistry of the terrain and, in the case of lakes, by the form of the basin, which affects the nature and rate of inflow and outflow. In general, the inorganic composition of the water of an **open lake** (one whose waters are drained by effluent channels) reflects the nature of the influent waters. The composition of such lakes is approximately that indicated in Table 11.1. The inorganic composition of **closed lakes** (those lacking significant effluent discharge) is, in contrast, greatly modified by precipitation and by the concentration of solutes (such as salts) by evaporation. Fluvial waters are considerably more varied than standing waters, but the larger streams, and especially major rivers, are generally similar in inorganic composition to open lakes. As fresh water passes through estuarine systems it undergoes various changes as a result of the interaction between the riverine and the marine aquatic systems; in terms of dissolved solids, estuaries are complex environments, and estuaries as a group exhibit a wide range of chemical characteristics. Marine waters, finally, are nearly constant in terms of their chemical composition and in terms of the proportions and activity of their major inorganic solutes.

TOTAL DISSOLVED SOLIDS

The total concentration of dissolved substances or minerals in natural waters is a useful parameter in describing the chemical density as a fitness factor, and as a general measure of edaphic relationships that contribute to productivity within the body of water. One measure referred to as either **total dissolved solids** or **total filterable residue** is determined by simply evaporating a quantity of filtered water at low temperatures. The dried residue contains both inorganic and organic materials. Ignition of this residue at high temperature eliminates volatile substances, usually organic in nature, and decomposes bicarbonates with the loss of carbon dioxide. The residue following ignition therefore contains the total inorganic solids, the difference between it and the original residue being termed the **loss on ignition**. The amount of residue in each operation and the loss on ignition is expressed as the proportion of the original water sample in terms of parts per thousand (‰) or parts per million (ppm) depending upon the concentration, or as mass in metric units (mg) per liter. The **salinity** of fresh waters is defined as the total concentration of the ionic components. Although common in marine sciences, this term has not been widely used in freshwater research. Because of the relatively small quantities of total ions generally encountered, freshwater salinity is usually expressed in milligrams per liter.

TABLE 11.1 Mean percentage composition of North American inland waters[a]

Ion	Percentage
$CO_3^=$	33.40
Ca^{++}	19.36
$SO_4^=$	15.31
SiO_2	8.60
Na^+	7.46
Cl^-	7.44
Mg^{++}	4.87
K^+	1.77
NO_3^-	1.15
Al_2O_3 Fe_2O_3 }	0.64

[a] Data from Clark, 1924.

A measure of the total amount of ionized materials in water can be obtained through determination of the electrical conductance of the solution. Commonly called **specific conductance**, this parameter closely approximates the residue in solution and may be correlated with salinity. By definition, conductance is the reciprocal of the resistance measured between two platinum electrodes separated by 1 cm and having a cross section of 1 sq cm. The conductivity is usually expressed as micromhos (reciprocal megohms) per centimeter at 25°C (see A.P.H.A., 1971, and Ellis, Westfall, and Ellis, 1946, for details). In general, the range of specific conductance of natural waters should approximate that of total dissolved solids given below. The very clear water of Bunny Lake in the Sierra Nevada, listed in Table 11.2, has a total dissolved solid concentration of 8.2 ppm at mid-depth in July, the specific conductance being 8.9 micromhos (Reimers, 1958). In Hot Lake, Washington (a saline lake), at 3 m depth in August, the total dissolved solids amounts to around 391,800 ppm, and the specific conductance is 60,440 micromhos.

Most lakes occupying open basins have a total dissolved solid concentration of between 100 and 200 ppm. Evaporation from lakes in closed basins raises the concentration of dissolved solids, in some cases, to over 100,000 ppm. These lakes with extremely high concentrations are considerably more saline than the sea, the latter averaging about 35,000 ppm.

Organic compounds such as the organic states of phosphorus and nitrogen, sugars, acids, and vitamins are known in natural waters. We are very much lacking, however, in our knowledge of the formation and role of many of the organic substances.

Although we shall relate dissolved substances to community metabolism in more detail in later chapters, it is well to keep in mind the idea that the quantity and quality of dissolved solids often determine the variety and abundance of plants and animals in a given aquatic situation. In a most general sense the limiting nature of dissolved solids is essentially two-fold. In the first place the chemical density of the environment of aquatic organisms is a function of the total dissolved solids. According to laws of osmosis and diffusion the water balance, or osmoregulation, of plants and animals is governed by this environmental density factor and by the physiological adaptations of the organisms to it. As we shall see later, osmoregulation is a central factor in restricting marine and freshwater organisms to their respective habitats. The second way in which dissolved solids influence the nature of the community relates to supply of nutrients and otherwise important materials. The only source of nutritionally important ions available to phytoplankton is the reservoir of matter dissolved in the water. The nature of the animal community is, of course, dependent upon the kinds and quantity of phytoplankton. In another sense certain animals may be directly limited by the availability of a given dissolved substance, animals bearing carbonaceous shells being an example. In highly acid waters some mollusks may be entirely absent, while under mildly acid conditions mollusks may occur but their shells may be reduced in thickness as compared with those inhabiting waters of higher pH value.

TABLE 11.2 Ionic composition (in parts per million) of water from several sources[a]

Ion	Rainwater	Silver Springs, Florida	Purcell's Pond, Nova Scotia	Bunny Lake, California	Lake Erie (1958)	Soap Lake, Washington
Cl	0.5	7.8	7.2	0.2	6.6	5467.0
SO_4	2.0	34.0	9.5	1.3	9.3	6240.0
B	0.01	0.01	NR[b]	0.01	NR	NR
Na + K	0.43	5.1	5.3	0.9	4.9	13,002.0
Mg	0.1	9.6	0.4	0.9	5.7	8.2
Ca	0.1 to 10	68.0	0.8	0.7	23.4	20.6
$N \cdot NH_3$	0.5	NR	NR	0.03	NR	NR
$N \cdot NO_3$	0.2	0.2	NR	0.2	NR	8.7
HCO_3	0.0	201.0	0.0	6.0	NR	5209.0
pH	4.5	7.8	3.9	5.8 to 6.3	8.2	9.4 to 10

[a] Data from Hutchinson, 1957; Ferguson et al., 1947; Gorham, 1957; Reimers, 1958; Hough, 1958; and Anderson, 1958.
[b] Data not reported.

With these general concepts of the importance of dissolved solids in community function and structure in mind, let us now consider some of the properties and activities of the more abundant and better known inorganic and organic substances in natural waters.

SOLIDS IN SOLUTION IN LAKE WATERS

In terms of their significance in the ecosystem, the most important dissolved solids in freshwater environments (under natural conditions) are the compounds of calcium and magnesium, sodium and potassium, nitrogen, phosphorus, iron, sulfur, and silicon. A few other elements occur in small, but significant, amounts; these will be considered toward the end of this section.

CALCIUM AND MAGNESIUM

Together, these two alkaline earth metals constitute the most abundant ions in fresh waters. The chemical activity of the two elements is similar, particularly in the formation of carbonate salts, and both may limit biological processes in streams and lakes. Magnesium is an important component of the chlorophyll molecule. Of these two ions, calcium is usually more abundant. In soft waters, containing less than 50 mg/liter dissolved solids, calcium makes up, on the average, about 48 percent, and magnesium about 14 percent, of the total cations present. In average hard waters the proportion of magnesium to calcium increases, approaching approximately 53 percent calcium and 34 percent magnesium. The increase of these ions in hard waters takes place at the expense of two alkali metals, sodium and potassium, to be considered later. The nature of the lake basin is reflected in the ionic proportions of the waters in open lakes (those with considerable inflow and outflow). In this type of lake, calcium averages about 63 percent; magnesium, on the other hand, makes up slightly more than 17 percent of the total cations.

The wide variation in calcium content and the corresponding correlations shown in hardness, stratification, and biological productivity have stimulated many attempts at classification based on this ion. The following scheme was proposed in 1934 by a German limnologist, W. Ohle:

< 10 mg Ca/liter	"poor"
10 to 25 mg Ca/liter	"medium"
> 25 mg Ca/liter	"rich"

Since calcium normally occurs in combination with the carbonate anion, the above classification should correspond generally with the designations of soft and hard waters given previously. Since they are concerned with the ionic content, both sets of designations are equally applicable to streams and lakes.

The distribution of calcium in stratified lakes generally follows a more or less characteristic pattern. In soft-water lakes of northern Wisconsin, the calcium content ranges from 0.7 to 2.3 mg/liter (poor) and during summer stagnation little stratification of calcium exists. Lakes of medium hardness and medium calcium content typically contain a moderately increased calcium concentration in the hypolimnion during stratification. Such has been shown for lakes in southern Wisconsin which ranged from 21.2 to 22.4 mg Ca/liter. The hypolimnion of hard-water (rich) lakes characteristically contains a greatly increased load of calcium. In some of these lakes the curve of calcium concentration shows a steady increase of the ion from the surface to the upper region of the hypolimnion, the hypolimnial content being on the order of twice that at the surface, and rather uniform throughout the stratum. Under other conditions the epilimnial calcium may be relatively uniform throughout, increasing rapidly with depth through the metalimnion, and with less rapid but nontheless increasing change with depth in the hypolimnion. Along with other lake substances calcium is redistributed throughout the lake during vernal and autumnal mixing. An exception to calcium stratification in calcium-rich lakes is seen in data from Cultus Lake, British Columbia. This lake contains 32 mg Ca/liter during summer but shows a somewhat uniform vertical distribution of the ion during stratification (Ricker, 1937).

Biologically, soft-water lakes usually contain less living matter per unit area than hard-water lakes. The total estimated biomass of plant substance in certain medium lakes of Wisconsin was found to be three to five times that of poor lakes, while the animal mass, not counting fishes, was as much as triple that of poor lakes. Although the total mass of organisms may be greatest in rich lakes, medium lakes often harbor a greater variety of kinds of plants and animals. In addition to the quantitatively meager mass of organisms, poor lakes generally support a unique assemblage of species.

Calcium Carbonate

We have already seen the interrelationships involving the carbon dioxide–bicarbonate–carbonate equilibrium, water, and the hydrogen ion concentration (Chapter 10). If the carbonate formed by dissociation of carbonic acid combines with calcium (or magnesium) ions in the water, calcium carbonate is produced by the reaction

$$CO_3^= + Ca^{++} \rightleftharpoons CaCO_3$$

Calcium carbonate, or lime, occurs as a white precipitate. Its formation is dependent upon the loss of carbon dioxide from the carbon dioxide–carbonate system. Recall that the addition of carbon dioxide serves to produce carbonic acid which dissociates and lowers the pH (increases the hydrogen ion concentration), thereby leading to acid conditions. The effect of carbon dioxide removal from the system is to split off CO_2 from HCO_3^-, which brings about the precipitation of $CaCO_3$.

In fresh waters, calcium carbonate deposits, in general, are referred to as **marl**, although some of this material may in fact include accretions of animal shells. In a more technical sense, the term "marl" refers only to the accumulation of calcium carbonate that has been precipitated directly from the water. This precipitation results from the removal of carbon dioxide from the water by physical loss to the atmosphere and by photosynthesis in green plants. It appears well established now that many species of algae and some of the so-called higher plants are capable of precipitating lime in the process of obtaining carbon dioxide for photosynthesis. In some of these plants the carbon dioxide is taken directly from bicarbonate after the free carbon dioxide content has been exhausted. Among the higher plants, *Elodea (Anacharis)* and *Potamogeton* often possess dense coatings of lime on their surfaces. In fact, it has been found that 100 kg of fresh *Elodea* can precipitate 2 kg of Ca CO_3 in a day with 10 hr of sunlight. Under hard-water conditions, green algae (Chlorophyta) such as *Chara* and *Cladophora* cause the formation of considerable quantities of marl. *Chara,* or stonewort, feels brittle to the touch and some 30 percent of its dry weight may be lime. *Cladophora* often occurs in limes balls littering the lake bottom. Some blue-green algae (Cyanophyta) and bacteria are also capable of forming lime.

Plant activity may not be the only process in the formation of lime deposits in fresh waters. It has been shown that carbon dioxide loss is also related to temperature, hydrogen ion concentration, partial pressure gradient across the water surface, and the nature of other substances in the water. Recently, it has been found that phosphate deficiency may be important in reducing the carbon dioxide content, thereby facilitating lime precipitation.

Calcium and magnesium enter into combination with anions other than carbonate, some of these being found in high concentrations in certain lakes. In the Swiss Alps several small lakes are rich in calcium sulfate ($CaSO_4$), or gypsum, derived from local deposits. Magnesium also combines with sulfate to form epsomite ($MgSO_4 \cdot 7H_2O$), which occurs in high concentration in certain saline lakes. The bottom of Hot Lake, a meromictic lake in Washington, is underlain by a layer of epsomite 4.5 m thick, from which epsom salts have been mined. A layer of gypsum is also found, lying below the epsomite stratum. Unlike the "typical" water we have been emphasizing, this lake lacks carbonates, but the concentration of sulfate in the deep zone (monimolimnion) is over 243,000 mg/liter. The same layer contains slightly over 700 mg Ca/liter, and nearly 54,000 mg Mg/liter. In other types of saline lakes magnesium may occur in combination with chloride as $MgCl_2$. Dolomite, a double carbonate of calcium and magnesium ($CaMg(CO_3)_2$), serves as an important source of both cations in water in various parts of the world.

The Calcium Cycle

From our study of lime relationships it is obvious that calcium and carbonates are derived almost entirely from sedimentary rock strata. The ions are dissolved out of these formations and carried in solution in lakes and streams. Biological activity such as shell construction, bone building, and plant precipitation of lime combines and concentrates the ions. Through streams, much of the calcium reaches the sea, where it becomes "locked up" in coral reefs and bottom deposits of animal shells. Here the calcium remains until geologic forces raise the sea-covered deposits and present them to the attacking forces of erosion and solution. The famous white cliffs of Dover, England, and the rich limestone regions of the Alps are examples of now-available sources of calcium in the perpetuation of the cycle.

SODIUM AND POTASSIUM

Where sodium and potassium occur in low concentrations the proportion of sodium is usually only slightly greater than that of potassium. As the total content of both increases, the concentration of sodium greatly exceeds that of potassium. In average soft waters the equivalent percentage of sodium is second to that of calcium. In hard waters the proportion is less, usually falling below calcium and magnesium. The most common form of sodium in natural waters is the halide NaCl. It may occur in high concentrations in saline lakes such as Great Salt Lake with its salt content ranging from 24 to 26 percent. (Seawater, by contrast, has a salt content of only 3.5 percent.) Certain lakes in California, and in other regions of the world contain sodium in the form of dissolved sodium tetraborate ($Na_2B_4O_7$), or borax. Sodium sulfate (Na_2SO_4) occurs in abundance in some Canadian lakes. In certain of the saline lakes in Nebraska, potassium accounts for about 23 percent of the total ionic composition. Because of analytical difficulties, sodium is frequently determined and reported with potassium (Table 11.2); it appears that the limnological importance and behavior of these elements are similar.

NITROGEN AND NITROGEN COMPOUNDS

Nitrogen is central to all ecosystems because of its role in the synthesis and maintenance of protein, which is, along with carbohydrates and fats, a major constituent of living substance. Derived originally from the atmosphere, nitrogen enters into a complex cycle involving plants and animals and several forms of the element. Nitrogenous compounds in natural waters may be derived from outside sources (allochthonous) or may be fixed within the body of water (autochthonous). The former category includes precipitation falling upon the earth carrying its own compounds (such as nitrate and ammonia), surface runoff containing terrestrial compounds of nitrogen (including pollution by human agencies), and inflow of ground waters as springs or seepages. **Endogenous nitrogen compounds** result from fixation processes carried on by certain bacteria and algae. However, the extent to which "in place" biologically fixed nitrogen contributes to the total supply is not yet known. In other words, we have yet to determine how much of the nitrogen content of a given lake is fixed within the lake by its own organisms and how much is delivered to the lake from the outside.

Once in the system, a great proportion of these nitrogenous compounds is caught up in a cycle of biological assimilation and decomposition, or becomes involved in inorganic processes in the economy of the ecosystem. Depending upon the efficiency of the community, certain quantities of these and other compounds will be lost. Lake, stream, and estuarine communities can lose nutrient materials through outflow, by incorporation of the substances in sediments, and through processes of **denitrification**, which cause the release of elemental nitrogen.

There are at least two possible sources of elemental, or **uncombined** nitrogen (N_2) in natural waters. One reservoir, and very likely the more important, is the atmosphere. The second, but poorly known, source of uncombined nitrogen is that produced by bacterial denitrification of ammonia. This uncombined nitrogen is rather inert; the only organisms capable of using it are certain microorganisms such as the nitrogen-fixing bacteria and algae.

The solubility of molecular nitrogen in fresh waters is related to temperature and pressure, the temperature relationship being an inverse one. Although little study has been made of nitrogen in lakes, it has been established that supersaturation can exist under certain pressure conditions at the air–water interface. During summer stratification, the vertical distribution of uncombined nitrogen follows inversely that of temperature. Mixing, during vernal and autumnal overturn, distributes the nitrogen throughout the lake.

The synthesis of inorganic substances into plant and animal tissues and the metabolic processes of protoplasm produce various compounds containing nitrogen. These **organic nitrogen compounds** include, for example, nitrogen in combination with carbon and other elements, animal and plant protein, and urea and uric acid as animal metabolic wastes. Of the total content of soluble nitrogen in filtered and centrifuged surface waters of lakes, it is probable that 50 percent or more is in the form of organic nitrogen. Some 60 to 80 percent of this organic nitrogen is composed of amino compounds such as free amino acids, polypeptides, and proteins. These substances are contained primarily in living plants and animals, and the presence of the compounds in water doubtless reflects the metabolic processes of living organisms as well as the decomposition of dead bodies. With respect to formation by living organisms, it has been established that many blue-green algae secrete extracellular nitrogenous materials, including polypeptides, amides and amino acids. The ecological importance of such liberated compounds is not fully known. In the algal species tested, the plants were not able to utilize their own excreted products as nitrogen sources.

The concentration of organic nitrogen may be expected to vary seasonally. Figure 11.1, for example, shows the seasonal distribution of various forms of nitrogen in Lake Mendota, Wisconsin. There is little evidence that much of the organic nitrogen is available as nutrient for plants and animals. The measure of the total organic nitrogen content is, however, a valuable indication of the productivity of the body of water, for certainly most of the substance will ultimately be transformed into states which can enter into production of living matter.

In addition to its occurrence in the uncombined state and in organic compounds, nitrogen is also present in natural waters in the form of **inorganic nitrogen compounds**, such as ammonia, nitrite, and nitrate. In most fresh waters the concentrations of these inorganic compounds are relatively slight; nevertheless they are very important in determining the productivity of a given community. All of the inorganic forms can be used by most green plants, particularly various algae in their role of primary producers of energy-containing mass that can enter the aquatic food web.

Nitrogen "locked up," so to speak, in organic compounds is returned to the environment through decomposition and, to a lesser extent, by excretion of nitrogenous wastes of animals. The end product of the first state of oxidative

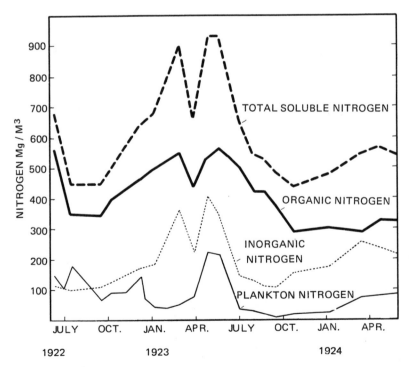

Figure 11.1   Seasonal variations in the quantity of various forms of nitrogen in Lake Mendota, Wisconsin, 1922-1924. (After Domogalla *et al.*, 1925.)

degradation of animal and plant proteins is mainly free ammonia (NH_3). Lesser amounts of ammonium compounds such as the base, ammonium hydroxide, and a salt, ammonium carbonate, are also released. The agents of the decomposition process are microbial organisms, *i.e.*, certain bacteria and fungi. The free ammonia content of natural waters is derived in part from this bacterial decomposition of proteins, and in part from **deamination** (the removal of an amino group (NH_2) from an amino acid), a process that also involves bacterial action.

Generally, in unpolluted waters ammonia and ammonium compounds occur in relatively small quantities, usually on the order of 1 mg/liter or less. With the uptake of oxygen, as in pollution, however, the concentration of ammonia may increase, in extreme cases to 12 mg/liter or more. The relationships between ammonia and its basic compound, undissociated ammonium hydroxide (NH_4OH), seem to rest upon the pH of the water. At a temperature of 18°C and pH6 the proportion of NH_4^+ to NH_4OH is approximately 3000:1, while at pH 8 the proportion is nearer to 30:1. Under highly alkaline conditions the concentration of NH_4OH may reach toxic levels. Free ammonia in concentrations over 2.5 mg/liter in neutral or alkaline waters is apt to be harmful to many freshwater species. The common ammonium salt of fresh waters, ammonium carbonate, is usually present in small amounts, but in concentrations of about 20 mg/liter or more under alkaline conditions it is also toxic to certain animals.

Although somewhat variable, the ammonia content of the upper, lighted zones of unpolluted lakes is generally low, particularly in stratified lakes in summer. During summer stagnation the hypolimnion of eutrophic lakes may become considerably enriched with ammonia from decomposition in the tropholytic zone. With the advent of autumnal circulation the hypolimnetic ammonia is distributed throught the lake, thereby increasing the content in the upper strata. Similarly, ammonia produced during winter stagnation is circulated throughout the lake during vernal overturn, producing another seasonally high concentration of the compound. Spring and summer growth of phytoplankton populations in the trophogenic zone, and the resultant increased biochemical demand, bring about a reduction of ammonia during summer and up to the time of fall overturn.

In an intermediate phase of oxidation of nitrogenous organic compounds, ammonia is attacked by certain of the nitrifying bacteria, such as *Nitrosomonas,* which absorb ammonia and release nitrite (NO_2^-) ions in the process. Energy is released in this activity and is used in the synthesis of carbohydrates by the bacteria. Nitrite may also be formed by reduction of nitrate (NO_3^-), and it is probable that more of the nitrite content of most natural waters is derived from this process than from oxidation of ammonia.

Nitrite nitrogen, as it is usually designated, occurs in very minute quantities (if at all) in unpolluted waters. In lakes in which nitrite nitrogen is present, seasonal variation in concentration of the compound appears to follow (except in early winter) that of the nitrate from which, as we have seen, the nitrite is probably formed. In Lake Mendota the lowest content was found during late summer, when nitrate was also at a minimum. It is now known that diatoms and certain algae, *Chlorella* for example, are capable of reducing nitrate to nitrite. Since all green plants require nitrate, the supply of this compound may become quite low toward the end of the growing season. It seems reasonable to assume therefore that nitrate reduction to nitrite would decline during the same period. During early winter, when nitrate and ammonia concentrations are low, nitrite content may increase; the reason for this is not known. Vertically, the concentration of nitrite is usually maximum between the oxygen-rich trophogenic region with high nitrate content toward its lower boundary, and the oxygen-poor tropholytic zone often high in ammonia. The ecological importance (if any) of small nitrite quantities is not fully understood. Large amounts of the compound usually indicate pollution by sewage.

In what might be considered the final phase in the ecologically important process of decomposition of nitrogenous substance, nitrite is oxidized by nitrifying bacteria to nitrate (NO_3), expressed usually as **nitrate nitrogen.** It is in this form that nitrogen is most easily taken up by green plants rooted in the substrate or floating in the water. To the bacteria involved in the process, *Nitrobacter* for example, formation of nitrate is of less importance than the energy released for their own utilization. To the community and its maintenance and productivity, however, nitrate is extremely important as a nutrient in supplying nitrogen for protein synthesis.

Nitrate nitrogen usually occurs in relatively small concentrations in unpolluted fresh waters, the world average being 0.30 ppm. In certain highly saline lakes (those of magnesium sulfate, for example) this form of nitrogen may be entirely absent at times. Lake Superior has a total dissolved solids content of about 60 ppm, making it, in this respect, the lowest of the Great Lakes. Nitrates account for 0.86 percent of the total in the lake, and this represents the highest concentration of all the Great Lakes. Under the influence of edaphic factors, particularly during flood times, and organic pollution, nitrate nitrogen content may be expected to increase significantly. Under normal conditions, however, the amount of nitrate in solution at a given time is determined by metabolic processes in the body of water, *i.e.,* production and decomposition of organic matter. As we have seen, nitrate is contributed to the ecosystem as a byproduct of bacterial nitrification. The compound is removed from solution through utilization by green plants, and through bacterial denitrification to uncombined nitrogen and reduction to ammonia nitrogen.

Seasonal fluctuations in nitrate nitrogen and other forms in Lake Mendota, Wisconsin, are shown in Figure 11.1. Annual cycles in lakes, generally, may be expected to vary as to time of maximum nitrogen production depending upon latitude, basin morphology and chemical nature of the drainage area substrate, and productivity. Vertical distribution of nitrate is apparently related to lake productivity. In some oligotrophic lakes that have been studied there is little evidence of nitrate stratification; in others the nitrate content of the upper zone decreases during summer from the amount present at spring overturn, with little change in deep waters. In eutrophic lakes, the nitrate concentration is typically decreased in the upper zones by plankton utilization and in the deepest regions by bacterial reduction; the result is a high content near the lower limits of the trophogenic zone.

PHOSPHORUS

In ecological thinking, phosphorus is often considered the single most critical factor in the maintenance of biogeochemical cycles. This extreme importance stems from the fact that phosphorus is vitally necessary in the operation of energy transfer systems of the cell, and that it normally occurs in very small amounts. The latter factor means that there is apt to be a deficiency of the nutrient, and this in turn could lead to inhibition of phytoplankton increase, resulting ultimately in decreased productivity in the system.

Phosphorus is known to occur in several forms, those of greater concern in natural waters being soluble phosphate phosphorus, soluble organic phosphorus, and particulate organic phosphorus of the seston. In water, phosphate may enter into combination with a number of ions—most conspicuously, perhaps, with iron and with the usually abundant calcium. The pH of the water determines to a great extent the nature of the phosphate compound. Under circumneutral and moderately alkaline conditions calcium phosphate is probably prevalent, while extremely high pH is associated with the presence of sodium phosphate. In acid waters phosphate attraction swings toward iron, to form ferric phosphate. Except for precise investigations of the activity of the ion itself or of biological assimilation of the various fractions, it usually suffices in general limnological studies to report total phosphorus, or phosphate, content.

The concentration of total exchangeable phosphorus in natural waters is determined primarily by four factors: (1) basin morphometry as it relates to volume and dilution, and to stratification or water movement; (2) chemical composition of the geological formations of the area as they contribute dissolved phosphate; (3) drainage area features in relation to introduction of organic matter; and (4) organic metabolism within the body of water, and the rate at which phosphorus is lost to sediments. Ground waters and flowing surface waters are typically richer in inorganic phosphate than are surface waters of open lakes, due mainly to less biological demand in proportion to water volume. In most open lakes assimilation by phytoplankton and bacteria serves to reduce the inorganic phosphate content. In closed basins of arid regions evaporation may result in very high concentrations of total phosphate. With respect to geological influence, it should suffice to say that waters in local regions of highly phosphatic substrate contain considerable quantities of the ion. It is doubtless safe to state that all bodies of water that support some plant populations contain a quantity of phosphate, albeit small, in some cases less than 0.001 ppm (frequently expressed as milligrams per cubic meter because of the minute quantities normally encountered). Variation in total phosphorus content of selected lakes is shown in Table 11.3. Note especially the difference in mean content of lakes in the highly phosphatic region of Florida and those of other regions. The mean total phosphorus content of most lakes ranges from about 0.010 to 0.030 ppm. Goodenough Lake, shown in the table, is a highly saline lake in an arid region.

The seasonal distribution of phosphorus in lakes is variable, being determined to a great degree by basin form, chemical composition of the surrounding terrain, land use, behavior of other substances in the particular lake, and the annual cycles of mixing. In relatively open basins receiving considerable influx of surface water, phosphorus concentration is often regulated by stream discharge, particularly following high rainfall. The Maumee River drainage of Ohio annually contributes on the order of 96 metric tons of soluble phosphate to the western basin of Lake Erie. Of this total, 68 percent is carried in seasonally during January and February, the time of greatest stream discharge. The Maumee load is apparently derived as leachings from the highly agricultural region, for other streams of the area transport proportionately less phosphorus. In lakes of more self-contained dynamics, seasonal variations in total content and vertical distribution of phosphorus are rather complexly related to other compounds and

TABLE 11.3 Concentration of dissolved phosphorus (parts per million) in lake waters from various regions[a]

	DISSOLVED PHOSPHORUS	
	mean	range
HUMID CLIMATE, OPEN LAKES		
Northeast Wisconsin	0.023	0.008-0.140
Connecticut		
eastern highland	0.011	0.004-0.021
western highland	0.013	0.007-0.031
central lowland	0.020	0.010-0.031
Japan	0.015	0.004-0.044
Austrian Alps	0.020	0.000-0.046
Sweden		
uplands	0.038	0.002-0.162
northern Sweden	0.024	0.007-0.064
Florida		
phosphate regions	0.290	0.100-0.660
other regions	0.038	0.000-0.197
ARID CLIMATE, CLOSED AND SALINE LAKES		
California (Owen's Lake)	0.078	
British Columbia (Goodenough Lake)	0.208	

[a] Data from Odum, 1953; and Hutchinson, 1957 (after various authors).

mixing cycles. In eutrophic lakes during summer stratification the phosphorus content in the hypolimnion increases significantly following oxygen depletion. The factors involved in this increase are not fully understood. One investigator has shown that this phosphorus is apparently released from a ferric iron–phosphorus complex which is insoluble in the presence of oxygen; in the absence of oxygen the ferric compound is reduced to a soluble ferrous form, thus liberating phosphorus. With lake overturn and the reintroduction of oxygen, the insoluble ferric phosphate is again formed and distributed throughout the lake. From another source has come evidence that in the presence of oxygen phosphorus is adsorbed on basic iron compounds in the oxidized microzone of the mud. Removal of oxygen brings about a reaction in which the ferric ion is reduced and phosphorus is released. While it is possible that both mechanisms operate, it seems certain that the presence or absence of oxygen is a critical factor. Soluble inorganic phosphorus in the upper waters of eutrophic lakes is usually low throughout the year, at times becoming depleted during summer. The concentration of total phosphorus (mainly the organic form) may increase in late summer.

The Cycle of Radioactive Phosphorus in Lakes

We are all aware of the astounding array of new tools of everyday living and instruments of research made possible by the twentieth-century exploitation of atomic energy. Some of the materials and techniques have already entered into limnological investigations, and many more are destined for future usefulness. Prior to the "radioactive era" our knowledge of phosphorus activity in natural waters was based on field and laboratory analyses and measurements of seasonal variations in concentration of the several forms of phosphorus, and the distribution of the substances in natural waters. Now through the use of radioactive phosphorus (^{32}P) we know quite a bit about the rate and processes of phosphorus circulation in at least certain types of natural waters. Even so, our knowledge at this stage is still meager, and much is yet to be learned. Actually, radioactive phosphorus is only one of many such materials presently available for use in new and exciting research in the dynamics of natural communities, be they aquatic or terrestrial. Recently, isotopes of calcium (^{45}Ca), strontium (^{90}Sr), and yttrium (^{90}Y) have been used in researches bearing on biological limnology.

Of the many questions asked in investigations of nutrient cycles, those of utmost importance concern the source of the nutrient, the forms in which the nutrient occurs throughout the cycle, the manner of transformation of the fractions, and the rate at which the nutrient is circulated. In the case of phosphorus, for example, the major forms such as we have just considered have been known for some time. How are these forms produced, and at what rates? We also know that when phosphorus fertilizer is dumped into a lake most of the compound disappears rapidly, the disappearance sometimes being accompanied by increased plant production and sometimes not. How do we account for this? Among other things, we are aware of the mud–water differential in phosphorus concentration, but, as indicated previously, we are not sure of the causes or of the rate of turnover of phosphorus from these and other sources.

If a quantity of radioactive phosphorus (^{32}P) is diluted to a desired strength of radioactivity (usually as counts per minute) and added to a body of water or used in laboratory experiments, the isotope enters into the system under study. The fate of the introduced ^{32}P can then be determined as radioactivity of inorganic and organic phosphorus in living and nonliving components of water and mud. The rate at which the isotope is transferred within the system may be measured also. Let us now review some of the findings from experiments designed along those lines.

In the first place, it was found that radiophosphorus, when added to a lake surface, decreased at a high rate, as had been observed for fertilizer. Within a few days between 70 and 90 percent of the ^{32}P was lost from the epilimnion. This led to the idea that phosphorus in lakes is held in a kind of repository from which the nutrient is withdrawn to maintain a steady-state equilibrium. In other words, there probably exists a sort of reciprocal exchange relationship between phosphorus in the water and its contents, and phosphorus in solids in contact with the epilimnion (recall that during stratification there can be little mixing of bottom-produced substance in the epilimnion). This state of equilibrium leads to the question of rate of exchange, or turnover time, between the two phases. Turnover time, in this case, refers to the rate at which atoms of phosphorus move from one phase to another, as for example from the lake water to the bottom mud. More precisely, it is the time during which as many atoms are transferred through a phase as there are atoms in that particular phase. The use of tracers permits recognition of atoms of a given phase. (Further details relating to concepts and methods are found in publications by Coffin *et al.*, Harris, Hayes, and Rigler cited in the Bibliography).

A cycle of phosphorus, including turnover times of the components, is shown in Figure 11.2. The exchange rate between mud surface and water, at the bottom of the figure, was determined from laboratory experiments rather than from studies of a lake—in which the rates are apt to be rather variable. Exchange rates in the water phase are probably constant within lakes generally. Of particular interest is the rate at which the introduced inorganic ^{32}P is taken up. Note that there is considerable competition among several factors for inorganic phosphate. Laboratory experiments indicate that over 95 percent of the original radiophosphorus may be taken up within 20 min by phytoplankton and bacteria, the turnover time of the inorganic fraction in the epilimnion being on the order of 5 min. Certain phytoplankton can convert inorganic PO_4 to the organic state in less than 1 min. As shown at the lower right (and this is probably the same substance as at the lower left) about 4 percent of the planktonic cells that take up inorganic PO_4 are sedimented to the bottom each day. Inorganic reactions involving the uptake of inorganic phosphate through the water–mud interface are given at lower center, these reactions proceeding when bacterial activity is inhibited. It is here seen that the rate of return of the phosphate to the water (15 days) is much slower than the turnover from the water (3 days).

The activity of bacteria in phosphorus utilization revealed in these studies is of great ecological significance. It was found, for example, that these microorganisms take up large quantities of inorganic phosphate either by assimilation into their own bodies or by conversion to the organic fraction, thus making the nutrient unavailable for use by green plants. In the absence of bacteria, the rate of uptake of ^{32}P by algae and rooted plants is very high. It appears that in the competition bacteria get a large share of the available inorganic phosphorus. This in turn could seriously limit production of animal mass in a community, although we do not know to what extent these autotrophic bacteria are utilized by consumer organisms in the food web. The problem of bacteria *versus* phytoplankton is difficult to assess, however, for we know very little about the phosphorus requirements in the maintenance of natural populations of the various algal species. Many studies have been made of this subject of plant food requirements, but in most cases laboratory-determined minimal concentrations necessary for growth have been higher than the average content of most natural waters. On the other hand, it has been shown that an excess of phosphorus inhibits plankton growth in laboratory cultures.

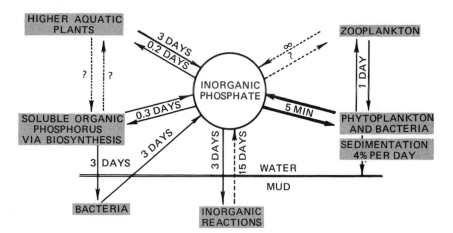

Figure 11.2   Phosphorus relationships in a lake. Turnover times are indicated for the various equilibria. The primary reaction, transferring phosphorus between the pool of inorganic phosphorus and floating bacterial and planktonic cells, is also the most rapid, occupying only about five minutes; the duration of later parts of the phosphorus cycle are measured in days. Two points are especially significant: first, the release of phosphorus from sediment through inorganic reactions is a very slow turnover process; second, turnover by some pathways (indicated by question marks) is immeasurably slow. The net effect is a continuing demand for allochthonous phosphorus. (After Hayes and Phillips, 1958.)

IRON

Iron is of particular interest because of its importance as a vital element in the respiratory pigments of many animal species and because of its part in many chemical reactions in water. We have already seen various aspects of iron in relation to oxidation–reduction systems, and we have examined the reactions of iron with carbonate and phosphate compounds. Whether or not these reactions are significant in any particular body of water depends upon the presence of the compounds other than iron and the momentary form in which the iron occurs. Iron is found widely in nature, usually as either bivalent Fe^{++} or trivalent Fe^{+++}. The bivalent, ferrous, state is soluble, but only under anaerobic conditions. In the presence of oxygen the trivalent, ferric, form is present as a colloidal complex in combination with other inorganic ions and simple decomposition products. With oxygen depletion, as for example in the hypolimnion in summer, the ferric form is reduced to ferrous, the latter going into solution. As a result of the breakdown of the ferric complex the concentration of silicate, phosphate, bicarbonate, or iron is often increased, depending, of course, upon the original chemical nature of the waters. It might be pointed out here that the mere absence of oxygen will not bring about the transformation from ferric to ferrous iron. The depletion of oxygen is a result of organic decomposition which also forms organic compounds that reduce the ferric iron. In some instances, that of ferrous bicarbonate for example, high carbon dioxide content and near-neutral or acid conditions are necessary in addition to absence of oxygen and presence of reducing substances of organic origin.

The vertical distribution of iron in lakes is primarily a composite picture of several forms of the element influenced by the solubility factors just considered. In view of the chemical conditions prevailing in the epilimnetic regions of most lakes the iron content, maintained by several forms, is quite low, usually less than 0.2 ppm. In the deeper zones the iron concentration is a function of oxygen; as in oligotrophic lakes, the iron is normally contained as the nonsoluble ferric complex. If oxygen becomes deficient at the mud–water interface, ferrous iron may go into solution. In eutrophic lakes the hypolimnion typically contains iron in solution during the later stages of summer stagnation. Even at this time the content is normally low, generally a few parts per million.

One of the forms of iron which appears to be most readily available to phytoplankton is ferric hydroxide ($Fe(OH)_3$). This form is the usual product of the oxidation of ferrous iron in waters containing dissolved ferrous salts. It normally occurs as an amorphous mass of a ferruginous organic complex. In the utilization of the organic component of the complex by bacteria, ferric hydroxide is precipitated as a by-product. This process also leads to the formation of lake ochre, the "parent" of iron ore. In the filamentous bacteria of the genus *Leptothrix*, iron hydroxide is left in sheaths surrounding the microbe as it metabolizes the organic material. The species of *Gallionella* excrete ribbon-shaped strands of ferric hydroxide that attach to objects in the water. This and such forms as *Ochrobium, Siderocapsa,* and others probably derive energy from the oxidation of ferrous iron.

SULFUR, SILICON, AND OTHER ELEMENTS

We have seen that compounds of a relatively small number of elements account for the major part of the dissolved solids found in natural waters. The most important of these have been discussed above. A small number of other elements are also of some importance in lacustrine ecology. Of these, the most important are sulfur and silicon.

Sulfur

The most frequently encountered forms of sulfur in fresh waters are (1) the anion sulfate ($SO_4^=$) in combination with the common cations, and (2) hydrogen sulfide (H_2S). Sulfate enters bodies of water with rain, and through solution of sulfate compounds in sedimentary geologic formations in the drainage area. With respect to the latter source we should add that sulfate is not highly soluble. Surface waters are generally low in sulfate, except in regions locally rich in the ion, and in closed lake basins where evaporation raises the concentration. In some geographic regions rain may be the major source of the ion. As shown in Table 11.1, the average content of sulfate in North American waters is about 15 percent. Although this quantity places the ion second in abundance, it is also less than half that of carbonate. Some idea of the range of sulfate concentration in natural waters can be gained from Table 11.2: the low sulfate content of Bunny Lake is probably characteristic of youthful lakes (geologically speaking) formed in regions of weakly soluble terrain; Purcell's Pond is in an area of granitic rocks which normally contain little sulfur in proportion to the alkali and alkaline earth metals. However, this lake has a large peat bog along one side, and, since *Sphagnum* is known to accumulate sulfur from the atmosphere and release it as sulfuric acid, this contribution would augment that from rainfall. Nevertheless, the content of 9.5 mg SO_4/liter for this pond is quite small compared with that of Hot Lake, Washington, which amounts to 103,680 mg SO_4/liter at the surface and 243,552 mg/liter at the bottom.

Sulfate is ecologically important in natural waters in several ways. It is apparently necessary for plant growth; short supply of the material can in inhibit the development of phytoplankton populations, and thereby reduce production. Sulfur is important in protein metabolism and is supplied to the organisms originally as sulfate. As we shall presently see, under anaerobic conditions sulfate is utilized in chemosynthetic processes of sulfur bacteria, the sulfate being reduced to hydrogen sulfide. Sulfur in this reduced state may have a parallel reaction with iron, resulting in the liberation of phosphates and other nutrients held in ferric complexes.

Seasonal pulses in sulfate concentration have been demonstrated in certain lakes. In some, the sulfate reaches a high in spring followed by decreasing content to a low in autumn. Bicarbonate variation in the same pond behaves in the reverse fashion. It has been suggested that the sulfate decrease is due to reduction to sulfide which becomes taken up in the bottom mud; the bicarbonate fluctuation apparently results partly from carbon dioxide removal by photosynthesis and partly from sulfuric acid activity in the winter oxidation of sulfide.

The presence of hydrogen sulfide and some aspects of its occurrence have been considered in Chapter 10. In relation to the sulfur cycle, however, we should note here the activity of bacteria in maintaining the cycle. The reactions involving H_2S are essentially between oxidation and reduction. Hydrogen sulfide can be oxidized by a number of so-called "colorless sulfur bacteria," among which *Beggiatoa* is probably the best known. The species of this genus are large filamentous forms often said to be more closely related to blue-green algae than to bacteria. Under normal conditions the cells of the filaments include granules of elemental sulfur accumulated only when both oxygen and H_2S are present. When H_2S becomes exhausted from the environment, the sulfur in the granules is metabolized by the organism and sulfuric acid is released. The cellular oxidation exhibited by these bacteria apparently proceeds as indicated in the equations

$$2H_2S + O_2 \rightarrow 2S + 2H_2O + \text{energy}$$
$$2S + 2H_2O + 3O_2 \rightarrow 2SO_4^= + 4H^+ + \text{energy}$$

Another group of bacteria, particularly the genus *Thiobacillus,* is obligately chemotrophic and can oxidize elemental sulfur to release sulfuric acid. Certain other bacteria, including the euphoniously named *Desulfovibrio desulphuricans,* can reduce sulfate (and other sulfur compounds), resulting in the release of H_2S. With this activity the cycle of consumption and production of H_2S is completed.

Silicon

More than 60 percent of the rock and soil occurring at the earth's crust consists of silicon dioxide (SiO_2). In view of this proportion we should expect to find the compound in most unpolluted waters. As is the case with many other biogenic substances, silica is particularly concentrated in certain sedimentary rocks and therefore generally occurs in higher concentrations in waters in regions of such rocks. Even within a given region, however, the silica content of waters may be quite variable, depending upon the nature of the host rock. For example, in Florida, a limestone area in which sedimentary soils and rocks predominate, the concentration of SiO_2 ranges from less than 1 mg/liter to over 30 mg/liter. Unlike many other minerals, silica does not appear to be important in the composition of animal or plant protoplasm. It does occur, however, in the "shells," or frustules, of certain algae (the diatoms), in cysts of yellow-brown algae, and in skeletal spicules of some sponges. Population development of such diatoms as *Asterionella, Melosira,* and *Tabellaria* is limited, at least partially, by concentrations of from 0.5 to 0.8 mg SiO_2/liter. The utilization of silica by one or more of these organisms takes up the mineral from lakes and streams, and often during blooms of diatoms the quantity removed may be great.

Other Elements

In addition to those elements previously considered, natural waters may contain others in various proportions. Manganese is commonly found, and, although it is less abundant, it has a role similar to that of iron. Aluminum, zinc, and copper are usually found in natural waters in varying quantities. Traces of molybdenum, gallium, and nickel have been detected on occasion. Uranium appears to be common in small amounts in most lakes and streams, and industrial activities may be a strong environmental factor influencing future concentrations of this element.

ORGANIC SUBSTANCES

Organic matter is present in water in solution and in the form of the **seston**, the fraction of plankton and organic debris suspended in water. Organic substances are also found, often abundantly, in bottom sediments. The total content of these materials in lakes is derived essentially from two sources: (1) incoming waters (and wind and wave action) at the shore, which introduce allochthonous matter into the lake to account for a portion of the total—the nature of the substances brought into the lake from the outside being quite varied, ranging from leaf litter of the shoreline to upland soil leachings and organic pollution; (2) autochthonous matter produced in the lake by living organisms and by decomposition of plant and animal bodies. The relative contribution of each of these sources to the lake total is of considerable importance in maintaining productivity. It would, therefore, be of interest to measure each, but at present we do not have precise, readily usable methods for doing so. Hutchinson (1957) has reviewed and commented upon several techniques.

Data on the organic composition of a large number of lakes in Wisconsin have shown that the total content and the proportions of various fractions of the total organic concentration vary widely in lakes (Birge and Juday, 1934). In organically poor lakes the carbon content ranges from 1 to 2 mg per liter, and, of the total organic content, carbohydrates comprise slightly over 73 percent and proteins about 24 percent. The richer lakes contain up to about 26 mg carbon per liter, with nearly 90 percent of the organic matter being composed of carbohydrates and about 10 percent of proteins. It appears that the protein content generally decreases with increasing total organic content. In lakes poor in organic matter (little autochthonous substance) the seston organic material averages about 16 percent of the total. In richer lakes seston matter comprises about 4 percent of the total organic content. The data from these studies indicate that the particulate organic matter of lakes increases with increase in dissolved organic substance up to a point; beyond this point additional dissolved organic matter does not cause an increase in the seston. This relationship would seemingly indicate that autochthonous organic materials do not contribute greatly to the productivity of lakes. In fact, it has been shown that the protein content generally is indeed low in the allochthonous portion of lake substances. See Table 11.4 for data concerning these relationships.

Lake color may be correlated as a rule with the source of organic matter. Allochthonous material derived from peat or shallow-water plant debris is mainly responsible for the dark-brown color of bogs. In such waters the protein fraction of the total organic content is low in proportion to carbon. In the waters of lakes receiving little allochthonous material the protein component is high in proportion to carbon (Table 11.4), the autochthonous matter being formed mainly by plankton decay. These waters are relatively uncolored. Color and source of organic matter are apparently associated with plankton production also. In highly stained waters which contain considerable quantities of allochthonous substances, plankton may comprise as little as 3 percent of the total organic substance. In unstained lakes receiving little allochthonous materials, over one fourth of the organic matter may be plankton.

The total organic content of most ''average'' lakes is as much as three or four times that of open seas. Sea water generally contains about 2 mg carbon and about 0.2 mg organic nitrogen per liter. These differences are of interest, but probably bear little on the composition of estuaries. Most estuarine basins receive considerable quantities of allochthonous materials from stream inflow and drainage from surrounding marshes or other land features.

TABLE 11.4 Nature and amount of organic matter from water of Wisconsin lakes, with reference to total organic carbon[a]

Carbon[b]	Organic seston[b]	Organic solutes[b]	C:N ratio	Water color[c]
1-2	0.6	16	12	
5	1.3	13	15	2.4
10	1.9	9	20	3.2
15	2.3	7	22	4.3
20-25	2.2	4	29	4.3

[a] Data from Birge and Juday, 1934, as reproduced in Hutchinson, 1957.
[b] Quantities given in mg / liter.
[c] Water color given in platinum-cobalt units (see page 139).

An impressive array of organic compounds has been isolated, identified, or otherwise recognized in fresh and marine waters, bottom sediments, and soils (see Vallentyne (1957) for a comprehensive review of the subject). At present we know very little about the role of most of the organic compounds in the overall ecology of freshwater or marine communities. There is sound evidence that certain bacteria can utilize dissolved organic substances as an energy source. Beyond this it has not been established that multicellular animals make use of such materials, even though a considerable controversy over the matter has existed for some 50 years.

Fats, in the form of simpler molecules of fatty acids, are known from freshwater and marine seston and water. Acids and certain waxes have been recovered from sediments. Some attention has already been given to organic nitrogen in natural waters. Other compounds such as free amino acids, tryptophan, glycine, glutamic acid, tyrosine, and others, are known from water, seston, and sediments. These have been recovered upon hydrolysis. Although studied but little, some of these amino substances may be important to organisms which are unable to synthesize certain of the acids, the organisms obtaining the items directly from the environment. Hydrolysis of seston and sediment components reveals a number of sugars and sugar-like compounds, the best known being glucose and galactose. In estuaries a compound having the qualities of a reducing sugar has been shown to influence the water-pumping rate of oysters.

Yellow carotenoid substances with fluorescent qualities have been extracted from shallow marine waters, and from fresh waters. The marine compound may not, however, be the same as yellow acids of lakes.

Fresh and marine waters contain biologically active compounds such as vitamin B_{12}. The seston of both environments has yielded also vitamin A and vitamin D. Thiamin (B_1), biotin (H), and niacin (B_6) are known from lakes, the last-mentioned being present in amounts up to 0.89 mg/m^3. Enzymes are present in natural waters and probably enter into various biological reactions in the environment.

It is now known that some organisms, particularly algae, produce metabolites that serve in various ways to inhibit, stimulate, or generally regulate the development of natural populations of other algae. The chemical nature of the particular metabolites is not known, although the possibility of hormone-like substances and fatty acids has been suggested. The effects of these and more specifically toxic compounds will be considered further in Chapters 12 and 16, in relation to populations.

DISSOLVED SOLIDS IN STREAMS

The reactions and behavior of the various ions and organic substances in streams are governed by chemical and biological processes and conditions such as we have considered in the preceding pages. The main differences between lake and stream characteristics with respect to dissolved solids usually pertain to relative concentration, composition, and longitudinal distribution of the substances. Seasonal variation in rainfall and surface runoff, and the geochemical nature of the drainage basin, strongly influence the composition of waters of small streams, thereby imparting considerable individuality to streams even within a restricted region. Variability becomes the key word in any attempt to categorize small streams. This variability is seen not only when comparing individual streams; it is also observed along the gradient of a particular stream, being markedly influenced by edaphic factors, human endeavors, and channel morphology. Waters of large streams, on the other hand, typically exhibit general uniformity of composition, so much so that a quantitative expression of average content becomes meaningful. The mean composition, as percentage proportions, of the major cations of river water has been given as: calcium 63.5, magnesium 17.4, sodium 15.7, and potassium 3.4. Note that the order of prominence here differs slightly from that of North American waters generally, as given in Table 11.1. In fact, the average composition of rivers is quite similar to that of open lakes, as might be expected.

The calcium content of surface waters varies with pH, substrate composition, and temperature. Acid swamp streams and streams on igneous rocks contain little or no calcium, but streams in the region of certain sedimentary formations often contain high proportions of the ion. Springs issuing from limestone frequently contain on the order of 70 ppm calcium. This concentration is similar to that of many large North American streams; throughout much of its length the Missouri River contains, on the average, about 60 ppm of calcium. Much of the ion in large streams is carried to the sea, where it enters into the calcium cycle.

As in lakes, there also appears to be a correlation in streams between quality and quantity of algae, and calcium carbonate concentration. Under proper conditions calcium in streams may be precipitated as the monocarbonate by lime-encrusting plants, and some is taken up by shell-building animals. The magnesium content of flowing waters is also highly variable, ranging from trace quantities to nearly 500 ppm in warm mineral waters. Although calcium and magnesium react similarly, particularly in entering into carbonate combinations, the proportion of calcium to magnesium in freshwater streams varies from about 2:1 to 10:1 or more.

Nitrogen compounds normally exhibit conspicuous seasonal fluctuations and pronounced variations along the gradient in small streams. The amount of nitrate nitrogen and nitrite nitrogen is influenced to a great degree by surface runoff and associated stream level and discharge (see Figure 11.3). Nitrogen in ammonium compounds is released into streams through decomposition of organic debris. In unpolluted streams the concentration is small, normally less than 1 ppm. Pollution raises the concentration of ammonium compounds, and this, within limits, increases biological productivity. In excess, certain compounds of ammonium can be toxic to stream organisms. High concentrations of nitrogen occur during times of winter and spring flooding, when plant populations are minimal. Interestingly, the floods that serve to bring in nitrogen may

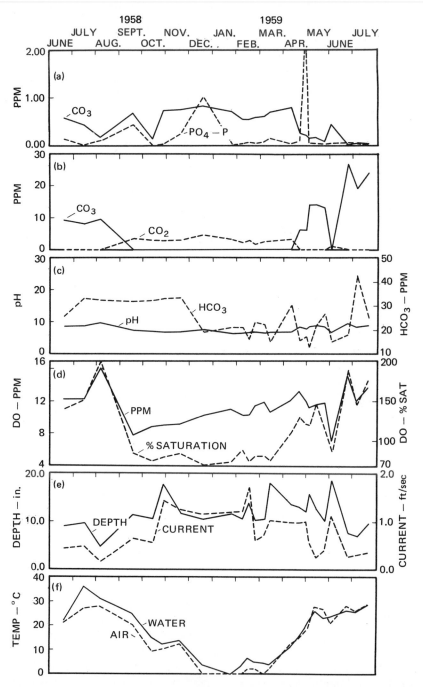

Figure 11.3 Seasonal variation in certain physical and chemical aspects of a segment of Stony Brook, a small stream on the New Jersey Piedmont near Princeton. The graphs of the various data are presented one above the other in order to indicate the correlations between different factors. (After Woods, 1960.)

also scour algae from rocks in the stream bed, thus minimizing consumption during the temporary increase in nitrogen. We might add that rapid decrease in nitrogen follows closely behind the flush of flood water. In streams in more stable watersheds, the nitrogen content may be lowered in spring and summer by plant utilization, with continued decrease to late summer. The major source of nitrogen compounds in many streams is pollution by drainage from agricultural areas and from sewage. These factors contribute large quantities of the substances regionally and have noticeable effects on stream communities.

The phosphate compounds of streams are derived from biological and chemical processes along the stream course. During summer the concentration of inorganic phosphates may increase somewhat, due to biological activity. Increased surface runoff contributes to the phosphorus content of streams by introducing allochthonous phosphorus-containing substances (Figure 11.3). Although there is considerable longitudinal and seasonal fluctuation in content, there does not appear to be any great depletion of stream phosphorus such as we have noted in surface waters of lakes during the growing season. This is due partly to mixing and partly to the normally greater proportion of water to phytoplankton. In fact, as we shall see later, there is little, if any, true plankton in streams of fair velocity.

Iron occurs naturally in streams, although in relatively small proportions. In unpolluted flowing waters this ion usually takes the ferric form because of continual aeration and the presence of oxygen. Where organic decomposition is great, oxygen depletion may result in the transformation to the ferrous state and the precipitation of iron hydroxide. Similarly, in stagnant stream pools, particularly following subsidence of flood conditions, iron bacteria may grow rapidly and form masses of ferruginous substance in the pools.

Sulfur, as sulfate or hydrogen sulfide, occurs in varying quantities, depending upon the composition of the substrate, the area of the catchment basin (its relation to rainfall), and the course of the stream. Obviously, the importance of the substrate rests on the proportion and solubility of minerals present. The stream course determines to a great extent the occurrence of hydrogen sulfide. If the course permits slow flow through organically rich swamps or marshes, the gas may be formed by decomposition or, as we have seen, by the reduction of sulfate.

A number of additional ions, including those discussed in the preceding section, are present in streams. Their abundance is usually associated with land forms and chemistry, and the ecological role of many of them is poorly known. The importance of climate, geochemistry, and physiography in stream composition has not been studied sufficiently. We can see some of the interrelationships in a few summarizing statements relating to certain of the major ions:

1 The waters of rivers whose headwaters lie in semiarid plains are generally high in sodium, sulfate, and chloride, but low in calcium and carbonate.

2 Streams draining youthful granite mountains typically contain high concentrations of silica, and the total dissolved solid content is low.

3 Valley streams in granitic mountain regions show an increase in total dissolved solids.

4 Rivers draining extensive plains in humid, temperate regions are characterized by high sulfate and carbonate concentrations.

DISSOLVED SOLIDS IN ESTUARINE WATERS

The chemical composition of estuarine waters, and the distribution of ions in an estuary, represent the net result of the meeting of waters from two quite different aquatic environments, the freshwater and the marine ecosystems. The preceding section indicated something of the variety we can expect to encounter when we study the chemistry of freshwater streams. All of these streams, moreover, can be expected to differ chemically from the relatively uniform seawater into which they flow. Table 11.5 suggests the range of this difference by presenting data from two North American streams that flow into estuaries, one on the Atlantic coast and the other in the Gulf of Mexico; the table compares the chemistry of both streams with that of seawater. Without even taking into account the effects of channel and basin morphology, tides, and currents, it is obvious from what has gone before that the chemical features of any two estuaries are likely to be quite different.

Of the cations in sea water, sodium is by far the most abundant. Streams typically contain more calcium than any other element, and calcium and

TABLE 11.5 Ionic composition of natural water from several sources[a]

Ion	Delaware River at Lambertville, New Jersey	Rio Grande at Laredo, Texas	Seawater
Na	6.70[b]	14.78	30.4
K	1.46	0.85	1.1
Ca	17.49	13.73	1.16
Mg	4.81	3.03	3.7
Cl	4.23	21.65	55.2
SO$_4$	17.49	30.10	7.7
CO$_3$	32.95	11.55	0.35[c]

[a] Data from Clarke, 1924; and Harvey, 1957.
[b] All figures indicate percentage composition.
[c] Includes dissolved HCO$_3$.

magnesium together constitute a greater fraction in fresh water than does sodium in seawater. In streams the anion sulfate characteristically occurs in greater proportion than does chloride, and in many cases carbonate does also, but compare the carbonate and sulfate percentages for the Delaware and the Rio Grande in Table 11.5. Offshore and away from stream influences, seawater exhibits a nearly constant content. Sodium predominates among the cations of seawater, making up nearly one third of the total salts content, and chloride is the most abundant anion, comprising over one half of the total ionic composition. Thus the chemical composition of any estuary must represent the dual contribution from fresh and salt water, in addition to which local drainage may play a part.

SALINITY

Numerous references have been made in preceding sections to salinity. These references have shown the influence of this factor on various estuarine parameters such as density and stratification, temperature, and dissolved gases. Let us now give attention to a more precise definition and to consideration of salinity in its more or less true light, that of its chemical aspects.

The ions shown in Table 11.5, together with strontium, bromide, and boric acid, contribute more than 99 percent of the total salts in solution in full-strength sea water. The weight in grams of these salts dissolved in 1 kg of sea water is termed the **salinity**. In the process of evaporating seawater in order to weigh the salts, carbonates are decomposed to oxide, and bromine and some quantity of chlorine are liberated; these transformations and losses are taken into consideration in the definition. Salinity is usually expressed as **parts per thousand** (‰), although **percent** (%) and **grams per kilogram** are also used. Since the salts of the metallic ions, the conservative elements, occur in very uniform proportions, it is possible to determine the salinity by measuring the chloride content, or **chlorinity**, and applying it in the relationship

$$\text{Salinity (‰)} = 0.030 + 1.8050 \times \text{chlorinity (‰)}$$

Chlorinity is rather easily and accurately measured by silver nitrate titration, using potassium chromate as an indicator. Salinity can be determined by electrical conductivity, and from measurements of density obtained by the use of hydrometers.

The salinity of open seas generally ranges between about 33 and 38‰, with the average being near 35‰. Since the average salinity of soft fresh water is 0.065‰ and of hard fresh water 0.30‰, it is apparent that high estuarine salinities are derived almost wholly from seawater, while the diluting effect of the influent serves to reduce the concentration of dissolved salts. In general, the proportion of dissolved salts in estuaries resembles that of seawater, while the total concentration is variable along the axis of the estuary. In some instances, however, the concentration of inflowing fresh waters may be such as to modify the normal ionic relationships in estuaries. Where this occurs the result is usually an increase in the ratios of carbonate and sulfate to chloride, and of calcium to sodium over those of average seawater. The momentary salinity may be regarded as a function of the quantity and quality of inflowing and outflowing waters, rainfall, and evaporation. Since these factors may vary with season (in some instances rather drastically), the general structure of the estuary also shifts. Therefore, attempts to fit estuaries into schemes of classification are often difficult. We can, however, recognize that certain estuaries are typically, or on the average, more or less saline than others. For example, the average salinity of the estuary-like Laguna Madre of Texas nearly always exceeds that of seawater. The waters of the James River estuary of Chesapeake Bay, on the other hand, grade from fresh to a salinity of about 17‰ near the mouth. On the basis of salinity characteristics such as mean and range, estuaries can be classified under various systems. One of these, the 1958 "Venice System," classifies marine waters according to certain approximate salinity zones:

ZONE	SALINITY, ‰
Hyperhaline	> 40
Euhaline	40 to 30
Mixohaline	(40)30 to 0.5
Mixoeuhaline	> 30 but $<$ adjacent euhaline sea
(Mixo-) polyhaline	30 to 18
(Mixo-) mesohaline	18 to 5
(Mixo-) oligohaline	5 to 0.5
Limnetic (fresh water)	< 0.5

According to this scheme the James River would be classified on the basis of salinity range as mixohaline, while its mean salinity of about 10‰ permits a designation of mixo-mesohaline. Laguna Madre would obviously be termed hyperhaline. The salinity range of Great South Bay (Figure 5.2) indicates mixo-polyhalinity.

Salinity is a very critical factor in the distribution and maintenance of many organisms in the estuary. To these plants and animals mean salinity is probably of less importance than the rate and magnitude of seasonal and tidal fluctuations in salinity. A bottom-dweller in the upper part of the Tees Estuary (Figure 11.4) may be subjected to a change of 12 to 13‰ during a winter tidal cycle, and seasonal variations in other estuaries can be much greater. The ability of organisms to withstand such changes is primarily related to their osmoregulatory adaptations, which will be considered in Chapter 12.

The horizontal and vertical distribution of saline waters in an estuary is closely related to the form of the basin and channel (Chapter 5), to tides and currents imposed by sea and stream (Chapter 9), to relative water volume contributed by the two sources, and to evaporation. These processes operate to varying extents to create a diversity of often complicated estuarine patterns. Generally, however, we may recognize two basic types of estuary. The first is the **positive estuary**, in which the influx of fresh water is sufficient to maintain mixing and a pattern of increasing salinity toward the mouth of the estuary, usually in the mixohaline range. This type of estuary is characterized by low oxygen concentration in the deeper waters and considerable organic material in the bottom sediments. Vertical distribution of salinity in normal estuaries may range from top to bottom homogeneity, through oblique layering, to complete vertical stratification, depending to a great degree upon depth of the basin, morphology of the bottom and especially the mouth, and relative temperatures and magnitude of currents in the conflicting waters. Great South Bay is a shallow, mixing estuary-lagoon in which the vertical salinity is nearly uniform, and in which the horizontal pattern is one of increasing salinity toward the mouth (Figure 5.2). The positions of lines connecting masses of similar salinity (isohalines) indicate deflection of inflowing currents of high salinity toward the left side of the estuary (facing toward the mouth), and outflowing masses of low salinity toward the right side. This configuration results from rotational effects of the earth (Coriolis forces) in the Northern Hemisphere. The profile of salinity distribution in the estuary of the River Tees, England (Figure 11.4), shows a vertical stratification maintained during complete tidal cycles. As depicted in the figure, the entire water mass shifts vertically and (slowly) horizontally with tides. As the surface layers of fresh water move seaward, salt water flows

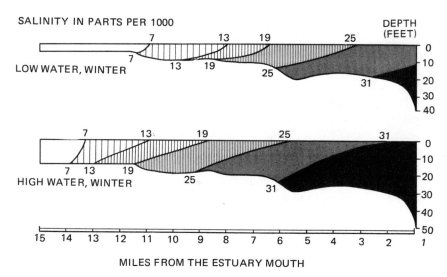

Figure 11.4 Longitudinal section of the Tees Estuary, England, formed where the Tees River flows into the North Sea. Winter salinity gradients are indicated in parts per thousand at low tide (upper figure) and high tide (lower figure). (After Macan and Worthington, 1951.)

upstream. The average time for all masses to move down the estuary is about six days during the dry season, and between two and three days during the wet period in winter.

It appears from the two examples of positive estuaries that salt water may flow into an estuary either as a sheet below a freshwater layer, or as something of a current moving along one side of the basin. This is essentially true, except that patterns of inflowing seawater may be complicated by movements ranging from the advection of a large salt-water wedge to eddy diffusion resulting from shallow-water tidal currents.

The second major type of estuary based upon salinity is the **negative estuary**. In arid regions, in particular, evaporation from an estuary may exceed the inflow of fresh water. This usually results in increased salinity in the upper part of the basin, especially if the mouth of the estuary is restricted by shoreline features that inhibit tidal flow. The resultant pattern of salinity distribution is therefore reversed with respect to that of positive estuaries. Negative estuaries (and lagoons) are typically hyperhaline, but possess a moderate oxygen concentration at depth. Bottom muds are generally poor in organic content. The Laguna Madre, previously mentioned, is an example of a negative lagoon. High salinity, sometimes up to 90‰, occurs in the upper segment. The salinity distribution pattern changes often, due to highly saline (more dense) masses being blown about in the lagoon by shifting winds.

We have given considerable attention to the more conspicuous and general reactions and relationships of the major ions in fresh water. These factors and processes have been studied very little in estuaries. However, as in freshwater research, much is known about certain phases of the chemistry of seawater. The chemical activities of calcium, magnesium, sodium, and potassium in salt water are basically the same as in fresh water, being tempered somewhat by salinity and by variation in relative proportion, and bearing certain relations to chlorinity which can be modified by the introduction of fresh water. These aspects of marine chemistry, along with fuller treatment of salinity, are found in Harvey (1957) and Sverdrup, Johnson, and Fleming (1942). In the estuary, therefore, we might expect the essential processes to be complicated by the meeting of chemically variable fresh water with relatively uniform seawater over what is often a considerable linear distance.

Similarly, the activities of nutrient elements and processes of nutrient cycles in estuaries are not greatly different from freshwater dynamics. The organisms involved, such as green plants, animals, and bacteria, may differ in name, but they perform similar ecological functions in both fresh water and estuaries. Estuaries may contain greater concentrations of certain nutrients than the adjoining sea, due to the introduction of the substances from terrestrial sources. Phosphorus, for example, may be found in higher concentration in the upper reaches of an estuary than near the mouth. Phosphorus has also been found in the deeper, unlighted and anaerobic regions and in surface waters at times when photosynthetic activity is low; this corresponds to what we have found in fresh waters. Similarly, the content of nitrogen (in its various forms) and silica is often greater in estuaries than in sea surface waters. With respect to most of these, the estuarine concentration typically increases linearly toward that of the contributing fresh water. Surface–bottom differentials in the concentration of nitrite, nitrate, phosphate, silica, and others have been observed in stratified estuaries. In some, the concentration of silica is highest in fresh waters flowing over the surface, while the greater content of nitrite and nitrate is in the deeper water. The seasonal cycles of nitrate, phosphorus, and silica, as nutrients in estuaries, are similar to those of lakes and of the sea, insofar as they are known. The pattern is essentially one of thorough mixing in winter due to uniform temperatures and winds, followed by spring and summer growth of phytoplankton, and depletion, or near-depletion, of nutrients in the upper regions during summer stratification (if present). During summer there may be little circulation of nutrients through the thermocline. Mixing and replenishment of the nutrient supply in autumn complete the cycle.

DISSOLVED SOLIDS IN THE OCEANS

The concentrations of dissolved solids in marine waters are very nearly constant; in this respect the marine environment is strikingly different from any freshwater habitat. Aside from biogenic salts, such as nitrates, phosphate, and those of silicon, iron, and (to a slighter degree) arsenic, manganese, and copper, the ratio of one ion to the others is constant, as are the absolute concentration values. For this reason analysis for most dissolved solids is rarely carried out, since it is possible, and simpler, to calculate the values for most materials on the basis of the chlorinity of the water.

The salinity of the oceans is fairly constant, usually in the range of 35 parts per thousand, but regional peculiarities can be observed. Some of these regional differences may be so persistent as to be considered permanent features of certain bodies of water. For example, salinity is reduced in great bays which are fed by river or glacial waters. Salinity is 30 in the Gulf of Saint Lawrence, and about the same in the Gulf of Chihli, near the mouth of China's Yellow River (all salinity values are expressed in parts per thousand). Tropical rains account for depressed salinities in equatorial seas (34.85 in the Pacific; 35.67 in the Atlantic). In the middle of the North and South Pacific Oceans, evaporation raises salinity values to 35.5 in the north and 36.0 in the south; in the Atlantic, values reach 37.5 in the Sargasso Sea. Especially high salinity —38 to 41—occurs in the Red Sea, while polar seas, in contrast, are adjacent to melting ice caps and exhibit markedly depressed salinity values. For virtually all marine waters, however, salinity falls in the range between 33 and 38 parts per thousand.

One consequence of this limited variation in seawater salinity is that the electrical conductivity of marine waters is nearly constant throughout the world. Thus **conductivity** is often used as a reliable measure of salinity in chemical oceanography. This would not be possible in estuaries, where the proportions of ions, and the proportions of ionized and nonionized substances, are more likely to vary. The most abundant ions, of course, are those of sodium and chlorine. In order of decreasing abundance, the other major ions are magnesium and the sulfates, calcium, potassium, boron, and bicarbonate. Each of these tends to occur in a constant, predictable concentration, as compared with the chloride ion. These ratios are so constant, in fact, that chlorosity alone can often be used to calculate the abundance of ions of other elements. In similar fashion, the abundance of one ion can be used to estimate the abundance of another, by relating the abundance of each to the (constant) frequency of the chloride ion (Sverdrup *et al.,* 1942).

Calcium is of special importance because of its involvement in the biological processes that form the skeletons of vertebrates and the calcareous shells and structures of many other species. Chemical measurements have shown some reduction in calcium concentration in surface waters, presumably a result of its incorporation in biological processes. Differences in Ca values have been detected at different depths, and the role of calcium in the deposition of limestone (as in coralline reefs) is of great interest in marine ecology. In recent decades, strontium, which is present only in low concentrations, has been of concern. The long-lived radioactive isotope ^{90}Sr, a by-product of the nuclear tests of the 1950s and early 1960s, is incorporated into living tissue along with calcium. In this way a dangerous radioactive material enters the major aquatic food chains and becomes a permanent element in the biogeochemical cycle.

Silicon, another biogenic element, is an essential component of diatom cell walls. It usually occurs in low concentration in surface waters. Plankton measurements indicate that the abundance of diatoms is reduced in the open seas, as compared, for example, with the dinoflagellates. One reason for this decreased abundance may be the paucity of silicon, which acts as a limiting factor.

Nitrogen compounds occur in three inorganic forms—nitrates, nitrites, and ammonia. Nitrates are most abundant at a depth of several hundred meters, while nitrites, for unknown reasons, seem to occur in a thin stratum just above the thermocline; little is known about the behavior of marine ammonia. Phosphorus occurs in a number of forms, both organic and inorganic, and, like nitrogen, it is most abundant at a depth of several hundred meters. Together with nitrates, phosphates are implicated metabolically in the limited plankton population of the tropical seas. Organic compounds such as yellow caretenoids and vitamins occur in small quantity, but are significant mostly near shallow waters; in open water such materials are barely detectable, as they are rapidly eliminated by biological activity. Major rivers serve to carry tremendous amounts of dissolved solids into marine waters; the discharge of the Amazon River can be detected at a distance of more than 300 km from the coast of South America. Fortunately, the river-borne solutes that are the product of natural biological processes can usually be integrated into the normal biogeochemical cycles of the ocean and the rest of the ecosphere. The same cannot be said, unfortunately, of the industrial and other wastes that result from human activity. These man-made materials are frequently incompatible with the ecology of the marine environment; often they cannot be used in the ecosystem, and no natural agent is able to cause their decomposition into biologically useful compounds. The result is an accumulation of nondegradable wastes, with negative, and possibly severe, impact on the environment.

PART FOUR

Organisms and Communities in Aquatic Environments

The Ecology of Species and Communities

Up to this point we have been considering, for the most part, only the geological, physical, and chemical features of different aquatic habitats. Now we add living things themselves—the **biota**. In this chapter we will see first how aquatic species are specially adapted—by structure or function—to life in fresh or salt water. We will also examine the special relationships through which groups of individuals, of the same or different species, are linked into ecological patterns and associations. Finally, we will see in Chapter 13 how it is possible to predict the effects of change in one or more factors in an ecosystem, either natural or intentional, by defining those factors in mathematical terms and using the techniques of systems analysis.

We will thus approach our topic—aquatic ecology—from three rather different points of view: that of the single **species**; that of the **community**; and finally that of the entire **ecosystem**. The first of these is essentially the approach taken by general biology. The second, exploring the ways in which groups of different species coexist in nature, leads us into the area usually thought of as ''basic'' ecology. The third, which has been important in recent ecological studies, involves broad, integrative concepts and principles. These three approaches may also be referred to respectively as **autecology**, **synecology**, and **ecosystem ecology**.

ECOLOGY OF THE SPECIES—AUTECOLOGY

The biology of the individual (or the species) may be thought of as the study of how the structure and function of each fits it into its environment. Evolution gives us two very important working principles that apply here. First, those organisms best fitted to an environment tend to survive, while others are eliminated. Second, the survivors which develop advantageous mutations become better and better suited to survival under specific environmental conditions. The characters that especially fit a species to a particular habitat are its **adaptations**. Among aquatic organisms, the major forms of adaptation may be grouped conveniently under six headings: whether the organism is attached to a substrate or is free-floating; the organism's usual (or alternative) modes of nutrition; its metabolic processes; its homeostatic mechanisms; its life cycle and mode of reproduction; and, finally, the way the organism responds to such environmental factors as heat, salinity, or the presence of other organisms.

ATTACHMENT TO SUBSTRATE

Among aquatic organisms (as among terrestrial plants) a major consideration is whether, and how, individuals are attached to the substrate. An organism that is attached to rock or sediment, for example, is thereby "locked into" a particular habitat and mode of existence. Body shape, manner of nutrition and reproduction, and many other aspects of adaptation are closely linked to this single factor; in fact, an experienced ecologist will be able to predict much about a species' general mode of life simply by observing its mode of attachment (Figure 12.1).

Modes of Attachment

Many modes of attachment are observed among aquatic plant and animal species. Rock and sediment are the usual substrates, but many organisms attach themselves to individuals of other species in various symbiotic relationships, and a wide variety of other attachment relationships can be found in aquatic communities.

Attachment to rock Some protozoans, such as the gobletlike ciliate *Vorticella,* have a stalk with a sticky end that attaches the individual to rock. Some bacteria and blue-green algae (such as *Calothrix*) produce a sticky capsule that attaches to rock in a similar fashion. Seaweeds and freshwater algae often have a cushion of filaments that adhere to rocks, and a similar attachment is found in some aquatic mosses. Among animals the snail *Littorina* has a muscular suction pad and the familiar barnacle *Balanus* attaches to rock with a cementlike glue; the immature states of many insect species, such as the dragonfly *Libellula* or the dobsonfly *Corydalis*, cling strongly to minute irregularities in rock surfaces. Such streamlined **torrential** forms, as they are called, are able to remain firmly attached to the rock surface even in extremely turbulent streams.

Many of these organisms are able to survive only as long as they remain attached to the substrate. Others, however, will survive if the attachment is broken; the individual may then prove to be adapted to a different role in a different habitat. Those that are circumneutrally buoyant may drift with the current, becoming part of the plankton or neuston; heavier forms may settle and grow on the bottom, becoming part of the epibenthos. Sea lettuce—the green

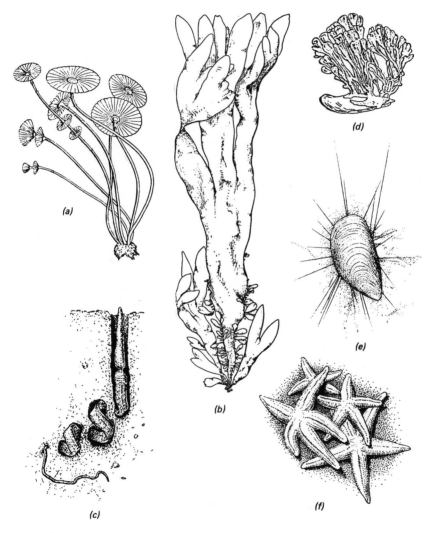

Figure 12.1 Representative aquatic species that live attached to their substrate. *(a)* The green alga *Acetabularia crenulata* is a marine species that occurs widely on tropical seacoasts. *(b) Rhodymenia palmata* (dulse) is a red alga of temperate seacoasts. *(c)* Acorn worms, Phylum Hemichordata, live in mucus-lined burrows in the sea bottom. *(d)* The redbeard sponge *Microciona* grows as an encrustation on rocks and mollusk shells in shallow waters from Cape Cod south to the Carolinas. *(e)* The edible mussel *Mytilus edulis* attaches itself to coastal rocks with strong threads secreted by the foot. *(f)* Starfish, such as *Asterias rubens,* are usually found clinging to the bottom, where they are often found feeding on beds of mussels or oysters.

alga *Ulva*—often breaks off and grows to large size as both a neustonic and an epibenthic form. The wedge-shaped cells of the colonial marine diatom *Licmophora* regularly break loose with age, becoming part of the phytoplankton.

Attachment in sediment Rooting into sediment is a common mode of attachment, and is nearly universal in such higher plants as the cattail *Typha* and eelgrass *Zostera*. The tropical green seaweeds *Caulerpa* and *Penicillus* are common on coral sediments, where they are held in place by rhizoids. The freshwater clam *Anodonta* burrows into bottom sediment with a muscular foot, and the lugworm *Arenicola* digs out a U-shaped burrow in which it spends its life. Here again, most organisms cannot survive if their attachment to the sediment is broken, but some can adapt well to the change. In the coontail *Ceratophyllum* the attachment is easily broken and the plants survive well as they are swept along by currents. In water milfoil, *Myriophyllum humile*, flower formation can occur only after segments of the plant have broken loose. In general, sediment-attached animals do not survive as well as plants if the attachment is disrupted.

Other attachment relationships Many species achieve attachment by entanglement. Among these are the seahorse, which coils its flexible tail around a host plant; the seaweed *Asparagopsis*, whose special hooks catch onto other algae; and the crinkly, wirelike green seaweed *Chaetomorpha linum*, which entangles firmly among the prongs of Irish moss and similar algae. Worms are commonly entwined in plant branches. Thus many species function as attached forms without having any special mode of permanent attachment.

A number of animal species conceal themselves in holes in rocks or other structures. The lobster *Homarus* and the moray eel *Muraena* are familiar examples. Some organisms digest holes in the substrate. These include the blue-green alga *Hyella*, which penetrates the shells of mollusks; the sea urchin *Echinometra*, which etches out pockets in limestone of coral reefs; and the ubiquitous marine shipworm *Toredo*, which burrows in great numbers into submerged wood. Many forms live attached to other organisms; these are referred to as **epiphytes** (attached plants) or **epizoites** (attached animals). In these cases the host organism is used only for support, and no parasitism is involved. More complex relationships are involved when one species lives within another. The polyps of corals, for example, are inhabited by photosynthetic dinoflagellates—zooxanthellae—in a relationship of **obligate mutualism**; the dinoflagellate and the polyp each use the waste product of the other as raw material, and neither can survive without the other. More will be said in a later section of these relatively complex relationships.

Most higher aquatic animals, such as fish and seals, are not attached to any substrate; this is true, too, of the free-living plankton of the open sea (Figure 12.2). Surface forms are grouped together as **neuston**. The lighter-than-water forms—the **epineuston**—stand above the surface; these include the spiderlike water strider *Gerris,* the whirligig beetle *Dineutes,* and the Portuguese man-of-war *Physalia.* Neustonic forms living at the surface film include the water flea *Daphnia,* the backswimmer *Notonecta,* and the duckweed *Lemna.* The term **hyponeuston** is used to indicate the species, such as the rhizopoid protozoan *Arcella,* which live just below the surface film. Drifting organisms of nearly neutral density are grouped together as **plankton** (although it should be noted that many planktonic species seem to exercise some control over the depth at which they drift). Finally, the term **nekton** is applied to all animal forms that have the ability to swim effectively against, or independently of, the current; the nekton includes all free-swimming pelagic forms, such as the jellyfish *Aurelia,* the arthropod forms mentioned above, and virtually all the vertebrates.

For planktonic organisms the **settling velocity** is an important environmental factor. As explained in Chapter 6, settling velocity is a function of organism density, water temperature, and viscosity (Stoke's equation), and it is significantly modified by **form resistance** (Oswald's equation); details are given on page 122. Cell density depends on cell-wall structure (bethnic diatoms, for example, have heavy, glasslike walls), on the composition of cytoplasm, and on gas content. In many planktonic species the cell form obviously retards the sinking velocity; threadlike and needlelike shapes are common, as are extended sheets of wings and elaborate spines (Figure 12.2).

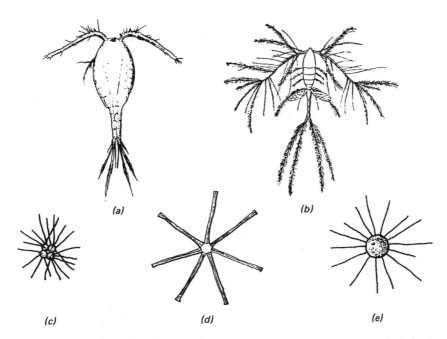

(a) *(b)*

(c) *(d)* *(e)*

Figure 12.2 Planktonic species. Species *a* is a marine copepod; *b, c,* and *d* are diatomaceous green algae. *(a) Oithona; (b) Micractinium; (c) Asterionella; (d) Golenkinia.*

NUTRITION

A second major adaptation of any species is its mode of nutrition. The manner in which an organism obtains the food necessary for its metabolism is fundamental to many aspects of ecology, since it can determine where the species fits into a food chain, and whether it will contribute to the complete recycling of some strategic material. For our purposes, all organisms may be placed in one of two groups: **autotrophic** species (those capable of independent nutrition) and **heterotrophic** species (those requiring an outside source of food).

Autotrophic Nutrition

Autotrophic organisms are able to use carbon and hydrogen to synthesize carbohydrates. The most important autotrophic species do this through **photosynthesis**, obtaining the carbon and hydrogen from CO_2 and water and releasing oxygen in the process. Other autotrophs are **chemosynthetic**. Rather than carry on food production using light energy (as do the photosynthetic species), chemosynthetic species obtain energy from oxidation-reduction reactions.

A few chemosynthetic species are of ecological significance. *Nitrosomonas* is a nitrifying bacterium which oxidizes ammonia to nitrites in soil and water; *Nitrobacter* is another, which oxidizes nitrites to nitrates (Carpenter, 1967). Sulfur bacteria (genus *Thiobacillus*) are common in acid bogs, where they oxidize the sulfur in H_2S to form sulfuric acid, and the hydrogen to synthesize carbohydrate. With few exceptions, however, the important autotrophic species utilize photosynthesis, which is a highly efficient method of capturing solar energy and holding it in the form of chemical bonds. In most ecosystems, photosynthetic plants are ultimately the only significant source of both food and free oxygen.

Heterotrophic Nutrition

Organisms that cannot synthesize their own food from inorganic sources must naturally rely on outside sources to furnish them with needed organic nutrients; they are thus **heterotrophic**. Most animals are **holozoic**—they ingest large amounts of complex organic material, usually the bodies of other plants or animals. Further distinctions are made among various classes of such heterotrophs. **Carnivores** and **herbivores** are distinguished, of course, by their feeding exclusively on animal or plant tissues, respectively. **Scavengers** are heterotrophs that feed on dead animal bodies, while **predators** generally hunt and kill their own prey; the two categories, however, are not mutually exclusive. Other distinctive modes of feeding include **grazing**, as when certain whales ingest large amounts of plankton and krill unselectively; **filter feeding**, in which water is pumped through the organism (such as certain mollusks) and the organic matter extracted; and **detritus feeding**, or the harvesting of bits of partly decomposed organic matter. We will examine some of these modes of feeding in a later section, when we study the **food webs**, or feeding relationships, that link the various members of natural communities.

A few other modes of heterotrophic nutrition are also of some interest. Most bacteria and fungi are **saprotrophic**; they digest dead plant or animal matter outside of their own bodies, and absorb the broken-down organic compounds. A number of heterotrophs are also photosynthetic or chemosynthetic, and thus combine the two major modes of nutrition. The aquatic nonsulfur purple bacteria (Athiorhodaceae) are photosynthetic, but must obtain certain vitamins from the breakdown of other living tissue. The common intestinal bacterium *Escherichia coli* and the infectious bacterium *Streptococcus faecalis* are chemosynthetic, but they require supplementary polypeptides, which they absorb from (digested) organic fecal matter in the human intestines.

METABOLISM

The sum total of chemical processes in an organism constitutes its **metabolism**. The processes and products characteristic of its metabolism have much to do with the place of each species in the natural ecosystem, since metabolic needs determine, ultimately, whether a species can compete successfully in a given environment. Similarly, the waste (and other) products of metabolism may have striking effects on the adaptive success of other species.

Enzymes

Enzymes are the organic catalysts (proteins) involved in digestion and assimilation of nutrients. Enzyme studies are essential to biochemistry, and are finding increased use in aquatic ecology. It has been suggested, for example, that it may be more useful to measure nitrogen-reductase activity in a water column (*i.e.,* enzyme activity) than to apply the usual chemical analyses for inorganic nitrates. The amount of enzyme activity indicates how rapidly nutrients are being utilized by organisms; simply measuring the concentration of a nutrient does not indicate how much is being taken up for use in metabolism.

Nearly all biological processes are enzyme-catalyzed. In general, any metabolic process involves a series of chemical changes, each catalyzed by a different enzyme. Incoming materials are passed along the system, changed somewhat at each step, and ultimately used or excreted as an end product. The compounds that are formed briefly during the intermediate steps do not usually accumulate in detectable amounts. An analogy may be drawn between such a process and an automobile assembly line: parts are fed in sequence to the workers, who add them one by one to finally produce the complete car. Unless the assembly line is disrupted, none of the separate parts accumulates in large quantities. In a factory, a worker may fail to do his job; in the organism, a catalyst may fail to function. In these cases, there is an accumulation of intermediate products. Some organisms, for example, lose the ability to synthesize some specific enzyme. They then become dependent on an outside source of the material usually produced by this enzyme.

An example is the common bacterium *Proteus vulgaris*, an organism that lacks the enzyme needed to obtain nicotinic acid (an essential metabolite) from the substrate. *Proteus* therefore can grow only if one of the intermediate metabolites is supplied by an outside source (Carpenter, 1967). If the metabolite is available from a host, the cell could survive as a parasite; if it is released during the decomposition of vegetable matter the organism could live as a saprophyte; if a potential host produces a surplus of the metabolite, *Proteus* might establish a **commensal** relationship, taking nutrients from the host but causing it no damage. Thus the distribution of important enzymes in the various organisms of a natural community could definitely affect ecological relationships.

Biological Oxidation

Foods are usually broken down, and their energy released, through **biological oxidations**. The energy obtained is retained in the cell temporarily in the form of high-energy **phosphate bonds**. These are two basic types of reaction. One involves the release of incompletely oxidized products and is called **fermentation**; the other involves (primarily) the release of gases and is called **respiration**. Fermentations usually involve the breakdown of carbohydrates and the release of ethanol or lactic acid. Respirations involve the uptake or release of gases, and are either **aerobic** or **anaerobic**. Aerobic respiration, the more common process in multicellular organisms, is usually based upon the uptake of oxygen and the release of carbon dioxide, although other gases (such as H_2S and NH_3) may be involved. Anaerobic respiration does not require free oxygen from the environment, but does release gas; the process is similar to aerobic respiration except that the oxygen used is chemically bound rather than free molecular oxygen. Anaerobic respiration is ecologically important because it is carried out by denitrifying and desulfofying bacteria, which release such gases as NH_3, CH_4, and sulfur compounds as end products. The bacterium *Desulfovibrio* releases H_2S. These reducing processes are common in **anoxic** sediments and waters (*i.e.*, those in which dissolved oxygen concentration approaches zero), and are especially noticed in marshes or in muds from considerable depths.

Nitrogen fixation Certain bacteria (such as *Azotobacter* and *Clostridium*) and blue-green algae (*e.g.*, *Nostoc, Aphanizomenon,* and the marine form *Trichodesmium*) carry on a form of respiration that results in the reduction of nitrogen to ammonia. This **nitrogen fixation** is a significant source of biologically available nitrogen in many aquatic habitats. The blue-green algae seem to be the most important agents of aquatic nitrogen fixation. It has been demonstrated that *Nostoc,* a symbiont with the water fern *Azolla,* fixes nitrogen efficiently (Brock, 1966). It has also been observed that nitrogen fixation in Sanctuary Lake, Pennsylvania, is greatest during the fall bloom of planktonic blue-green algae (Dugdale and Dugdale, 1965).

End products of metabolism A number of metabolic end products accumulate during the assimilation and dissimilation of nutrients (*i.e.,* during their intracellular syntheses or breakdown). Thus many compounds may accumulate, later being discharged as waste products or released by death and decomposition. Products of carbohydrate dissimilation include acids (formic, acetic, proprionic, and others), alcohols (such as ethyl and propyl), and such gases as hydrogen and carbon dioxide. Protein dissimilation releases products of decarboxylation (amines, many of which have putrid odors) and deamination (organic acids and ammonia). Other products include tannins, humic acids, carotenoids, amino acids, fatty acids, and uric acids. Fecal pellets contain a wide variety of substances. Some released products are of ecological importance. They may attract or repel organisms of other species, and they may be toxic to some. Some species of the marine dinoflagellates *Gymnodinium* and *Gonyaulax* release substances that act as potent nerve poisons; these species are abundant in the well-known "red tides," which may seriously harm fish and other animal life through the accumulation of toxins in the inshore food chain.

HOMEOSTASIS

The term **homeostasis** is often used to indicate a state of equilibrium among the interdependent parts of a cell, an organism, or a living community. Homeostasis is maintained by processes that are so regulated that there is a continuing balance between inflowing and outflowing materials; thus any sudden net change is avoided, because the change in one material is offset by adjustments in others.

Solute Uptake

Whether a dissolved solid will be taken up by organisms often seems to depend more on the organism and solute involved than on concentration of the substances. Sodium and chloride ions are concentrated in the ocean and are dilute or absent in fresh water, yet the concentration of these ions usually is the same in animals from either habitat. Potassium ions occur in the freshwater stonewort *Nitella* in concentrations up to 1000 times that in the surrounding water (Osterhout, 1933). Thus, certain ions are taken up preferentially by organisms, and various species accumulate and concentrate scarce ions according to their special needs.

Microorganisms with contractile vacuoles are able to excrete surplus ions along with water, and in higher animals the urinary system provides effective control of internal ion concentration. In species lacking such active mechanisms, such as blue-green and green algae, control of ion concentration results from passive cellular adjustments and from structural adaptations such as rigid cell walls and "leaky" cell membranes.

The principle that different dissolved substances are taken up at predictable rates is widely applied in aquatic ecology. Biogenic salts are usually taken up by a given species in proportions approximately equal to the proportions at which they occur in body tissues. These proportions are sometimes expressed as ratios, such as N:P and C:N, which give the proportion of nitrogen to phosphorus, and carbon to nitrogen, respectively. In marine plankton, for example, N:P is about 7.2:1 and C:N is approximately 5.8:1 (Fleming, 1940).

Many substances not utilized in normal metabolism may enter cells rapidly because of their nonpolar structure or their solubility in fats. In the case of toxic petrochemicals (gasoline and xylene, for example) uptake may be quite rapid and extremely harmful. On the other hand, important substances, such as vitamins, may enter slowly if they possess attached polar radicals. (**Polar radicals**, such as $-OH$ or $-COOH$, strongly affect the behavior of organic molecules by causing them to become polarized; **nonpolar radicals**, such as CH_2, do not.) Thus extremely large nonpolarized molecules, such as the fatty acids, will penetrate cell membranes much more readily than will much smaller polarized molecules. The principle of differential uptake of solutes finds an interesting application in retarding eutrophication in freshwater ponds. Mats of algae are introduced and allowed to take up the available nutrients. After the nitrates and phosphates have been depleted from the water, the algal mats are seined off and discarded, leaving relatively oligotrophic water in the basin.

Water Balance and Osmosis

Marine waters contain large amounts of dissolved solids in nearly constant amounts; while estuarine and fresh water is more variable in this regard, and usually contains less dissolved matter (Chapter 11). This difference has important implications for **osmosis**, the process by which water diffuses from a region of higher to a region of lower concentration (of water) through the differentially permeable membranes of the cells. A freshwater algal cell put in salt water **plasmolyzes**; the pure water in the cell vacuole diffuses out into the diluted seawater and the protoplasm shrinks away from the cell wall and collapses. A marine algal cell placed in fresh water often swells as pure water diffuses in, and the cell will usually rupture. Osmotic pressure is a colligative property of solutions (Brescia *et al.*, 1966). As such, it is related to the concentration of solute particles, and the subject is usually treated, therefore, in terms of solutes rather than of the solvent, water; but it is the movement of water that leads to ecologically important results (see page 123).

Some species do not suffer severe effects from changes in osmotic pressure. Organisms without vacuoles consist of virtually solid cytoplasm, and may not be affected at all. Animals with contractile vacuoles may pump out water rapidly enough to avoid bursting; this ability permits them to survive both inland fresh water and in dilute seawater. Plants with very strong cell walls can withstand great internal (**turgor**) pressure without bursting. In other species the cell membranes may be so permeable to both water and salts that a new equilibrium is established before damage results. The special features of each species determines its tolerance to change in solute concentration and, thus, the habitat in which it can live and reproduce.

Waste Disposal

The waste products accumulated by simple organisms usually diffuse out into the surrounding water. More highly specialized species have contractile vacuoles or even simple kidneys, which concentrate and eject metabolic wastes. In the "higher" invertebrate species, solid material is compacted and ejected as fecal material, which is commonly found in the periphyton (*aufwuchs*) community. The metabolic by-products in fluid and solid waste may be immediately useful to other organisms in the community. A wide variety of such wastes is found in any natural community; this may account in part for the complex and often obscure relationships and balances that link the species of every habitat. Within limits, even man's domestic waste can greatly increase the productivity of many aquatic habitats.

REPRODUCTION

It is essential for the ecologist to understand clearly the processes and patterns of reproduction of any species being studied. An organism lives for a limited period of time; thus its reproductive success determines how many representatives of its species will continue to inhabit a given region. (For this reason, reproductive success is also a major factor in evolution.) We might begin by considering the **life span** of a species. This is not so simple an idea as it might seem, and the relation of life span to reproduction may be rather complex.

Among microorganisms which reproduce by cellular fission, "life span" is generally taken to be the equivalent of one generation: the time between one cell division and the next. When an individual of such a species reproduces, it ceases to exist, but it does not "die" in the usual meaning of the word. In a limited sense, such organisms might be considered immortal. In this instance the concept of life span has a special, limited application.

Higher plant and animal species have specialized gamete-producing organs and can thus produce offspring without themselves ceasing to exist; many are able to produce several generations of young in a single lifetime. Species that produce offspring but live for only one season are called **annuals**; those that live for several or many seasons are called **perennials**. Most higher species are perennial. Many complex variations on this basic theme can be found. Amphibians, for example, go through the process of metamorphosis; during the course of a single lifetime an individual may be at one time exclusively herbivorous and at another exclusively carnivorous; it may be alternately aquatic and terrestrial. It will thus fill very different roles in the local ecology at different times, and it might be valid to consider the larval and the adult forms as having quite different life spans. Again, the salmon lives for many years, returning from marine waters to spawn in the stream of its birth; unlike most vertebrates, however, the salmon dies after spawning a single time. The major ecologically important aspects of a species' reproductive pattern are its **life history**—the stages of its life cycle—and its **dispersal of progeny**. The two are of course closely interrelated.

As suggested above, life histories vary greatly in complexity. They may involve nothing more complicated than simple binary fission, or they may involve several larval stages, spores of different types, relationships with a number of different hosts, and periods of "rest" (estivation or hibernation). The protozoan *Paramecium* or the blue-green algae seem to represent the simplest pattern. On the other hand the pelagic crustaceans, the aquatic stages of many insects, or such plants as the brown seaweed *Laminaria* have more complicated life histories. In *Laminaria* the large blade may produce filamentous **thalli** of several types, vegetative or sexual, diploid or haploid (Dawson, 1966); the complexity of the life history gives the species adaptive flexibility, permitting it to survive a wide range of adverse marine conditions. Some life histories, on the contrary, are highly restrictive, involving a series of stages, each of which in turn requires a particular habitat. Many marine animal species must spend the early part of their lives in an estuarine environment. For such species, the fitness of the environment is especially critical. Equally demanding are those life histories that require a several-years' migration between fresh and marine waters, which are characteristic of the eels (*Anguilla*) and the salmon (*Salmo*). In a surprising number of species the details of the life history are keyed to the particular requirements of the environment with remarkable precision.

The means whereby the progeny of different species are dispersed are extremely varied. In primitive species the cells, produced by fission, may sink toward the bottom or be carried along by currents. Species with gametes and spores are dispersed by the **motility** of these reproductive structures. Some heavier eggs or spores, such as those of the green alga *Oedogonium*, have **delayed germination**; spores remain dormant until a storm, or spring flood waters, disperse the progeny far downstream. The special feature involved here is a particularly long period of encystment. Many species have specialized flanges, spines, and the like, that aid in dispersal by adding to buoyance of water resistance. In the diatom *Fragilaria*, long filaments break apart into small fragments which disperse easily; such **fragmentation**, as it is called, is common among plants. Higher aquatic plants may have specialized terminal buds

—**turions**—which survive through the winter and germinate in spring. The seeds and spores of many aquatic plants may be eaten by birds, fish, or other animals and be deposited, still fertile, at a considerable distance with the animal's fecal matter. This often happens with some freshwater desmids (Proctor, 1962, 1966). Such a dispersal mechanism permits dispersal not only downstream, but also upstream, over barrier hills and mountains, and—when the animals involved are migratory birds—along the major flyways.

The few examples given here suggest the **adaptive** significance of modes of reproduction and dispersal in fitting a species to the conditions of its habitat. In later chapters we will be examining some of these in greater detail. Our point here is to stress the adaptive significance of the various modifications of structure and behavior that we will encounter in our later discussion. Some of these adaptations seem, at first, impossibly complex or even bizarre. Nevertheless, each such modification had its origin in the fundamental processes of mutation and natural selection, and each plays a part in the survival of the species.

RESPONSES TO ENVIRONMENTAL FACTORS

All aquatic organisms, from algae to the most elaborate plants or animals, respond to changes in the environment; such a response is part of the definition of living as opposed to nonliving matter. Two major types of responses may be distinguished: responses or reactions to particular, nonrecurring stimuli; and rhythmic (daily, seasonal, etc.) changes, which are usually related to recurring factors (such as tide or day length) in the environment.

Nonrhythmic Responses

Any of the processes described in the foregoing pages is affected by the individual's physicochemical and biotic environment. Certain factors have especially pronounced effects: light, temperature, dissolved oxygen content of the water, hydrogen ion concentration, nutrient availability, and solute concentration. If we were to measure the rate of each process (metabolism, nutrient uptake, etc.), as affected by each of these factors, we could obtain a set of curves to indicate the apparent effect of each factor in the environment upon each biotic process for any given species. One such curve is shown in Figure 12.3, indicating the effect of temperature upon metabolism in Irish moss (*Chondrus*). If the curve were extended the figure would show three **cardinal points**: a **minimum** value below which metabolism cannot be detected; an **optimum** range in which metabolism proceeds at its fastest rate, and a **maximum** value above which metabolic activity stops. A similar curve could be obtained for the apparent effects of most environmental factors. Great care must be taken, however, in interpreting any such data and curves, for in many cases the effect of one factor on a process changes when other factors are varied simultaneously, a condition known as **interactions**. For example, an increase in dissolved CO_2 will augment the effect of light on photosynthesis in algae.

It is important in ecological studies to determine which factors are most critical for any given species, and then to determine the cardinal values for each significant factor. The mutual interactions between different factors may make it

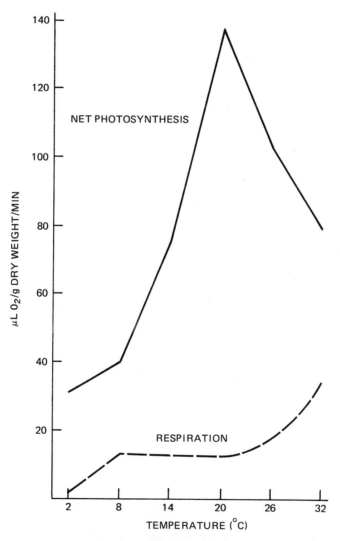

Figure 12.3 Effect of temperature on metabolism in Irish moss (*Chondrus crispus* Stackhouse). Note that photosynthesis rises higher, and peaks at a lower temperature, than respiration. At a slightly higher temperature, respiration would overtake photosynthesis. (After Mathieson and Burns, 1971.)

extremely difficult to interpret data collected by observation or experiment, and the difficulty is compounded because the organism's response to one factor, such as temperature, may alter its response to another, such as the presence of a solute in the water. It is often impossible to predict the results of a particular set of changes in the environment, even when using the most sophisticated mathematical techniques. In most cases the only way to determine how an organism will respond to various combinations of environmental changes is to conduct experimental tests involving all factors simultaneously.

Tolerance to features of the environment varies widely among aquatic organisms. In describing the ranges of tolerance of an organism, the prefix **eury-**, meaning wide, or **steno-**, meaning close or narrow, is added to a term for the particular feature. Commonly used terms include **euryhaline** (wide salt tolerance) and **stenohaline** (narrow salt tolerance); these conditions are shown in Figure 12.4. To these, we might add **eurythermal** and **stenothermal**, pertaining to temperature. Let us now consider some examples of tolerance relationships, for these are of great importance in regulating the abundance and distribution of organisms.

Within our provinces of fresh and marine waters, salinity constitutes a striking example of an environmental factor which limits organisms, primarily by its effect on physiological processes other than the uptake and utilization of nutrients. Although we have defined salinity in terms of total dissolved solids, some of which are plant nutrients, our reference here is directed toward chloride concentration in relation to density of the environment. It is this factor, and the various biological adaptations to it, that essentially distinguish marine and freshwater organisms.

The extent to which inhabitants of fresh or marine waters can invade the contrasting environments, and, indeed, the ability of certain euryhaline popula-

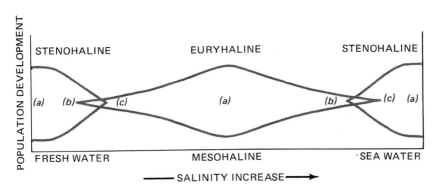

Figure 12.4 Schematic comparison of population development in stenohaline and euryhaline species. Tolerance to a salinity gradient, as indicated by growth, reproduction, and migration patterns, is indicated by the width of the figures. Optimum conditions are shown at *(a)*, minimum limits at *(b)*, and maximum limits at *(c)*. Stenohaline organisms may be tolerant to either a high or a low salinity, and population development is often affected by slight changes in salinity.

tions to occupy estuaries permanently, is related to the limit of tolerance of each organism to variation in salinity. Among animals generally, tolerance to salinity changes varies greatly, as do the mechanisms associated with adjustment. The salt concentration in the body fluids of most marine invertebrates is nearly the same (**isotonic**) as that of the environment. These forms are essentially stenohaline and are restricted to regions of relatively stable, near-seawater salinities. Freshwater organisms, on the other hand, inhabit a medium in which the concentration of salts is usually less (**hypotonic**) than that of the body juices; this is also true of regular inhabitants of brackish waters. Because of osmosis these animals continually take up water across exposed, water-permeable membranes. Thus, freshwater animals must possess some structure or mechanism for maintaining proper internal balance of salts and water as well as for ridding the tissues of excess water.

Adaptation to estuarine existence demands various modifications of these osmoregulatory organs, as a few examples will indicate. Contractile vacuoles are characteristic of freshwater protozoans, but not of their marine relatives. In certain flatworms (*Gyratrix*, for example) the flame-cell mechanism serving in osmoregulation and excretion is more complexly developed in freshwater forms than in estuarine ones; in marine flatworms, major parts of the system are absent. Freshwater crayfish possess long, well developed nephridial canals; in the estuarine and marine lobsters these structures are reduced and poorly developed.

Rhythmic Responses

The activities of many species follow definite cycles, which are apparently linked to the daily, tidal, monthly, or seasonal cycles of the earth. Frequently

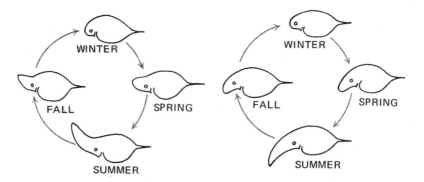

Figure 12.5   Rather extreme and conventionalized cyclomorphosis exhibited by two races of *Daphnia* from large lakes. The winter generations of the two races are nearly identical; individuals of summer generations were thought, at one time, to be of different genera than the winter forms. The internal mechanism governing cyclomorphosis is thought to be associated with environmental conditions, possibly available nutrients, temperature, or seasonal water currents. (After Coker, 1954.)

these cycles continue with little or no change even when the species is isolated from any direct influence of the natural environment. These internal rhythms, or "biological clocks", affect many of the complicated activities of higher aquatic animals, and are thought to be inherited.

It is well known that many planktonic species, plant and animal (such as dinoflagellates and copepods), migrate upward to the surface of the water and down again in a **circadian** (approximately one-day) cycle. The water flea *Daphnia* undergoes cyclomorphosis, changing structurally from one season to the next in a **circannual** (approximately one-year) cycle, developing an elongated "helmet" in the summer (Figure 12.5). In both of these cases the rhythmic change persists in the absence of any outside stimulus; the plankton species will continue their migrations in total darkness, and cyclomorphosis has been shown to occur under similarly controlled conditions. Many well-known examples of rhythmic behavior, especially migratory behavior, are related to reproductive activity; very little is yet known, however, as to the mechanism involved, and the search continues for factors that might trigger cyclic responses.

ECOLOGY OF THE COMMUNITY—SYNECOLOGY

Groups of like organisms constitute **species populations**; these populations form **communities**, the distinctive and connected aggregations, which are the subject of **synecology**. At this level we are concerned with how different species populations, each with its peculiar biological needs and attributes, share a habitat. The basic questions we ask here are: What species are present? What patterns of interaction does each species develop with the others and with the abiotic environment? How can we assess the major patterns—food chains, abiotic factors, territoriality, and distribution? Synecology is an attempt to find answers to these and related questions.

POPULATIONS AND COMMUNITIES

Community ecology begins with the reproductive processes in single species; reproduction increases the number of individuals in each species population. This leads in turn to new interaction phenomena among the species in the community. The most important of these are population growth, relationships and interactions, and ecological succession.

Natural Growth of Species Populations

Natural environments are populated with a number of different species, each with its own features and peculiarities. But, to understand the whole complex, we need to have a working knowledge of each one. Here, we are concerned with the growth and development of a species population. A population can be defined as a group of organisms occupying a given space at a particular moment in time. It may consist of a single species, and thus constitute a species population, or it may be composed of a number of species exhibiting such similar ecological traits as food requirements and breeding sites; it is then termed a **mixed population**.

Ecologists use a number of abstract concepts to explore the phenomenon of population growth. The central concepts are increase in number of individuals, in volume, in dry weight, in amino acids or protein content, and in genetic material. Numbers alone are often a misleading measure. It is of little value, for example, to compare the numbers of protozoans and the number of fish in a cubic meter of lake water. Even within a single species, numbers may be misleading when both young and old individuals are involved. **Volume** and **dry weight** are the most commonly used measures of species population growth, but these too must be used cautiously, because of possible error due to water or mineral deposits. The most reliable (but more difficult) measures will probably be those of proteins or DNA, which measure only living protoplasm.

In using numbers alone to assess growth, ecologists ordinarily use various mathematical expressions, in which the basic terms are N (number of individuals), N_n (number of new individuals), t (time), and r (growth rate). Using these terms, **natality**—or **birth rate**—is expressed as N_n/t. The derivative of this equation (d) is used to express predicted long-term growth rates: dN/dt. In general, population growth is expressed as $dN/dt = rN$. The **mortality,** or net death rate, would be expressed as a negative value for rN.) Another useful form of this equation is $dN/Ndt = r$. This form of the equation provides a measure of natility per unit of population (r). Another form commonly uses the **exponential** form of this equation based on the natural log e:

$$N_t = N_0 e^{rt}$$

The shapes of growth curves are often good indicators of population dynamics. If growth were unrestricted—that is, if there were no limiting factors and environmental resources were infinite—the offspring of each generation would accumulate at an awesome rate; many bacteria reproduce by fission every twenty minutes, doubling their numbers three times per hour. If such **logarithmic** (or **exponential**) growth continued unchecked, it would conform to the expression $dN/dt = rN$; when the data from such growth are plotted the result is the exponential **J-curve** (Figure 12.6).

No species, however, actually exhibits such a pattern of growth indefinitely. To use the bacteria as an example, this exponential growth would, in a few weeks, cause the entire ecosystem to be filled with bacteria! Clearly, some factor or factors intervene and limit growth very early in the process. Ordinarily, nutrient supplies become depleted or exhausted, waste products accumulate until they become toxic, or the population is limited by predation. Ultimately the increase stops and the population levels off at a more or less constant figure. The population is then said to have reached a **plateau**, which can be characterized by a formula that includes the value K, the maximum number of organisms of a particular species that can occur in the habitat:

$$\frac{dN}{dt} = rN \frac{K - N}{K}$$

Plotting this equation, we get a curve variously known as the **logistic growth curve**, the **sigmoid growth curve**, or simply the **S-curve** (Figure 12.7). The curve levels off at the value K, which is called the **asymptote**, and indicates the **carrying capacity** of the environment. If the rate of growth is plotted, rather than the numbers involved, a bell-shaped curve—the normal distribution

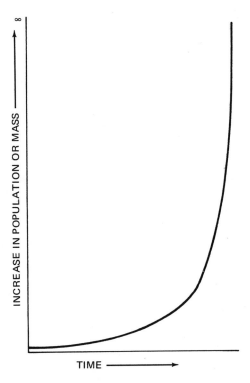

Figure 12.6 Unrestricted growth. Where conditions are favorable and nothing limits population growth except biotic potential, organisms accumulate in prodigious numbers in a short time, as shown in this typical *J-curve* or *logarithmic curve*. Populations of protists often show this acceleration phase for short periods.

curve—is produced (Figure 12.8). A species population may overshoot the asymptote for a while, then react to the limiting factor and drop below it; ordinarily, after several **oscillations**, the population will level off at the asymptote. The reduction of these oscillations is called **damping**. The mathematical interpretation and manipulation of growth data are treated in more detail later in this chapter. For the interested student, however, a number of works are available (Cutbill, 1971; Milsum, 1966; Patten, 1971, 1972; Walters, 1971) that offer information on mathematical simulation and modeling in general. These are listed in the bibliography at the end of the book.

Data on **age structure** can also tell us a great deal about the growth state of a population. If individuals of all ages are present in roughly equal proportion, then the birth, growth, and death rates are probably in balance—a healthy, stable condition. A high proportion of young or old individuals may suggest some abnormality in the reproductive cycle or it may reflect the ordinary survivorship pattern characteristic of the species being studied (Figure 12.9).

The ecologist is also interested in the **distribution pattern** of a species population. Knowing whether distribution in space is uniform or spotty is important in assessing field data. A basic contrast is drawn here between **random** dispersion (individuals evenly spaced) and **aggregation** (individuals occurring in clusters); sampling must be carried out with great care when an aggregated population is the subject of a field study. The distribution pattern of a species will be affected by whether a species is territorial or tends to occupy a restricted home range; aggregations will result from the clustering of offspring close to reproductively active individuals; or adverse environmental conditions may alter the way organisms are spaced out in the habitat. The statistical concept **degree of randomness** is of great use to the ecologist studying species populations. In most species dispersion shows some degree of regularity as a response

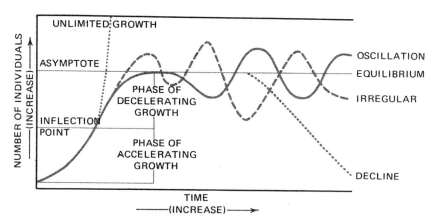

Figure 12.7   Growth curves involving limiting factors. The original logarithmic J-curve levels off toward an asymptote, but overshoots and subsequently oscillates around the asymptote. (After Allee *et al.*, 1949.)

to environmental or biological conditions; aggregation itself may be a limiting factor, affecting the distribution pattern.

The growth factors we have been discussing are summed up in the twin concepts of **biotic potential** and **environmental resistance**. Biotic potential is a species population's innate capacity to increase at a maximun rate—a rate which varies greatly from one species to another. Full biotic potential is almost never realized because environmental resistance intervenes and checks population growth, via such diverse factors as temperature, salinity, available nutrients,

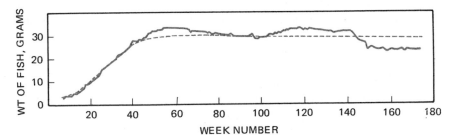

Figure 12.8   Growth in guppies (*Lebistes reticulatus*) in a laboratory tank reached its asymptote in about eight weeks, oscillating in roughly a 60-day cycle that approximates the logistic curve. (After Silliman and Gutsell, 1958.)

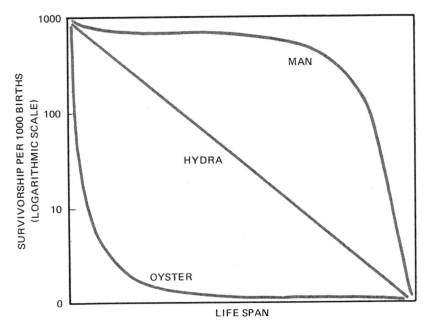

Figure 12.9   Mortality (survivorship) curves for three species. The curves are based on number of survivors per thousand births (log scales) and age in relative units of mean life span. (After Deevey, 1950.)

disease, and predation. Thus **potential natality**—the maximum birth rate theoretically possible for a species—is usually greater than **realized natality**. We can also distinguish similarly between **physiological longevity**—the theoretical maximum age—and the observed **realized longevity.** These ideas lead us back to the concept of life span (see page 271 and Figure 12.9).

Biological Communities

A natural community includes not one but a number of different species populations—the total of all species occupying a given habitat. The term **community** is applied broadly to groups of any size or complexity. The kinds of organisms that are present in a habitat depends on the history of the region (which species were available); the ability of various species to colonize the area; their tolerance for limiting factors; and the adequacy and moderation of the physical, chemical, and biological parameters. The species composition is the starting point in any study of community structure.

A hypothetical small pond will serve to illustrate the complexity of species composition in even the simplest habitat. The biota might include the green alga *Chlorella* and the diatoms *Suriella* and *Fragilaria;* the bacterium *Nitrosomonas;* the protozoan *Amoeba;* the quillwort *Isoetes;* the sponge *Spongilla;* and the midge *Chironomus.* Several of these species might occupy different regions of the pond simultaneously. The attached benthic forms would include *Isoetes* (rooted in the bottom sediment), bottom-dwelling *Chironomus* larvae and *Amoeba,* and *Spongilla,* which adheres to the bottom or to projecting debris by a specialized basal cell; the other four species would occur as unattached benthic (**epibenthic**) forms. Planktonic forms are absent, but during and after windy periods some of the microscopic epibenthic organisms would likely be swept up into the water column and functionally become plankton. Most of these species (all but *Isoetes* and *Chironomus*) could also be found living as epiphytes. Thus we see that in a simple community of only eight species, complex relationships already begin to arise among the component species. Synecology is concerned with identifying and assessing these relationships.

RELATIONSHIPS AND INTERACTIONS

When particular species typically occur in recognizable assemblages, the group is referred to as an **association**. The association may be purely fortuitous—the species may occur together because they have similar biotic processes or needs. More commonly, however, at least some of the member species are linked by some form of biological interdependence. Some species may be present because of the presence of another, which serves the first as food or as a host. In Chapter 6, for example, we referred to the abundance of birds and mammals in a region of the eastern Pacific Ocean that is particularly rich in upwelling nutrients and consequently supports a large population of crustaceans. In an inland pond a similar association may often be observed in the neuston. Floating duckweed often supports an association of smaller species of algae, diatoms, rotifers, protozoans, and so forth, all of them linked to one another and to their common host. In lakes and ponds in particular, biotic associations are often characterized by the depth at which they occur; it is usual to distinguish at least five major zones: the marsh, the emergent zone, a transitional floating-attached zone, the floating zone, and the submerged zone.

Symbiosis

Wherever two or more species share space intimately in nature, the opportunity may exist for interaction. Wherever interaction does take place, to the benefit of any of the species involved, we term the situation **symbiosis**. (Note that some writers restrict the use of the term to only those cases in which the benefit is mutual; in this text the term **mutualism**, as explained below, is used to indicate such a relationship.)

Symbiosis may take a number of different forms. Two species may be in **competition** with one another for resources; the result is ordinarily the partial restriction of population growth for both species. **Amensalism** refers to a relationship that affects one species adversely without benefiting the other; **commensalism** benefits one species without affecting the other adversely. **Parasitism** refers to any relationship that benefits one species at the expense of the other. **Mutualism** refers to those associations beneficial to both species involved. Mutualism is surprisingly common in the natural world. Lichens, for example, are organisms formed of two different species, a fungus and an alga; many species of freshwater *Hydra* contain photosynthetic algae, as do most reef-building corals. We shall have occasion in later chapters to refer to this principle. If two species share the same habitat but no interaction takes place between them, their relationship is referred to as **neutralism.** Symbiotic relationships are of particular importance to ecology, since they may sharply affect the analysis of food webs, pathways of energy flow, and the overall pattern of production and consumption.

The Niche

The ecological **niche**, or **econiche**, of a species is its role in the natural economy of the habitat. The niche thus includes all the factors of the environment that affect the species, including both spatial and trophic involvements. It is an established idea in ecology that only one species occupies a given niche at a time (''one niche–one species''), and this seems to be true in most cases; two similar organisms in virtually the same niche may compete for the same factors, and almost invariably one becomes dominant and eliminates the other. This tendency for competition to bring about separation of similar organisms by eliminating all but one is an aspect of the **competitive exclusion principle** (Hardin, 1960), or **Gause's principle** (Odum, 1971). Connell's study of intertidal barnacles along the Scottish coast (1961) is often cited in this regard. Connell noted that adults of two genera, *Chthalamus* and *Balanus,* occupied the high tide zone on the rocky coast. When both are present the *Balanus* species occupy a zone extending up to the mean high neap tide level and *Chthalamus* species are restricted to a zone slightly higher; when Connell experimentally removed one species or another, he found that in the absence of competition *Chthalamus* larvae would successfully colonize the upper portion of the *Balanus* region. By competing more successfully for resources, *Balanus* would normally exclude *Chthalamus* from the region above mean high water. The same sort of phenomenon has been observed in cultures of the water flea *Daphnia,* with *D. pulicaria* displacing *D. magna* when the two are in competition for resources.

A second point of importance is that an organism with a complex life cycle may occupy a different niche at different stages of the life history. The larva of the dobson- or caddisfly clings to rocks in oxygenated waters of turbulent streams, whereas the adult is a flying insect restricted to the air. Thus a species may be rather narrowly and firmly established in a given niche, or, with different adaptations, it may function in a number of niches during a life span.

Feeding Relationships

What an organism does determines its niche and the place of the niche in the environment. Feeding habits are a major aspect of the niche, determining how available food will be used and the patterns by which food reaches the members of the community. Nutrient materials are usually moved through the species of a community in a predictable sequence of **trophic** (eating) relationships. These are often described in terms of a **food chain**, in which the first link consists of autotrophic plants and the other links are heterotrophic plant and animal species, including decomposers. The sequence of species in a food chain usually is based on body size, but it may vary, depending on mode of obtaining and metabolizing food. In a typical pond food chain, bacteria and unicellular algae are grazed upon primarily by microscopic protozoans and crustaceans; these herbivores are eaten by larger invertebrates, usually scavengers; the invertebrates are eaten by small fish; and the small fish are eaten by larger fish. Such a stepwise movement of nutrients is the basis of support for virtually all animal life. It should be noted that these "chains" involve a funneling effect, by which great masses of algae and bacteria ultimately support a much smaller number of larger organisms; at each stage (or link in the chain), the number of organisms becomes smaller.

The movement of pollutants through food chains has become a matter of intense concern. The finding of extremely high concentrations of wastes such as DDT, lead, and radioisotopes in organisms high in food chains has led to the recognition of **biological magnification** as a serious threat to living things, including man. The small amounts of such pollutants picked up by producers (algae and aquatic plants) is not dangerous; but as vast numbers of producers are fed upon by herbivores, the pollutants are concentrated in their tissues. This leads to successively higher concentrations of these non-biodegradable pollutants at each successive trophic level. For predators at the higher levels, such as gulls, fish, crabs, and man, the concentrations may be dangerous. Many freshwater habitats that once served as valuable food sources (Lake Erie has been a notorious example) were later condemned and became valueless. Eniwetok Atoll in the Marshall Islands (western Pacific Ocean) was evacuated and used for nuclear bomb testing from 1948 to 1956. Over the past 20 years the atoll has become largely free of radioactive pollution, and some of the former Marshallese residents are returning. Nuclear pollution persists in the adjacent food chains, however; for example, the coconut crabs—once a local delicacy—are still highly radioactive.

Most heterotrophic organisms feed on more than one species, although their feeding behavior will show distinct preferences for various foods. Thus, describing a food chain may involve recognizing a large number of interconnections and crossconnections, to take into account the alternative foods used by the different species. The ecologist often finds the concept of a **food web** to be more useful than that of a simpler "chain." A complete diagram for a community, indicating which animals eat which plants and other animals, would in fact resemble a complicated web. An example of a simplified food web is given in Figure 12.10.

Competition

Feeding relationships are one important element of interspecific **competition**. Where two (or more) species depend upon the same food source, species populations may increase until there is not enough nutrient material to support both. The same is true in the case of any environmental factor—space, light, etc.—for which competition is possible. If the species are evenly matched in their ability to utilize the environmental factor, both would be affected by the shortage and their numbers would be reduced equally. The scarcity of vegetation and plankton in oligotrophic waters may reflect such a relationship. Frequently, however, one species has a competitive advantage over others, perhaps due to some biotic factor such as growth rate, nutrient-uptake rate, or photosynthetic efficiency. In such a case, the better-adapted species becomes dominant, and reduces or even eliminates the others. Competition between predator species can be an active struggle; competition between plants or sessile animals is a more passive process, involving the progressive crowding of the less efficient species out of the habitat. In either case, the result is often competitive exclusion, as discussed earlier.

Competition and dominance   In any community, one or more species may have features which determine or control environmental conditions for the others. Such a species is referred to as a **dominant** or, in the case of more than one, a **co-dominant**. Many ecologists distinguish **dominance** from **predominance**, a term which implies larger numbers or size but no special control. In plankton associations in lakes, the diatoms *Asterionella* and *Fragilaria* are commonly predominant, but there seems to be no evidence that they dominate in the sense of controlling the community. By contrast, the inland benthic *Myriophyllum* usually replaces most other waterweeds, and is thus properly referred to as a dominant.

Allelopathy   In many species, competition involves the secretion or excretion of minute amounts of chemical materials to which other species may react

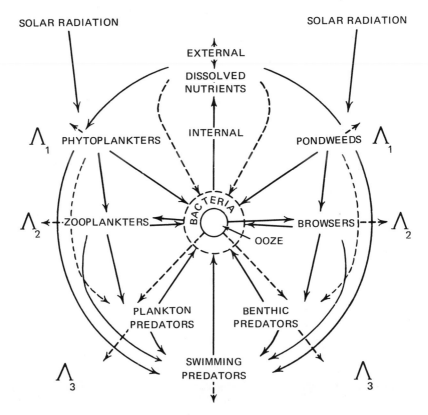

Figure 12.10   Generalized scheme of food relationships in a lake. Efficiency in utilization of food increases in the higher consumer levels. (After Lindeman, 1942.)

violently. Some investigators see in this form of competition a sort of natural "chemical warfare." In general, we speak of **allelopathy** when a small amount of a material released by one organism affects another adversely. **Ectocrines** are special substances, probably including a number of vitamins (B_{12}, biotin, uracil), that have a stimulatory or regulatory effect on other organisms. There is evidence, obtained in bioassay tests using the freshwater alga *Selenastrum,* that filtered water from Worden Pond, Rhode Island, had a growth-stimulating effect if it was collected from the pond after May, or during the summer (Ventura, 1973). It is presumed that a plant exudate or ectocrine is involved in stimulating the spring growth of plankton algae and benthic plants.

Material exuded by one species or organism often inhibits growth or life processes in another, a phenomenon generally called **antibiosis**. The displacement of other water plants such as *Potamogeton* by the dominant *Myriophyllum* seems to involve antibiosis. Considerable interest has developed in testing paired algae species, one of which ordinarily displaces the other in nature. It has been suggested (Pratt, 1966) that the flagellate *Olisthodiscus* dominates the diatom *Skeletonema* in Narragansett Bay by antibiosis. This phenomenon is by no means limited to plants; the skin secretions of the pickerel frog *Rana palustris* are lethal to other species of frogs, and thus serve as an extremely effective mechanism for securing feeding and breeding territory.

Competition and species diversity Competition is an important factor in determing the **diversity of species** in a given habitat. It has been suggested that the more intense the competition for resources, the greater the species diversity, and vice versa. High diversity is found in eutrophic ponds and bays, for example, where a great many species utilize the available food intensely; diversity is low in oligotrophic systems. An explanation may be found in the checks and balances maintained within the community: species potentially capable of competing very successfully (*e.g.,* by rapid reproduction) are held in check by symbionts, especially predators. Thus more species get a chance to survive. A recent film by the oceanographer Jacques Yves Cousteau showed how heavy harvesting of seals and whales near Tierra del Fuego has resulted in a tremendous increase in the numbers of krill, the small oceanic crustaceans that are the basic food of these species. The proliferation of krill sharply reduced the kinds and numbers of other plankton crustaceans. Removing the mammals, and thereby ending predation of the krill, reduced competition among plankters, which in turn led to low species diversity. Species diversity is a matter of concern because low diversity is often linked to some problem of environmental pollution; untreated municipal and industrial waste, when dumped into a stream, can sharply reduce the diversity of the benthic biota for hundreds of kilometers downstream (Wilhm, 1967).

ECOLOGICAL SUCCESSION

Communities of plant and animal species undergo changes of various sorts. In aquatic environments, as we have seen, the species composition of a habitat may change in a periodic fashion that is related to the changing seasons; such changes are recognized as a normal **progression** of species. Other changes are not periodic, and involve the displacement of one species assemblage by another over a longer period of time; the term **succession** is applied to a sequence of such relatively permanent changes.

Seres and Seral Stages

In many well-studied cases succession is an orderly process. Assemblages of living things succeed, or displace, each other until a stable community (the **climax**) is established; at that point succession stops. The pattern of succession is often fairly constant in any given region, and the sequence of communities is called a **sere**; each component community, that is, each step in the sequence, is called a **seral stage**.

Seres may be observed on a large or a small scale. An example of a microscopic sere is shown in Figure 12.11; here the sequence of alga groups inhabiting a leaf in spring is considered to be a sere leading toward a summer climax. Another well-known succession is the classic "bog hydrosere" shown in Figure 12.12; this sere is usually initiated in an oligotrophic lake or pond. A

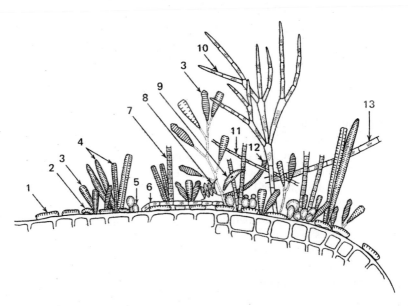

Figure 12.11 Succession in aufwuchs community: the *Cocconeis-stigeoclonium* association epiphytic on a leaf of *Sagittaria* in a hard, freshwater spring. Succession is from left to right: (1) *Cocconeis*, (2) *Achnanthes*, (3) *Gomphonema*, (4) *Synedra*, (5) *Xenococcus*, (6) *Pseudoulvella*, (7) *Lyngbya*, (8) *Scenedesmus*, (9) *Cymbella*, (10) *Stigeoclonium*, (11) *Melosira*, (12) *Fragilaria*, and (13) *Microspora*. (After Whitford, 1956.)

floating sedge mat may begin covering the margin of the pond, accompanied by a drop in pH. The hydrosere then continues as the mat covers the pond surface and sphagnum moss and such ericaceous shrubs as leatherleaf (*Chamaedaphne*) and Labrador tea (*Ledum*) become established. Ultimately a sequence of tree species, in this case tamarack (*Larix*) followed by a complete cover of *Picea* spruce forest, leads to the climax (Gates, 1926).

Succession implies the replacement of one association of species by another, and the conditions which initiate or accompany such replacements involve both physicochemical and biological factors. A physicochemical factor might be a change in light intensity or quality, affecting the distribution of species. The development of bacterial slime on submerged rocks, which appears essential for the growth of attached benthic microflora, is largely a biological factor (Hedgpeth, 1957). Succession involves the greater success of organisms which are increasingly tolerant of changed conditions; these remain after the more intolerant ones—which initiated the changes—have departed. When no further species are able to invade the habitat, the community becomes stable at climax.

What constitutes a climax in an aquatic community is a problem which is by no means resolved. In the past, little attention was given to aquatic climaxes because many are only temporary in character (Dice, 1952). Open-water habitats

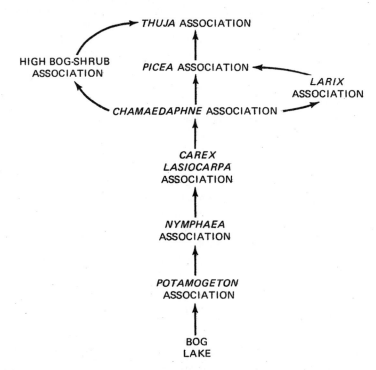

Figure 12.12   Hydrosere in bog lakes, Michigan. The sedge *Carex lasiocarpa* initiates the mat, which expands from the shoreline toward the center of the lake. (After Gates, 1926.)

are particularly unstable due to constant exposure to winds and waves, and shores of many bodies of water are subject to erosion as well as deposition of material. Further, according to Clements and Shelford (1939), only a few fresh-water communities are able to control the environment sufficiently to reach a climax stage, although some relatively stable aquatic climaxes can be recognized along rocky marine shores. In recent years, however, various workers have observed climax-like conditions developing in cultures and among microscopic organisms in the periphyton and epiphyton (see, *e.g.,* Margalef, 1963). It may be that aquatic synecologists have sought patterns that would be directly comparable to succession in the terrestrial environment, and have missed the possibility that sequence of climaxes may occur more rapidly in aquatic habitats.

Climax Communities

In theory, succession continues until conditions become stable, and thereaf-ter the self-regenerating climax community persists without major change. This concept is well established as a part of terrestrial ecology, from which it is taken. The concept of an aquatic climax is useful, but the idea of a climax community has never been applied with equal success to aquatic ecosystems, largely because equilibrium conditions are unlikely to persist. As earlier chapters make clear, change itself is an important parameter of most aquatic environments.

Climax stability   A climax is the stable community that becomes established at the end of a sere. In microscopic communities such as the epiphytic algal community described above (Figure 12.11), climax may be reached rapidly and then rapidly eliminated by other successions. In macroscopic communities in-volving perennial species, however, climax may be reached only after many years and may persist for long periods. An important feature of climax com-munities is the manner in which severe damage to the biota initiates a sequence of cause and effect that reestablishes the climax biota at a faster-than-normal rate. For instance, a swath cut through a mass of benthic plants will originally disrupt the community. At the same time, however, the change increases avail-able light and thus increases the rate of germination and photosynthesis, which in turn augments the growth rate; in a relatively short time the climax vegetation and fauna are replaced, without repeating the earlier stages. For this reason climaxes are said to be **self-repairing** systems, with a high degree of homeo-stasis; the homeostatic feedback mechanisms might be said to operate in a manner comparable to those found in an individual organism. More will be said of this in Chapter 13.

Historical determinants   The history of the organism in a habitat— when and how each species entered the region—is of great and obvious impor-tance in determining the subsequent distribution of species populations. When we are concentrating on more abstract ecological problems, it is easy to overlook this obvious factor. Thus we find that certain species, such as the cattail *Typha* and frogs of the genus *Rana,* are ubiquitous in shallow-water marshes along North American rivers. When we look at the rest of the biota, however, we soon notice that a particular species of snail or amphibian might be abundant in one watershed and entirely absent from the streams of an adjacent watershed (Storer, 1951). This may be true although all physical and chemical factors (and even most biotic factors) are similar. Ultimately, the presence or absence of a particular species in a habitat is a result of this **historical factor:** by biological accident, one particular species of salamander or snail simply did not invade or colonize an availible habitat. In any community, such historical factors must be considered in assessing the species composition.

Ecological equivalents In different regions, different and often unrelated species will often occupy similar ecological niches. The two species may be quite dissimilar genetically and physically, but if they perform the same function in the two ecosystems they are referred to as **ecological equivalents**. The salt-marsh grasses *Spartina townsendii* and *S. alterniflora* are often cited as examples of equivalence, the former occurring widely in Europe, and the latter in North America. Similarly, the European frog *Rana temporaria* and the North American wood frog *R. sylvatica* occupy virtually identical niches on the two continents, and even resemble one another rather closely. Ecological equivalence is a common phenomenon. When widely separated species assemblages, often consisting of quite different species, function ecologically in nearly identical ways, they are referred to as **physiognomically** *(i.e.,* structurally) similar associations.

Ecotones

The region between two distinct climax communities, where some members of both communities occur, is referred to as an **ecotone**. Often the number of individuals in an ecotone, and occasionally the number of species, is greater than in either of the climaxes. This phenomenon is called the **edge effect**, and is easily seen at the marshy edge of a small pond. The edge effect does not occur universally, however; the sandy beach might be considered an ecotone between the terrestrial and the marine environments, yet its biota is rather sparse. Similarly, an estuary might be regarded as an ecotone, but there too species tend to be few in number. In view of such exceptions, we might conclude that the value of the ecotone concept is greatest when it is applied to regions between somewhat more similar habitats, *e.g.,* two freshwater or two marine environments. Thus we might regard as ecotones the regions between each of the horizontal communities of littoral grasses and marine algae as we move through an estuary from land to deep water. In these more narrowly defined ecotones the edge effect might be more easily observed.

Succession in Applied Ecology

The principles of succession are of great importance in forming a policy for managing aquatic resources. When we recognize that the existing community in a pond, stream, or estuary represents a seral stage, we realize that ongoing physical, chemical, and biotic processes must be considered in evaluating the productivity and possible uses of aquatic resources. No practical course of action can be selected until, as policy makers, we understand the dynamics of aquatic ecosystems. On a large scale, succession thus becomes a very practical idea, an essential biological concept that must be considered in planning such projects as weed control in ponds, biological control of pest species, importation of parasites, dredging of harbors, or opening of breachways for coastal ponds to the sea.

Another important point follows inevitably: the successional patterns can be upset by any drastic change. For example, a landslide that forms a lake or cuts off a stream channel may change all the seres in the affected basins. The shifting of sand in a stream may alter the watercourse and disrupt the benthic seres completely. Catastrophic changes such as drought, flood, or avalanche will have obvious effects; not so obvious, but often of equal importance, are such changes as disease-linked mortality or unusually heavy predation on a key species. At the present time most alteration of natural seres can be traced to human activities, such as damming or channelizing waterways, harvesting game or food species, or, often, polluting or depleting natural waters.

AQUATIC CLIMAXES

From the point of view of the ecologist (particularly one concerned with applied ecology), it may not be essential to know the exact nature of the climax community. More important is a knowledge of the pattern of succession involved. For this reason the ecologist concentrates, in general, on the community that currently occupies a particular habitat, and on the ways in which the species associations are likely to change.

Aquatic climaxes are distinctive in that they are so often controlled directly by outside (*e.g.,* geological) factors rather than by internal biotic processes. Aside from such long-stable climax communities as the algal turf of windward reefs of atolls and certain of the algal belts in marine intertidal zones, most aquatic communities, especially freshwater communities, appear to be in a transitory state. Deeper ponds are silted in and their communities change continually, until the shallow pond in turn gives way to a marshy–terrestrial succession ending in a climax forest. In rivers, too, the riffle and torrential communities persist, with seasonal progressions and fluctuations, only until the river's continuous cutting and sedimentation eliminate the habitats or displaces the communities.

Eutrophication

As we saw in earlier chapters, an important distinction is made between **oligotrophic** and **eutrophic** (''nutrient-poor'' and ''nutrient-rich'') lakes. Eutrophic lakes are characterized by an abundance of available biogenic nutrients, a scarcity of hypolimnetic oxygen (page 199), heavy growth of planktonic algae, especially the ''nuisance'' blue-greens, and an acceleration of the normal rate of ecological succession. In recent years, **eutrophication** has been used as a negative term, to describe a situation where a natural body of water becomes clogged with algal scum as a result of organic pollution. In fact, the term is more properly applied to any change in an aquatic habitat that is correlated with an increase in available nutrients. Eutrophication is closely associated with both the biological and geological elements of ecological succession, and it often occurs without human intervention.

Natural eutrophication   Some degree of eutrophication normally begins as soon as a natural basin is formed. This **natural eutrophication** can be observed as soon as the erosion of rock and soil releases minerals into standing water and the water becomes capable of supporting life. Continued erosion brings in silt and nutrients, gradually or more rapidly filling up the deeper regions of the basin and supporting aquatic organisms which, as they die, further contribute to the bottom sediments. An interesting relationship between succession and eutrophication in a lake is shown in Figure 12.13, which reveals evidence that succession is not necessarily steady and continuous. Rather, eutrophication is but one phase in the process of lake or pond succession, which ultimately leads to

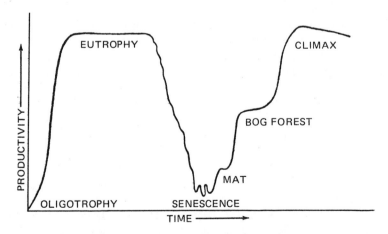

Figure 12.13   A concept of succession in a deep lake. The hydrosere develops from an oligotrophic state, through a eutrophic bog stage, to a climax forest. Note changes and stepwise levels of productivity. Change in productivity is irregular rather than steady, indicating variations in eutrophy of the lake. (After Lindeman, 1942.)

senescence and finally to replacement of the aquatic habitat by terrestrial vegetation. So many lakes have developed eutrophication problems in recent years that the cause is sought in man's activities. Human activity does often affect the process. But comparable trends are found in isolated lakes in pristine areas. Thus it also seems probable that many lakes formed after the retreat of the Pleistocene glaciers have been following this trend at somewhat comparable rates, with eutrophication becoming apparent only recently. This provides an interesting problem for investigation, but in any case it is apparent that, once filling in begins and biological systems get a start, the trend toward senescence and disappearance of the body of water is almost inevitable.

Cultural eutrophication Eutrophication of lakes is a natural phenomenon locked together with succession; but the process can be greatly accelerated by man. However, it is uncertain whether the biotic changes accompanying accelerated, or **cultural**, eutrophication of a lake parallel those of the natural process. An interesting point is whether eutrophication can be reversed. This idea is not consistent with the concepts of ecological succession discussed earlier. In some lakes, however, the effects of eutrophication have been significantly reversed after the introduction of such corrective measures as sewage diversion.

Human sewage and industrial wastes are significant sources of the nutrients that contribute to lake eutrophication. Other sources include runoff from farmland (much of which has leached through manure spread on frozen ground) and drainage from urban areas (which is rich in phosphate and nitrate). Significant amounts of nitrates are released in the smoke from fossil fuels and washed down in rains. Unexpected factors in cultural eutrophication are a number of vitamins, growth substances, amino acids, and trace elements which are synthesized during the biological treatment of sewage.

While excessively rapid eutrophication is a cause for concern, we should not lose sight of the possibility that chemical enrichment can be desirable. As mentioned in Chapter 11, nutrient enrichment has proven to be of value in increasing the production of fish and in increasing the productivity of valuable aquatic plant life.

Vulnerability of Aquatic Systems to Pollution

Aquatic environments are, in a number of ways, more vulnerable to change than are terrestrial ones. Likens and Bormann, summarizing current research findings, have pointed out how pollutants in the biosphere move downhill from terrestrial into aquatic ecosystems (Likens and Bormann, 1974). (Human settlements have always made use of this fact by letting their wastes wash downhill into streams, ponds, lakes and oceans.) Thus the wastes produced by cultural activities anywhere in a watershed move into the water system of that watershed, often with detrimental effects. This is true not only of industrial wastes, but also of any changes (*e.g.*, forest cutting, clearing roadways, clearing new land for settlement) that tend to cause increased erosion and consequent runoff of soil and detritus into streams and lakes. Substances used in agriculture (such as pesticides and fertilizers) also find their way into inland water systems, either directly as runoff or indirectly by being carried some distance in the atmosphere and later precipitated out with rain. Thus we see that aquatic environments are especially vulnerable. They receive and concentrate virtually every natural and cultural contaminant.

It has been noted that eutrophication may proceed at variable rates rather than at a single constant rate; periods of recovery are occasionally noted. But natural or cultural events may place too great a burden on the homeostatic capabilities of any aquatic system, leading to drastic changes in succession and accelerated eutrophication. A well-established example is the history of Lago di Monterosi, a lake on the Via Cassia, north of Rome. The lake was formed about 26,000 years ago, and was apparently oligotrophic for some 20,000 years. Later, when the Via Cassia was built in about 171 B.C. the lake changed drastically, becoming strongly eutrophic as a result of increased runoff from cleared land. Today the lake remains somewhat eutrophic, although the artificial eutrophication process had been partially reversed after about 800 years (Hutchinson *et al.*, 1970).

Until recently ecologists have assumed that diversity in an ecosystem provided some guarantee of ecosystem stability. Recently, however, a number of researchers have suggested that complex systems of high diversity may in fact be at least as fragile and susceptible to disruption as simpler ones (Likens and Bormann, 1974). According to this controversial view, rivers and estuaries may be better able to recover from environmental insult than lakes and coastal marine communities.

Other factors that may disrupt aquatic ecosystems are sulfur and nitrogen oxides (released during the combustion of fossil fuels), which accumulate in watersheds and finally in standing waters in such forms as sulfuric acid. In soft, poorly buffered waters, this can have the effect of abruptly lowering the pH and damaging acid-sensitive species, with widespread damage to the system. In addition, metals such as zinc, copper, and nickel are also released in many industrial processes and precipitated from the air. Mine tailings (waste) can be an especially critical problem in many areas otherwise largely free of environmental contamination.

ECOLOGY AND WATER RESOURCE MANAGEMENT

As we have seen, aquatic communities usually consist of a large number of species populations, each with its specific physicochemical and biotic needs. Each species is linked more or less directly to others in the community, and the community as a whole changes constantly through the slow processes of ecological succession. The susceptibility of aquatic ecosystems to disruption suggests some of the difficulties faced by planners in trying to make optimum use of natural aquatic resources, or in trying to ameliorate damage already caused by natural or cultural processes. Environmental problems are currently being attacked by using such advanced techniques as ecosystem analysis (discussed in the following chapter), biome studies, and mathematical (computer) modeling. Despite this impressive array of tools, progress is often slow, because we still have little knowledge of the interactions among many significant variables. There are innumerable links between a local ecosystem and a whole range of external vectors, some of them geological (landforms, available water, mineral substrate), some meteorological (precipitation, suspended aerosols, available light), and others biological (animal migration, biological concentration of contaminants, extinction of other species).

Intact terrestrial ecosystems develop "tight" biogeochemical and nutrient cycles, and lose only small amounts of nutrient to the aquatic ecosystems. But when there is major disturbance, the output of these materials is often grossly accelerated. There is thus a crucial balance between the output of terrestrial systems and the input into aquatic systems. To this should be added the observation that aquatic habitats tend to adjust slowly, with long-term lag, to many changes, including corrective measures. Thus corrective measures that seem, in the short run, to be quite effective might often have long-term results that are dangerously harmful. It is largely in order to better predict the outcome of future environmental changes that ecologists have developed the ecosystem modeling techniques that are discussed in Chapter 13.

SUGGESTED FOR FURTHER READING

Hynes, H. B. N. 1970. *The ecology of running waters*. University of Toronto Press, Toronto. The most recent and authoritative textbook on stream ecology.

Likens, G. E., ed. 1972. *Nutrients and eutrophication: the limiting-nutrient controversy*. Allen Press, Lawrence, Kansas. Papers presented at a symposium in 1971 sponsored by the American Society of Limnology and Oceanography.

Margalef, R. 1968. *Perspectives in ecological theory*. University of Chicago Press, Chicago. Brief lectures on advanced topics in ecology; useful primarily for graduate and advanced undergraduate students.

National Academy of Sciences. 1969. *Eutrophication: causes, consequences, correctives*. National Academy of Sciences, Washington, D.C. A collection of papers presented at a symposium on eutrophication, held in 1967.

Odum, E. P. 1971. *Fundamentals of ecology*, third edition. W.B. Saunders Company, Philadelphia. A comprehensive, updated presentation of the basic principles of ecological science. Clearly and concisely written, with emphasis on analysis; very strong in quantitative models.

Russell-Hunter, W. D. 1970. *Aquatic productivity: an introduction to some basic aspects of biological oceanography and limnology.* The Macmillan Company, New York. A timely and readable book, with an unusual breadth of coverage.

Whittaker, R. H. 1970. *Communities and ecosystems.* The Macmillan Company, London. An excellent coverage of ecological concepts from the viewpoint of a community ecologist; suitable for undergraduate reading.

13 | Ecosystem Ecology

The problems raised at the end of the previous chapter point to the need to develop methods for quantitative analysis of ecological **systems.** Great steps have been made in this effort, many of them using techniques borrowed from mathematics and systems analysis. In this chapter we turn from the concerns and terminology of the descriptive ecologist to the more abstract concerns and terminology of the functional ecologist. Our aim here will be to describe ecology in terms of **integrative concepts.**

THE APPROACH TO ECOSYSTEM ECOLOGY

The approach used in ecosystem ecology introduces a new way of analyzing biological communities as systems. Communities are not described by lists of associated species, but instead are analyzed in terms of functions, processes, compartments—values which can be measured **quantitatively.** Using such measurements, mathematicians have developed equations for analyzing communities, determining rates of change, and assessing the effects of a change in one factor on various other factors or processes. The methods of computer simulation and systems analysis have been adopted to provide further tools for these analyses. Thus, the substantial importance of the methodology of ecosystems analysis stems largely from its success in providing a way to gather and use hard quantitative data, often to very practical ends.

THE ECOSYSTEM

An ecosystem can be described as the entire complex of interacting physicochemical and biological activities operating in a fairly self-supporting community. Few ecosystems are truly independent in nature, and the size of individual ecosystems can vary greatly. An entire watershed, a pond in a tributary, or a mixed culture in a laboratory flask may all be examples of more or less distinct ecosystems.

The operation of an ecosystem can be viewed as components or, in ecosystems terminology, **compartments.** The functional biological compartments of an ecosystem can be grouped into three basic categories: **producers, consumers,** and **decomposers.** In addition to these biotic components, the system includes **abiotic** substances, and perhaps the entire array of physicochemical factors. In addition, we can think of energy as a distinct, fifth, category.

PRODUCERS

In ecosystems ecology, the **producers** compartment refers to the total of all organisms involved in synthesizing food from inorganic substances. The list is long and includes autotrophic bacteria (both photosynthetic and chemosynthetic) and endophytic and epiphytic photosynthetic organisms as well as phytoplankton and macroscopic seaweeds and plants. It is not considered necessary to know all the organisms in a community to understand the system; instead, integrative ecologists use such values as amount of producers or the results of their activities. The amount of producers is measured by totaling biomass or caloric content of autotrophs (Table 13.1) or by measuring the concentration of photosynthetic pigments (usually chlorophyll a); production is measured by determining the rate of synthetic processes.

CONSUMERS

The **consumers** compartment includes all organisms which actively feed on other organisms, and includes phagotrophic pigmented microorganisms such as

TABLE 13.1 Caloric equivalents of selected biologic categories[a]

	CALORIC CONTENT (GCAL/G)	
Biologic Category	dry weight	ash-free weight
Algae	4900	5100
Invertebrates	3000	5500
Insects	5400	5700
Vertebrates	5600	6300

[a] Data from Golley, 1961; E. P. Odum et al., 1965; and Cummings, 1967.

Euglena and *Cryptomonas* as well as unpigmented plants and animals that ingest food. In order to isolate consumers of particular interest, they are grouped by trophic levels into herbivores, primary carnivores, secondary carnivores, and so on, until the series ends with the **terminal** carnivores. The consumers are measured by totaling biomass or caloric content of all constituents, or indirectly from respiration measurements.

DECOMPOSERS

The **decomposers** compartment comprises the total of all organisms that release enzymes which break down dead organisms. Examples of decomposers are the bacteria and fungi which grow on organic matter in sediment and detritus. Total decomposition can be measured in various ways, such as by total respiration of ecosystem minus that of the producers. More frequently, decomposers are grouped with consumers as heterotrophs, and can be measured as total ecosystem biomass minus that of autotrophs.

OTHER COMPONENTS

Functionally, the living system can be represented by the three categories above; but it is sometimes useful also to distinguish the living and nonliving, or the **biotic** and **abiotic,** components of the system.

The biotic component includes the entire living community—producers, consumers, and decomposers. It is expressed as **standing crop,** or **biomass.** Standing crop is measured in number, volume, or weight. Biomass is usually measured in dry weight per unit area or volume. Both values can be expressed in caloric content, measured by calorimeter or calculated from values of caloric content per gram for the particular type of organism.

The abiotic component includes nonliving materials available in water or from adjacent structures (sediment, interstitial surfaces, air-water interface, inflow). Of special interest are the biogenic elements and ions such as Fe, NO_3, PO_4, and S. The abiotic component is measured as total solids, or total filterable solutes, or by measuring each in mg/liter or other suitable units.

ENERGY FLOW AND COMMUNITY METABOLISM

Energy flow is an integrative concept encompassing the entire productive and metabolic complex in a community. Food might be the important central theme in an ecosystem, but energy is even more general and less difficult to measure. In actuality, the two are nearly synonymous. Energy flow is the term designating the total energy budget of the system; it encompasses how energy enters the system, what it does, and how it leaves. The original trapping and conversion of light to chemical energy is included as **assimilation.** The food chains and food webs are included as **pathways** of energy flow. The loss of energy by each level of users, as in respiration, is considered an **energy sink.** The accumulation of energy in large organisms is **energy storage** (*i.e.,* accumulation or concentration). Thus, the movement through the pathways of photosynthesis, respiration, metabolism, storage, and energy release as heat, motion, etc., is included in this one integrative concept, energy flow (Figure 13.1). The sum total of the processes of living things is the **environmental or community metabolism.**

Since food is the carrier of potential energy for living things, energy encompasses the details of the food chains and webs. On the other hand, the kinetic energy—light, heat, electrical, and mechanical energy—is not considered, unless it is involved in a **work gate** (see description below) in the ecosystem energy flow; only those kinetic energies which have been converted to potential energy in certain organic compounds (foods) are available. Energy flow involves producers, consumers, and decomposers, listed above as the major ecosystems components. In addition, energy flow allows for the loss of kinetic energy not retained in the biotic component.

PRODUCTIVITY

Productivity is the sum total of energy-trapping processes, comprising all the photosynthetic and chemosynthetic processes in the system. Production is usually measured by the amount of dissolved oxygen released or carbon dioxide taken up, by chlorophyll, or by increase in dry weight. However, all the organisms in the system undergo respiration, and therefore what is measured is actually the **net production.** Respiration is measured in the dark; the respiration, added to the net or apparent photosynthesis, gives the **gross production.** It is usually expressed in grams of dry weight or calories produced per m² of earth (or bottom) surface over a measured period, usually a year. This figure is designated **gross primary production;** that value, less the amount used in respiration and other metabolic processes, is **net primary production.** This net production represents the amount of energy remaining for other purposes.

Representative gross primary production values for aquatic habitats have been given by Ryther (1969). Expressed in kilocalories (1 kcal = 1000 cal) per square meter per year, these values are on the order of 20,000 kcal/m²/yr for estuaries and coral reefs, 6000 for regions of upwelling nutrients (as in the eastern Pacific), 2000 for coastal zones, and 1000 for open oceans. Since production is usually expressed in grams, it would be helpful to convert these to a more universal form. Using an average value (which has been determined from many samples) of about 4.7 kcal per gram of dry weight of aquatic organisms, and adjusting these for variations during the season, the production for a summer

day might be about 24 g/m²/day in estuaries, 7.2 for upwelling areas, 2.4 for coastal zones, and 1.2 for open oceans (see Odum, 1971). When compared with the **maximum probable production** for intensely cultivated agriculture of up to 50,000 kcal/m²/yr or 60g/m²/day, the maximum values approached in nature are

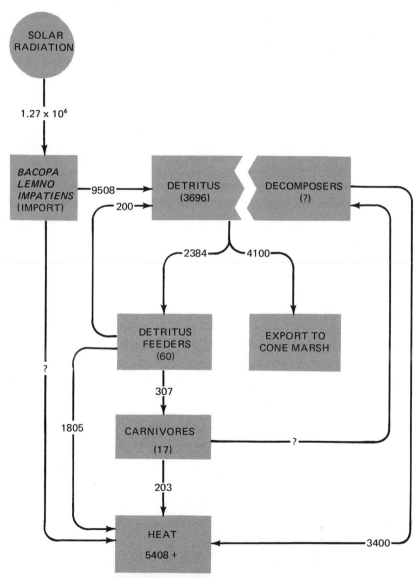

Figure 13.1 Energy flow diagram for Cone Spring, Iowa. Values are monthly averages of standing crop expressed in kcal/m². Energy flow values per year are indicated on the arrows. (After Tilly, 1968.)

TABLE 13.2 Productivity values for Long Island Sound[a]

Measure	kcal/m²/year
Gross primary production	5700
Net primary production	2500
Net community production	approaches zero

[a] From Riley, 1956.

rather high in estuaries and coral reefs, while the values for oceans is very low by contrast. Productivity values for Long Island Sound are given in Table 13.2.

TROPHIC STRUCTURE

Trophic structure is a concept that deals with food chains and webs without becoming burdened with details. Instead of considering which species feed on which others, we focus on the functional compartments described earlier. Linking these compartments are the pathways showing how energy (food) is passed from one level to the next. Major pathways are those of the herbivores, detritus feeders, aggregate feeders, and carnivores (Figure 13.2). These categories involve animals that eat plants, those that live on decaying

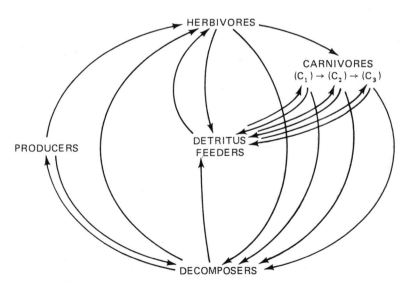

Figure 13.2   Pathways among the producers and consumers of an ecosystem.

organic matter (detritus), those that eat other living animals, and, finally, a category only recently recognized, those that obtain nourishment from inorganic particles (especially calcium carbonate) or bubbles which have adsorbed dissolved organic matter to their walls. Aggregates and bubbles may adsorb fatty acids, proteins, or carotenoids; certain fish, especially those of coral habitats and possibly in deeper oceans, appear to thrive on such "bubble detritus".

The energy flow in each pathway is determined by measuring the caloric content of each compartment, then calculating the difference to determine the energy change. It can also be done by measuring the rate of the process. The energy **pathway** can be established by labeling nutrients with radioisotopes and, after a measured period of time, testing the consumers to determine which have taken up the labeled food.

Another way of picturing trophic structures is by the use of **trophic pyramids.** Here, the same components are used, but are arranged in layers (levels) following the sequence of energy flow. Thus, producers are on the bottom, supporting the entire pyramid. The advantage of this system is that an idea of the biomass and energy involved, as well as that lost from one level to the next, is vividly clear (Figure 13.3).

Two aspects of the significance of trophic structure are illustrated by the concepts of ecological efficiency and abbreviated food chains. **Ecological effi-**

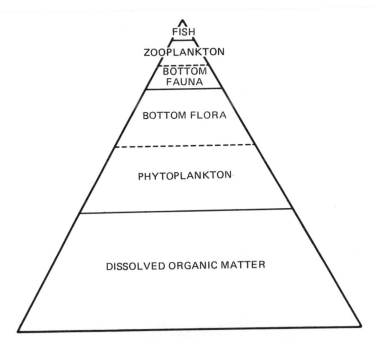

Figure 13.3   Pyramid of biomass and dissolved organic matter in Weber Lake, Wisconsin. Proportional areas of the triangle are based on the weight relationships among the various components. (After Juday, 1942.)

ciency is a measure of the proportion of energy lost at a step in a trophic pathway. For gross photosynthesis, this is the proportion of light received relative to that energy actually fixed in photosynthesis. At each subsequent level, ecological efficiency measures the percent of energy transferred to the next stage. In aquatic habitats, the phytoplankton level is commonly about ten times as great as the herbivore level; thus, the ecological efficiency at this step would be level 2 divided by level 1 times 100 $L_2/L_1 \times 100$ or 10 percent. Notice in Figure 13.3 how efficiency increases at the higher trophic level, rising to as much as 25 percent in some carnivores.

Efficiency is also involved in the concept of **abbreviated food chains.** Since energy in a food chain is lost at each trophic level, less energy would be lost by reducing the number of levels. A spectacular example is found in the grazing by baleen whales, the secondary carnivore level, on krill (euphausiid planktonic crustaceans, which in turn live on phytoplankton). In contrast to most higher-level carnivores, which function at the end of a rather long food chain, the plankton grazer gets more of the original assimilated energy.

An **ecological pyramid** is a diagrammatic representation (Figure 13.4) of the relations between different trophic levels. It indicates the biomass, the energy content, or the number of individuals needed at each trophic level to support the population at the next higher level. Since food is funneled into the system, the

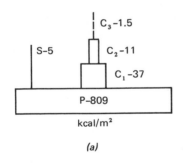

(a)

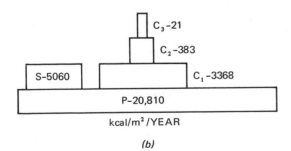

(b)

Figure 13.4 Ecological pyramids for Silver Springs, Florida, indicating standing crop (a) and biomass (b). P = producers; C = consumers (primary, secondary, tertiary, and terminal); S = saprophytes or heterotrophs. (After H. T. Odum, 1957.)

number of individuals usually decreases at each level, causing the diagram to resemble a pyramid. The terminal consumer (often man) is at the peak of the trophic pyramid. **Inverted pyramids** occur in oligotrophic waters when a large organism dies and scavengers (and to some extent the decomposers) enter the food chain and support a large number of carnivores on the single carcass.

BIOGEOCHEMICAL CYCLES

Biogeochemical cycles involve short- or long-term use and regeneration of biogenic solutes. On an integrative level of thinking, these involve compartments, or pools, which exchange solutes at measurable rates. A pool exists in solution in the water column, initial uptake is by autotrophs, and the autotrophs transfer the food to heterotrophs which release it directly or indirectly back into the water column. Depletion of solute pools is an important factor, for at various points in a cycle some of the material may become trapped in a nonsoluble form, lost in deep water, or swept away in soluble form. Depletion is common in actively draining ponds and lakes by materials being washed away through the outlet. The loss of certain nutrients in this way may be the causative factor in cases of natural recovery from eutrophication or pollution. The pathways can be traced by radioactive isotopes, and turnover rates can be determined by measuring the amounts of the solute in the various compartments.

A detailed knowledge of such cycles is an important consideration in ecological relationships and ecosystem ecology, since it gives insight into what can be expected to happen in the future as well as what might happen if certain changes occurred or were introduced. In aquatic environments iron, for example, occurs alternately in soluble and insoluble form, depending on dissolved oxygen concentration and redox potential (page 241). Thus the usual seasonal cycle of soluble iron can be depended upon only so long as the cycles of these other parameters continue. Ecologists and systems analysts must take into account the possibility that the cycles of important biogenic nutrients may be disrupted, especially since changes may occur quite suddenly. The cycles of most of the major biogenic solutes were considered in Chapter 11; here we will be concerned with the ways these cycles are measured and used in ecosystem ecology.

BIOGEOCHEMICAL CYCLE DYNAMICS

Five major concepts are of special value in considering biogeochemical cycles from the ecosystems point of view: turnover rate, turnover time, residence time, standing state, and fallout. **Turnover rate** is the fraction of a nutrient pool released per unit of time. If, for example, 100 mg/liter of a solute were present, and 4 mg/liter were used and replaced by biotic activity every hour, the turnover rate would be 4/100, or 4 percent, per hour. **Turnover time** is the reciprocal of the turnover rate, and is equal to the time required to replace the total amount in the system. In the example given below, turnover time is 1/0.04, or 25 hours.

Residence time is the time a material remains in a given compartment of the system, whereas turnover time refers to the whole system. **Standing state** refers to the amount of a solute available in the system. Note that rapid turnover rate by small organisms keeps the nutrient active; accumulation in large, long-lived organisms ties it up. **Fallout** from the atmosphere brings nitrogen (fixed in atmosphere), sulphur (from volcanic dust), and silicon (from dust), as well as carbon dioxide. In addition, fertilizers, pesticide sprays and dusts, radioactive materials from bomb testing, spores, bacteria, and other propagules may accumulate. Few biogeochemical cycles show short-term steady states, but, over entire seasons or periods of a few years, an average balance can be detected.

AMOUNTS AND RATES IN BIOGEOCHEMICAL CYCLES

One of the important tasks of this decade has been the evaluation of biogeochemical cycles. How fast are the materials released, and how long does it take them to be regenerated? These values are sought in terms of amounts actually present at one time. Some idea of these values can be found in the case of radiophosphorous (Hutchinson, 1957) which moved into water with a turnover time of 5.4 to 7.6 days, and moved into the sediment in 37 to 39 days. These figures can be recalculated to give turnover rates of about 0.185 to 0.13 per day. Fallout is generally proportional to rainfall and natural NO_3 from atmospheric oxidation and man-made sources, so representative rates are highly variable. However, for ^{90}Sr from megaton atomic blasts, fallout in millicuries is about 200 $mCi/mi^2/day$, which, when expressed on a more comparable basis, amounts to $6.66 \times 10^{-4} \mu Ci/m^2/day$.

SYSTEMS ECOLOGY

Systems ecology involves using techniques borrowed from the burgeoning field known as **systems analysis,** and applying them to ecological problems. Systems analysis arose primarily from the approach taken by engineers to solve practical problems. The basic practice was to reduce intricate operations to a form in which only key factors needed to be studied. Then only a relatively few measurements need be taken, and from that point data could be handled by equations. This type of approach to complex problems has found applications in a wide variety of modern-day areas such as the operation of hospitals, businesses, and subway systems, economics, sociology, and, of course, ecology.

PURPOSES OF SYSTEMS ECOLOGY

Systems analysis may be applied to ecology to help understand how a community or climax population functions, to help detect the more important parameters, or to eliminate the ones which are of little functional significance. Fundamentally, systems ecology permits the reduction of descriptive concepts to quantitative values that are suitable for mathematical analysis. In addition, it offers a way of predicting effects of change in a way more reproducible than the opinions of experts (though this is very important) and in cases where the situation is too complex to permit a reasonable opinion.

Quantification

As is usual in the evolution of different sciences, ecology began as a purely descriptive activity and later progressed to the stage of **quantification.** At the beginning, ecology sought such things as what kinds of organisms occurred in a habitat, what biotic and other factors were involved in the system, and how these were distributed and related. As data accumulated and understanding deepened, it became possible to synthesize a wide variety of data and to develop a body of ecological theory. In recent years quantitative study of natural systems has grown substantially, until now there seems to be sufficient background to permit the definition of ecological principles in mathematical terms, and the use of data from field and laboratory investigations. The development of this last phase has been greatly accelerated by the recent availability of practical computers. A great deal of quantitative data are already available in articles and books, although much of the material presented so far is conflicting or inadequate. At this point much work is still needed in merely measuring various aspects of communities and environments. Other workers have moved on ahead and examined ecological principles and relationships in terms of formulas and equations, putting their ideas in mathematical form. Thus, the quantification of the data and the concepts of aquatic ecology are changing the discipline, making it more precise and giving it a higher degree of predictive power.

Predictive Value

The predictive value of systems ecology is becoming more widely established. With quantitative data, mathematical equations, and computer processing techniques, it is possible to "try out" various combinations of control methods before actually applying the conclusions in nature or on the job. Such simple models as a quantitatively complete food web permit prediction of the effect of removing one food pathway or source. More involved mathematical models have also proved applicable. Riley (1953) had considerable success in predicting seasonal variation in phytoplankton populations for several regions by taking into account just six variables, five of them physiochemical and one biological (Figure 13.5). Slobodkin (1962) has expanded on this, applying mathematical models to the study of growth and regulation in animal populations. Such predictive theory is at the heart of much current higher level environmental decision-making today; often such an ecosystem analysis forms the most impressive, and the most easily accepted, portion of environmental impact statements.

BASIC TOOLS OF SYSTEMS ECOLOGY

The basic tools of systems ecology are few—a knowledge of a natural system, an understanding of its functional aspects, data on amounts of each important parameter and rate of change, and an awareness of the techniques available for constructing problem-solving models. This approach to ecology may move the investigator from the field into the world of computers and desk work. However, to obtain data, workers must often get back into the field. Often a field team investigating the habitat and its problems can solicit the aid of a systems analyst. Thus there is constant exchange between fieldwork and model-building. The concepts are rather specialized, so we will first examine mathematical models, then consider how they can be built, and finally observe how engineering language is used in studies of energetics (the study of energy involvements in a system) and flux (nutrient flow).

Mathematical Models

Models can be of such various types as graphs, diagrams, or verbal descriptions. Recently, the type of model which has excited the greatest interest in aquatic ecology is the **mathematical model.** A very simple mathematical model, discussed in Chapter 12, is the equation for population growth:

$$\frac{dN}{dt} = rN \qquad \text{or} \qquad N_t = N_0 e^{rt}$$

As we have seen, this equation gives the rate of growth, r, and the number of individuals, N. A second common model is growth under some limiting condition

$$\frac{dN}{dt} = rN\left(\frac{K-N}{K}\right)$$

in which K, the asymptote, indicates the carrying capacity, and $(K - N/K)$ causes the exponential curve to become a sigmoid curve (see pages 278, 279). As we have seen, both of these models are useful in describing growth rates and patterns, and would be useful in predicting growth of many aquatic organisms so long as the rate, r, the asymptotic value, K, and the original number, N, are known.

Wright (1965) studied the population of several species of *Daphnia* in relation to chlorophyll concentration of the samples (a measure of phytoplankton), phosphate concentration, zooplankton standing crop, and the predator *Leptodora kindtii*. From his data on these values, as well as number of eggs, duration, birth rates, and mortality, he was able to deduce an equation

$$N_t = N_0e(0.025 \text{ chl} - 0.088 - 0.0095L)t$$

where N_t = number of *Daphnia* at a selected time, N_0 = number at start, chl = μg chl/liter, e = base of natural logarithms ($= 2.718$), L = number of *Leptodora* per liter, and t = elapsed time. This model produced a calculated curve that matched the observed curves fairly well, and moreover enabled the investigator to point out areas and causes of discrepancy. The lack of agreement between predicted and observed results often provides clues to unexpected, but significant, parameters.

A more complex model is that developed by Riley (1963) for phytoplankton dynamics, which was given in the following form:

$$\frac{dP}{dt} = P \left[\frac{pI_0}{kz}(1 - e^{-kz})(1 - N)\frac{z_1}{z_2} - R_0e^{rt} - gZ \right]$$

where P = phytoplankton population

t = time

p = empirical constant relating photosynthesis and light

I_0 = intensity of light at depth 0

k = empirical exponential constant (extinction coefficient)

z_1 = depth at first reading

z_2 = depth at second reading, at water mixing

N = depletion degree of phosphorus

R = respiration; R_0 = respiration at 0°C, R_T = respiration at selected temperature

T = temperature

r = exponential constant for effect of T on respiration

g = linear constant for grazing rate

Z = density of zooplankton

Using this model, Riley was able to introduce values for the various constants, and predict plankton populations for such environments as Georges Bank, the New England coastal waters, and Uson Harbor in Korea with almost phenomenal accuracy (Figure 13.5). The derivation of this equation involves an interesting number of steps, and would be well worth looking up by students interested in mathematical applications to ecology.

There are several approaches to model building, and three of these can be distinguished as (1) the compartmental system approach, (2) the electronic analog approach, and (3) the experimental components approach. The first seems best for descriptive models which do not have to be especially realistic, while the last is best for precise, realistic models dealing with interactions and making predictions. The second depends more on electronic equipment arrangement, and seems less suited to ecologists, who generally lack the special engineering training needed.

The first step in developing a model is to define the components of the system. The second is to determine the **state conditions** of the various components (*i.e.*, measured values at a given time). Standing crop of producers or herbivores would be examples of this. The third step is to measure the rates of various inputs and outputs across the outer boundary of the habitat. This could mean input of light energy, and output of reflected light, heat, sinking, and export. The fourth step is to measure the rates of transfer among the components; the fifth is to define the parameters and, as far as possible, to define them in terms of one another in order to reduce the number of variables to a minimum. Sixth, the mathematician can then locate or develop a differential equation which will express each of the components in terms of the transfer values. Finally, the seventh step is to change a factor such as input experimentally, getting data on all variables under three values of input. These steps are illustrated by Collier *et al.* (1973) for a generalized model. In this model, only three compartments were considered: environment *(E)*, producers *(P)*, and herbivores *(H)*. The boundary

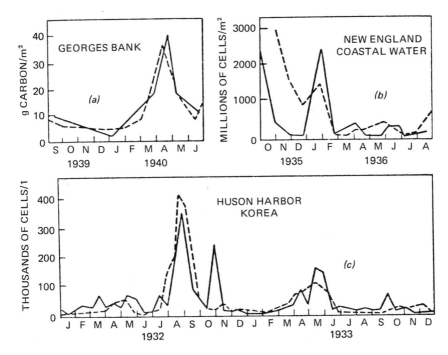

Figure 13.5   Applicability of a mathematical model in predicting natural populations of marine plankton. Dotted lines are predictions based on Riley's equations (see text); solid lines are the observed seasonal plankton. (After Riley, 1965.)

transfers were inputs *(a)* and outputs *(z)*; transfers between compartments were treated as processes. The energy transfers were coded according to the compartments involved. Thus, a_E was the energy input into the environment, while *EP* and *PH* were transfer functions (environment to producers, and producers to herbivores, respectively). Three simultaneous differential equations were generated:

$$\frac{dE}{dt} = a_E + 5P + 5H - 10EP - 2E = 0$$

$$\frac{dP}{dt} = 10EP - 5P - 10PH = 0$$

$$\frac{dH}{dt} = 10PH - 5H - H^2 = 0$$

The three-equation model attempts to explain change in each compartment, over measured periods of time, in terms of changes in the other compartments. The values of the different factors would of course be determined by actual data obtained in the field. We should emphasize that at this time, because of the tremendous complexity of natural ecosystems, mathematical methods are not yet available to handle full-scale models. For that reason, most researchers in ecosystem analysis concentrate on one segment or portion of an ecosystem, and design the model to answer only the exact question that is being asked. This would usually involve the environmental components approach, and only those variables would be considered that are immediately involved in the function.

Energy Flow Diagrams

Simple expressive diagrams are especially important in developing models of aquatic ecological conditions. These diagrams are useful in displaying the functional aspects of community metabolism. Two types are popular—the component **energy flow diagram** and the engineering circuitry energy flow diagram. Whereas energy is the most often used common denominator in ecosystem analysis, the flow of materials can also be shown. One convenient form of an energy flow diagram is the Y-shaped pattern, in which the stem of the Y represents the energy lost in respiration, one arm is the producer component, and the other arm is the consumer component (Figure 13.6). Adding caloric values to the diagram makes it useful for quantitative analysis. It is helpful to consider this particular example in greater detail.

Root Spring is a small spring in Massachusetts, about 2 m in diameter and 10 to 20 cm deep. A study of this small ecosystem showed that the producer compartment consisted of benthic algae and duckweed (*Lemna*). Approximately 50 species of animals inhabited the spring, most of these being herbivores or detritus-feeding forms. Community respiration and net production were measured monthly by "light and dark" glass cylinders pushed into the bottom. Energy transformation and the composition and size of the community standing crop were also determined. Inflow and outflow of organic matter (including the emergence of insects) were computed. From these data and the calculation of respiration rates of the various levels and the total community, the energy flow could be diagramed as shown in Figure 13.6. The energy contained in each level is shown as kilocalories per square meter per year in the numbers in the compartments of the diagram. It is interesting that primary production (algae) accounted for little of the energy of the higher trophic levels; 76 percent of the energy transformed by the animals entered the system as allochthonous detritus, while only 23 percent of the energy was derived from photosynthesis within the spring. Of the total annual energy inflow, 71 percent was transformed to heat, 28 percent was deposited in the community, and 1 percent was lost through insect emergence.

Among the important aspects of community metabolism illustrated in the Root Spring data, the distinction between biomass and energy flow is especially evident; the most abundant herbivore species (in terms of biomass) was third in amount of energy assimilated and transformed. Unlike the cyclic flow of basic nutrients in the ecosystem, energy flow is usually unidirectional; energy flows through a community and is used only one time by the components. Because of the incompleteness of our present state of knowledge, a number of assumptions must go into studies of community metabolism. Nevertheless, we do gain considerable insight into the subject through functional analyses of various types of aquatic ecosystems.

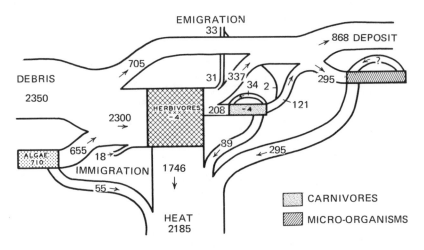

Figure 13.6   Energy flow model of an ecosystem, Root Spring, Massachusetts. Trophic levels are represented by boxes, the numbers indicating changes in standing crop. Energy is expressed in kcal/m² per year, and direction of flow is shown by arrows. (After Teal, 1957.)

Engineering circuitry flow diagrams, or **electrical analog circuitry models**, are often constructed to summarize this sort of data. These models use engineering or electrical symbols (Figure 13.7) to represent components and processes. In many cases these symbols are of great value as a step in mathemati-

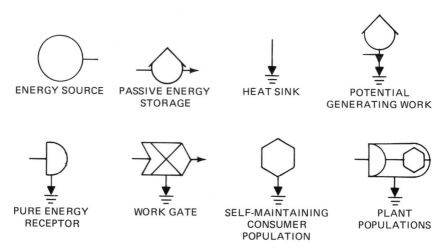

Figure 13.7 Engineering circuitry symbols useful in ecosystems analysis. (After E. P. Odum, 1971.)

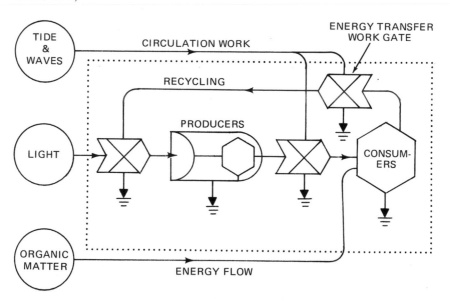

Figure 13.8 Engineering circuitry energy flow diagram of an estuary. Two main sources of energy, tides and wave currents, operate through work gates to augment production. Organic matter washed into the estuary is available to consumers, while light energy is fixed directly by producers. (After H. T. Odum et al., 1969.)

cal model making. The symbols are semipictorial, and in general they suggest visually the function they are supposed to indicate. The prime energy source is of course the sun, represented by a circle. The other symbols represent primary producers, consumers, energy storage, and heat (or energy) sinks, functions in which energy is lost to the system. One symbol, that for the **work gate,** needs further explanation. It represents in very general terms any process that facilitates another, different process. In electronics, this could mean a relay system where a low-voltage current activates a solenoid switch and thus permits the passage of a high-voltage current. In ecology, a work gate is usually some force external to the system—wind, current, or the like—which increases the flow of energy or nutrients into a compartment of the system. In the estuarine ecosystem diagramed in Figure 13.8, the circulation work of waves and tides is seen to affect the system by driving two work gates that permit energy to move "upstream" from the consumer back to the producer compartment, and a third that facilitates the movement of energy from plants to herbivores. Such a circuitry model can also be used (Figure 13.9) to describe the energy budget of an ecosystem, in this case a hypothetical marine system. Here the number above each symbol indicates the biomass in kcal/m²; numbers on or near the path lines indicate energy flow in kcal/m²/day. The diagram separates processes in the water column (upper path) from those that occur in detritus and sediment.

LAW OF THE MINIMUM

The concept of limiting factors as it affects the individual was treated in Chapter 12. Liebig's "law" of the minimum takes this concept further, by

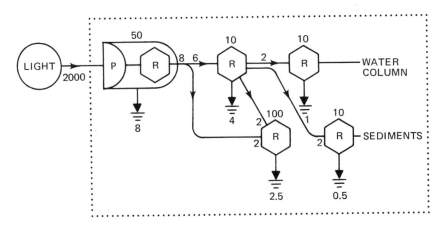

Figure 13.9 Engineering circuitry flow model of a hypothetical marine ecosystem. Numbers above symbols give biomass in kcal/m²; numbers on or beside path lines are energy flow in kcal/m²/day. The diagram separates actions in the water column (upper path) from processes that occur in sediment detritus (lower path). (Data from E. P. Odum, 1971.)

suggesting that the rate of any process, such as growth, is limited by the availability of the growth factor that is in shortest supply. This "law" was initially formulated to explain the relationship between growing plants and available nutrients, but the concept is now broadened to apply to many other factors. It is important to note, however, that the law is not strictly true, since interaction between factors usually makes the response to the limiting factor a gradual rather than a sharp, "cutoff" type of limitation. The integrative concept is interpreted by the overall effect of low (or excessive) concentrations as repressing production of other process. Since different organisms have different minima, low value of a factor could affect the sensitive ones first and cause a shift in the species community structure. Where the gradation is gradual and continuous, the series is considered a **cline**. A clinal shift in species composition is easily handled by synecological techniques, but would not be readily detected by systems ecological methods, since these are expressly involved with the functional aspects of a system rather than with its species composition.

SYSTEMS ANALYSIS OF SUCCESSION

The concept of succession has long been a special aspect of synecology, and ecologists have strongly emphasized descriptions of naturally occurring seres and climaxes. More work of this type is needed in aquatic ecology. However, the functional, decompartmentalizing, and mathematical approaches of systems ecology have brought a fresh start in this area. Four aspects of succession which could be selected to illustrate this trend are homeostasis, ecological succession, pollution, and resource depletion.

Homeostasis

The concept of **homeostasis**, as applied to ecosystems, is the integrative concept for the apparent tendency of a system, especially a climax community, to resist change. In such a community, producers receive an input of energy adequate to support the system, and consumers or decomposers (or both) utilize the excess production. In addition, climax systems seem to benefit from an extra work gate mechanism which increases production by the mechanical removal of what otherwise might be accumulating masses of vegetable matter. In the stonewort (*Chara*) community at the boil of Silver Springs, Florida, and the eelgrass (*Sagittaria*) community farther downstream, the rapid flow of water sweeps materials downstream as export or output. In *Rhizophora* (red mangrove) swamps and in beds of northern eelgrass (*Zostera*), daily tides function as work gates. Feedback mechanisms are effective in regenerating the community where portions are damaged, such as by grazing or intentional channel cutting. By contrast with such relatively stable communities, a *Typha* (cattail) marsh generally exhibits an excess of production over consumption and decomposition, with no well-developed export mechanism; thus input and output are out of balance and the excess accumulates, accelerating change, and thus instability, in the habitat.

Ecological Succession

From the viewpoint of ecosystem ecology, ecological succession can be seen to involve a series of stages of instability in which forces are not in balance. In communities in which production exceeds consumption, the excess net production is available to support additional organisms, and succession almost inevitably follows. It is not always easy to recognize when a community is still changing and when it has reached climax, but Margalef (1967) has noted a number of measurable features which he observed in succession in coastal marine waters. The major features can be listed:

(1) Productivity exceeds respiration in a growing autotrophic succession, and excess energy accumulates (P>R);

(2) Productivity is less than respiration in heterotrophic succession, and excess organic matter diminishes (P<R);

(3) When productivity equals respiration (at the climax), no energy accumulates, and change approaches stability (P=R);

(4) Production is high, but net production is zero (P/R=1);

(5) Pitment diversity, measured as the proportion of yellow to green, increases;

(6) Ratio of production to biomass drops (P/B↓);

(7) The frequency of individuals of each species decreases;

(8) Species diversity and total biomass are at a maximum in non-ecotone as well as ecotone regions.

Margalef has also recognized that a number of other factors reach their maximum values in climax communities. Thus extrametabolites occur in greatest concentration in climax communities; biogeochemical cycles are locked into the system to the greatest degree; and maximum values are observed for homeostasis, variation in light intensity, variety of niches, and stratification. It should be noted that no single one of these factors is an inevitable indicator of climax conditions; the pattern of many of these features occurring together is, however, a valuable quantitative measure of climax community functioning.

Pollution

The application of systems ecology to pollution problems is doubtless one of the major current developments in this field. Systems analysis offers a way to obtain data on present conditions, run a relatively modest number of tests to determine the effects of one factor, pollutant, or control measure on a water basin or stream, and use computations to provide a prediction in quantitative terms. Not only can ecological effects be quantitatively handled, but cost analyses can be done concurrently and a range of estimates can be offered for various degrees of completeness of controls. Environmental impact statements, now required before authorization can be given to undertake major changes in land or water resources that are of national or regional concern, are almost always supported by systems analysis reports predicting effects on the environment.

Resource Depletion

Resource depletion and resource management go hand in hand. With respect to depletion, one thinks at once in terms of trout fishing, esthetically pleasing rivers and lakes, clean water, commercial fishing, and the gradual disappearance of such species as whales, lobsters, or abalone. Recently, fisheries experts have developed effective mathematical models for managing such resource species. One problem that was solved by mathematical analysis was the belief that heavy harvesting of running alewives—which return from the sea to spawn in freshwater lakes—would have no effect on the number of spawn produced in the next generation, so long as the remaining breeding population maintained at least a certain minimum density. This proved to be true; any number of fish above that minimum population was in excess, especially since most adults die at the end of the run. The same type of model can be applied to determine permissible harvesting rates for other freshwater and marine species; the rates so determined may then be used to formulate public environmental policy. The models must, of course, be carefully designed to avoid focusing only on maximizing fish production and thereby adversely affecting other aspects of the environment and community. At this point in management modeling, the aquatic ecologist can play a particularly valuable role by ensuring that systems analyses take the whole ecological system into consideration.

SUGGESTED FOR FURTHER READING

Collier, B. D., G. W. Cox, A. W. Johnson, and P. C. Miller. 1973. *Dynamic ecology*. Prentice-Hall, Englewood Cliffs, New Jersey. A recent introductory ecology text that attempts to summarize ecological theory as it applies, in particular, to the functioning as well as the description of ecosystems. There is a good deal of descriptive material, providing the student with a clear introduction to the mathematical analyses that are a major feature of the text.

Krebs, C. J. 1972. *Ecology: the experimental analysis of distribution and abundance*. Harper and Row, Publishers, New York. As the title suggests, this book approaches ecology largely from the viewpoint of population dynamics. It is a strongly problem-oriented introductory text, relying heavily upon mathematical models of various sorts.

14

Monerans, Protists, and Plants of Aquatic Habitats

Representatives of virtually all major groups of organisms can be found in aquatic communities. In most communities the small algae are particularly important, since they form the basis of aquatic food webs. Other species are important because they decompose the remains of other organisms and thus re-release essential nutrients to the aquatic ecosystem. In many freshwater habitats larger algae and rooted plants provide food and cover for animals of many kinds. In this chapter we shall be concerned not only with plants, but with all those organisms that are not animals. If this seems an unexpected way to delimit our subject, the student should bear in mind that the question of how the different forms of life should be grouped has long been a matter of intense scientific speculation and controversy.

Traditionally, all species have been grouped two into kingdoms—plants and animals. A large number of organisms, however, do not fit very well the usual criteria by which a species was assigned to one or the other kingdom. For many years biologists have recognized that bacteria, flagellates, protozoans and other groups could not be accurately called either plants or animals. As a result, new systems have been erected to provide a way to classify such organisms. A recent suggestion (Whittaker, 1969) is to classify living things primarily according to function; on this basis we can recognize five distinct kingdoms (Figure 14.1).

This system accommodates nicely all the major types of living things. The system organizes the bacteria and bacteria-like organisms (without distinct nuclei, mitochondria, or plastids) into a separate kingdom, **Monera.** The tiny flagellated organisms, which although similar in structure and properties may lack or have chloroplasts, are grouped into kingdom **Protista.** Like all organisms except the monerans, these have distinct nuclei and mitochondria. From the protists, the other three types apparently arose: **plants,** the producers or autotrophs, with photosynthetic ability, cellulose-and-lignin cell walls, and fixed position; **animals**, the consumer heterotrophs, lacking photosynthetic ability, but with protein-and-lipid cell structure, and the ability to move; and **fungi,** the decomposers, heterotrophs which evolved penetrating filaments that release enzymes and digest organic matter. We shall follow Whittaker's schema, with one

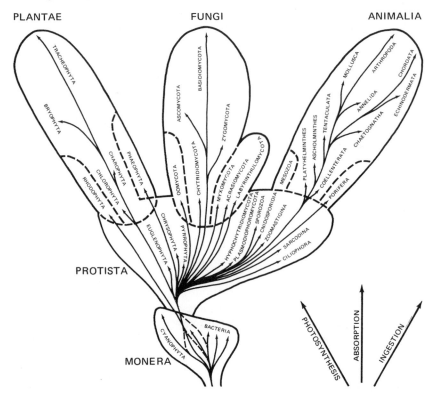

Figure 14.1   A current conception recognizes five different kingdoms, distinguished on the basis of (1) levels of organization and (2) mode of nutrition. Procaryotic organisms are grouped into kingdom Monera, eucaryotic unicellular organisms are in kingdom Protista, and more complex forms are placed in the other three kingdoms, as show. Within each level the major lines of descent are further distinguished according to the principal modes of nutrition: photosynthesis, absorption, and ingestion. (After Whittaker, 1969.)

important difference: we shall continue to treat the fungi as a distinct group within the traditional kingdom Plantae.

THE MONERA

The most primitive organisms are the monerans—the bacteria and bacterial-like forms. These organisms lack a nuclear apparatus, and thus they are said to be **procaryotic**. Pigments, where present in the monera, are dispersed throughout the protoplasm rather than concentrated in plastids. There are no mitochondria, and flagella (when present) are simple in structure, with only two fibrils. Aquatic monerans are predominantly of two types: bacteria and blue-green algae. Other moneran groups include the slime molds, the rickettsians, and the viruses.

Phylum Schizomycophyta (Bacteria)

Bacteria are extremely diverse organisms with respect to form, habitat, and function in the economy of inland waters. They range in size from about 0.5 microns (one micron, abbreviated μ, equals 0.001mm) to cover 50μ. Mostly, however, the size is less than 10μ. True bacteria occur in two primary forms: the spherical types, called **cocci**, and the cylindrical types, which include the **bacilli** and **vibrios**. Many secrete sheaths or stalks, thereby increasing the variety of body forms. Some of the true bacteria are capable of "swimming" locomotion. Another group, the Myxobacteria, distinct from true bacteria, move by gliding.

Within the aquatic environment bacteria inhabit the water phase, the bottom muds, and, of course, live on and in plants, animals, and detritus. As we have already seen, some (obligate anaerobes) exist under completely anaerobic conditions, others (obligate aerobes) require oxygen, while still others (facultative forms) function equally well whether or not oxygen is present. In terms of deriving energy and maintaining their own metabolism and synthesis, a few bacteria are capable of photosynthesis and are said to be **autotrophic**. Chemosynthetic bacteria obtain energy from inorganic substances. Most bacteria, however, are **heterotrophic** in that energy is obtained from environmental sources such as organic compounds in the surroundings.

The contributions of a number of bacteria to community maintenance have been considered in preceding pages. We refer to those forms so instrumental in deriving carbon (as carbon dioxide, ammonia nitrogen, phosphate phosphorus) and sulfur (as hydrogen sulfide) from complex organic substances. Additionally, bacterial activity can reduce certain of those compounds to methane, molecular nitrogen, hydrogen sulfide, or nitrites. We see here some of the important activities of certain kinds of bacteria in the ecology of natural waters. The development and maintenance of the various bacterial populations is, of course, dependent upon the availability of raw materials in a particular body of water. For example, a whole host of bacteria, the sulfur bacteria, is found only in waters containing hydrogen sulfide. Some of these, the green sulfur bacteria, do not normally appear until the hydrogen sulfide concentration exceeds about 50 ppm. Similarly, the bacterial assemblage in polluted waters is determined primarily by the chemical nature of the pollutant.

Marine bacteria are abundant, and they occur at all depths and in all major oceans. Also represented in marine waters are 15 species of yeast, 12 of mycobacteria, and two of actinomycetes. More bacteria occur in tropical waters, but cold-water species are able to use a greater variety of organic materials (McConnaughy, 1970).

Phylum Cyanophyta (Blue-green "Algae")

The reference to "algae" in the name of these organisms indicates that the designation is used loosely, for they are often considered to be more closely related to bacteria than to true algae. Furthermore, many of them are not blue-green in color, but rather various shades of blue, yellow, red, and green. Each body, or cell, is encased in a gelatinous sheath secreted by the organism. Some of the blue-greens are filamentous, the forms being made up of numerous cells attached end to end; others are simply masses, or colonies, of cells. The cells contain a number of pigments, including chlorophyll, carotene, phycocyanin, and phycoerythrin, which impart certain colors to the cells. Also present in the cell are false vacuoles, various granules, and the diffused nucleus. The pigments of blue-greens are not contained in discrete bodies, the plastids, as in true green plants, but are diffused throughout the cells. Some of these organisms, such as the Nostocaceae, are able to fix free nitrogen in a fashion similar to bacteria. Reproduction is largely asexual by cellular fission.

Cyanophytes most frequently occur in masses, either floating or attached to some object in the water. They are mainly inhabitants of fresh waters, and many are found in estuaries and in the sea. "Pond scum" or "water blooms" often seen in ponds and small lakes are sometimes formed by *Anabaena*, a filamentous form resembling a string of beads, or *Coelosphaerium*, a spherical mass of cells, or by other blue-greens. The color of the Red Sea is said to be caused by the presence of a red-pigmented cyanophyte, *Trichodesmium erythraeum*, which is a small filamentous species. The pigment phycocyanin, so common in these forms, is water-soluble, and its release into a lake through disintegration of an algal bloom may color the water. Although not necessarily "indicators" of pollution, blue-greens often thrive under such conditions, *Lyngbya* and *Oscillatoria* being notable examples.

THE PROTISTA

The protistans, the other kingdom of largely one-celled organisms, are far more complex than the monerans, but much simpler than the plants, fungi, or animals. These organisms, which have a nuclear apparatus, are said to be **eucaryotic**. Most protists are flagellated, with complex (2 + 9 internal fibrils) flagella. All have mitochondria, and chlorophyll pigments are found, when they occur, in plastids. In many respects, the protistans seem like "primitive" forms of plants, fungi, and animals. We shall limit our discussion here to the major phylum **Mastigophora**, which includes both chlorophyllous and non-chlorophyllous flagellates.

Phylum Mastigophora (Flagellates)

As a whole, this phylum includes a decidedly heterogeneous assortment of organisms. Most of the group possess flagella, or whip-like structures, that serve for locomotion (up to 300μ/sec), although some lose the structure in early development and move in ameboid fashion. Among the many mastigophorans, almost every known type of nutrition is to be found. Some are holophytic, carrying on photosynthesis; others are holozoic, ingesting particulate food; many are saprophytic, absorbing decay products through the cell surfaces; and quite a number are parasitic. Certain flagellates are capable of alternating modes of nutrition; *Euglena,* for example, can be photosynthetic or saprophytic. Within this phylum three classes are recognized.

Class Phytomastigina   This class includes a number of flagellates which are photosynthetic, some which are nonphotosynthetic, and others which may be either. The lumping of these diverse organisms into one class is based largely on morphological similarities. Ecologically, the contributions of the members of this class are varied. Photosynthetic forms such as *Euglena, Cryptomonas,* and others are important in the cycles of respiratory gases. Some, in the role of saprophyte, aid in the breakdown of complex organic compounds. Many serve as food for larger organisms and thereby enter into food webs in the community. In the sea and estuaries, photosynthetic members of this class and the dinoflagellates, to be considered next, share with certain of the true algae some fundamental roles as producers in energy cycles.

Class Dinoflagellata   This class includes forms which are holozoic, holophytic, and saprophytic. Most of the dinoflagellates are unicells, some possessing covering plates of cellulose while others are naked. *Peridinium* and *Ceratium* (Figure 14.2) are examples of the armored types. Both of these genera, as well as others, are represented in both salt and fresh waters. As shown in the figures, these dinoflagellates typically have one or more flagella lying in a groove about the body. The photosynthetic dinoflagellates possess chlorophyll *a,* chlorophyll *c,* carotene, four xanthophylls, and other yellow-brown pigments. Bioluminescence occurs in many of the organisms, and these are often responsible for "luminescent waters" of the sea and estuaries. Under certain conditions, the details of which are not fully known, the population of one or another of the flagellates in estuaries or open sea many increase rapidly, giving rise to a "bloom" which colors the water purplish or reddish-brown. *Gymnodinium breve* is responsible for the "red tides" that produce great mortality of fishes in the Gulf of Mexico. Other species of *Gymnodinium* may also discolor estuarine waters. In fresh waters, blooms of *Peridinium* or a host of others impart "fishy" odors to reservoirs or lakes. *Ceratium* in abundance produces a particularly obnoxious odor.

Class Zoomastigina This class is characterized by flagellated forms which are free-living or which have entered into various symbiotic relations, such as saprophytism or parasitism. Because of their nutrition, these are often placed with Protozoa. Some of these are pathogenic, others are obligate mutuals in the gut of cellulose-ingesting animals. The ecology of the free-living flagellates is not well known. In laboratory "wild" cultures, they typically represent a successional stage in a series of population changes, and in natural waters often show seasonal cycles of abundance.

PLANTS

In contrast to the vague phylogenetic relationships and the primitive features of the plant-like protists, the lineages, structure, and biochemistry of the "true" (green) plants are rather clear and indicate separation of these into a category, the Plant Kingdom. However, much controversy over this matter has prevailed for many years, and even today the classification of plants, animals, and "in-between" forms at the unicellular level is not settled. Reference to almost any textbook of zoology or botany will amplify this point. In the classification adopted here, plants include algae, fungi, bryophytes, and

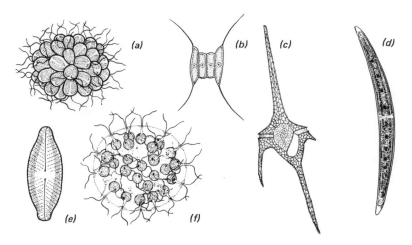

Figure 14.2   Some representative protists and algae of lakes and ponds. (a) *Synura uvella* Ehrenberg, × 200; (b) *Scenedesmus quadricauda* (Turp.), × 450; (c) *Ceratium hirundella* (O.F.M.) Schrank, × 250; (d) *Closterium* sp., × 125; (e) *Navicula* sp., × 675; (f) *Eudorina elegans* Ehrenberg, × 250. See text for classification. (Figure c after Smith, 1950.)

tracheophytes. All of these have representatives in fresh water; bryophytes apparently have not become adapted to marine existence and are, therefore, of limited distribution in estuarine and marine waters. In discussing plants, the term "division" (corresponding the term "phylum" in discussions of other kingdoms) is used to refer to the major taxonomic groups.

ALGAE

Algae are found in natural waters in an impressive array of shapes, sizes, biochemical characteristics, and ecological roles. Most of the vegetation in the sea consists of algae, for few tracheophytes (no ferns at all) inhabit saline waters. In fresh waters a great variety of algae and tracheophytes (including a large number of ferns) are found. Classification of algae is at present based upon chemical composition of food storage substances and of the cell wall, and upon the quality of pigments present. The algae and the plant-like protists constitute one major segment of the aquatic "pastures," the level in which radiant energy is fixed in protoplasm and transferred to nonautotrophic organisms ranging from zooplankton to fish to man.

Division Chlorophyta (Green Algae)

These plants occur widely wherever there is water. Many are found in moist soil and on tree trunks, and even in ice. They are especially common in a variety of body forms. Some of the greens are highly conspicuous, sheet-like (or often

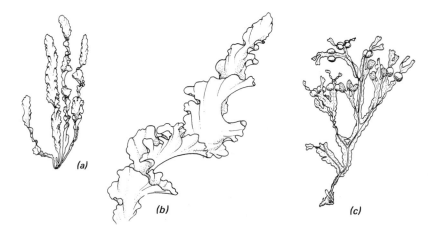

Figure 14.3   Green and brown algae of estuarine and marine environments. *(a) Enteromorpha linza* (Linnaeus) J. Agardh; *(b) Ulva lactuca* L; *(c) Fucus vesiculosus* L. (Redrawn from Taylor, 1937.)

tubular) organisms, such as *Ulva* and *Enteromorpha* (Figure 14.3) which inhabit estuaries and the sea. The thallus, or body, of other greens may appear as a continuous filament or as a partitioned, or septate, branching filament (*Cladophora* in Figure 14.4).

Single-celled forms such as *Chlorella* are sometimes common in lakes and ponds, often in numbers sufficient to color the water. *Chlorella* is also interesting in that it occurs in the cell body of some protozoans, *Stentor* for example, and in the tissues of certain coelenterates. The green hydra, *Chlorohydra viridissima*, owes its color to the presence of zoochlorellae in the animal tissues. It is

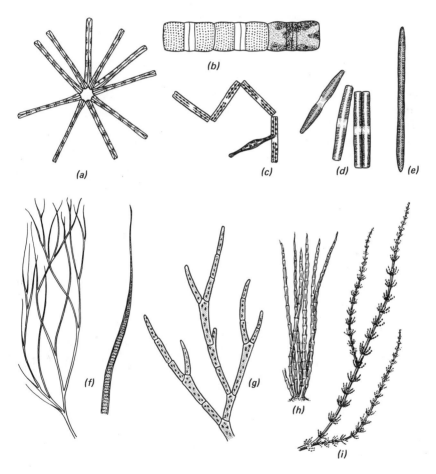

Figure 14.4   Algae of lakes and streams: *(a) Asterionella formosa* Hass. × 320; *(b) Melosira granulata* (Ehr.) Ralfs, × 600; *(c) Tabellaria fenestrata* (Lyngb.) Keutzing, × 250; *(d) Fragilaria capucina* Desmazieres, × 330; *(e) Synedra ulna* (Nitzch.) Enr., × 350; *(f) Compsopogon coeruleus* Mont., natural size; *(g) Cladophora crispata* (Roth) Kuetzing, × 80; *(h) Lemanea annulata* Kuetzing, natural size; *(i) Batrachospermum vagum* (Roth) Ag., × 15.

presumed that the alga-hydra relationship is a symbiotic one in which there is exchange of nutrients and respiratory gases. The details of the association are not fully understood, however, for both members can exist independently. Colonial green algae, such as *Volvox* and *Eudorina* (Figure 14.2) are often common in lakes and ponds. Desmids are green algae which occur as single cells or sometimes as small colonies and filaments of minute cells. These often exhibit intricately beautiful patterns of body form and surface markings. Desmids may be found on objects in the lake or freely floating as important constituents of phytoplankton. *Cosmarium* and *Closterium* (Figure 14.2) are common forms in lake or pond plankton. None occurs in marine waters.

Stoneworts, of which *Chara* is an example, are often placed in Division Chlorophyta. These large algae exhibit a form and habit similar to tracheophytes in possessing whorls of branchlets along a stem-like filament and holdfasts that anchor the plants in the substrate. The common name derives from the brittle, stony texture of some species, caused by deposits of lime frequently found on the plant surface in highly calcareous streams or lakes. *Nitella*, similar in form to *Chara*, often thrives in soft or acid waters. Under such conditions, *Nitella* generally lacks the hard texture.

Chlorophytes range in size from microscopic cells and filaments to large colonies several feet long and quite broad. This group typically contains chlorophyll *a* and chlorophyll *b,* together with various carotenoid and xanthophyll pigments contained in chloroplasts. Some green algae may not appear green due to the masking of chlorophyll by one or more of the accessory pigments. Manufactured food is stored principally as starch, this feature making the iodine starch test a useful aid in identification of the group. Cellulose is present in the cell walls.

Division Chrysophyta (Yellow-green Algae, Golden-brown Algae and Diatoms)

This is a rather heterogeneous assemblage of algae containing chlorophyll *a* and, in the golden-brown algae, chlorophyll *c*, beta carotene, and xanthophylls, including fucoxanthin. Starch is absent, the food substances being stored primarily as other carbohydrates or as oils. Silica, rather than cellulose, is the major component of the cell walls.

Among the chrysophytes, diatoms are probably of greater importance in the energy cycle of natural waters. These algae are frequently considered among the foremost of the photosynthetic producers, especially in the sea, and to a certain extent in the fast regions of upland streams. Diatoms are typically unicellular, although in some forms the individuals may be aggregated in variously shaped colonies such as the radiating colony of *Asterionella* or the chain-like colony of *Tabellaria* or *Melosira* as shown in Figure 14.4.

A most unique characteristic of diatoms is the enclosure of the cell in two siliceous shells, or frustules, which fit together in overlapping, or "pill-box," fashion. The frustules are frequently decorated with highly ornate and precise sculpture consisting of fine striations, pits, and shallow depressions. Reproduction is normally vegetative, and during the process the frustules separate; each new cell forms a new frustule that fits inside the original one. Thus part of a population is constantly being reduced in size. This diminution of size apparently continues until a certain minimum is reached, whereupon a form of sexual reproduction occurs involving the formation of an auxospore by two individuals. During this process the organism may form new, large frustules.

Diatoms are generally microscopic, but some reach a size of nearly 200μ. They occur abundantly as floating forms in plankton and on submerged objects such as stones, larger plants, and animals. In streams, the slick nature of stones may be due to dense diatom populations and their secretions. The "plankton" of streams is derived almost wholly from scouring of the substrate by stream action, and this plankton (tychoplankton) consists in large part of diatoms. It often serves as the principal food source for many small stream animals.

Division Phaeophyta (Brown Algae)

Of the thousand-odd known species of this group only a few inhabit fresh water. One rare kind, *Heribaudiella*, forms relatively large, dark, disk-shaped crusts on stones in rapidly flowing upland streams. Most of the brown algae are adapted to existence in salt water, the distribution of many extending into estuaries until limited by decreasing salinity. These algae include the familiar large kelps and rockweeds found in coastal and intertidal waters. Most are robust plants. All are multicellular. *Fucus* (Figure 14.3) is a common rockweed of cooler climates. The giant *Macrocystis* may grow to over 80 m in length. Most of the browns are attached to the substrate by holdfasts. One species of *Sargassum*, usually harboring a unique community of plants and animals, is a free-floating plant of the open seas. The floating of this form is made possible by morphological adaptations in the form of air-filled tissues which add buoyancy. In coastal waters and estuaries, brown algae in abundance provide numerous habitats in which populations of animals and of other plants develop and constitute a vigorous and productive community.

The pigments of phaeophytes include chlorophyll (*a* and *c*), carotene, and xanthophylls including fucoxanthin. The last-mentioned pigment predominates, and gives the characteristic color. Food reserve is maintained as a sugar, as mannitol, or as laminarin, the latter being unique to the brown algae.

Division Rhodophyta (Red Algae)

The members of this phylum are characteristically marine. A few forms, including *Batrachospermum* and *Lemanea*, shown in Figure 14.4, inhabit fresh water, mainly streams. *Lemanea* sometimes occurs in the upper reaches of estuaries where it receives washings in saline water. *Compsopogon* (Figure 14.4) is also a stream form, but it may be found growing on mangrove plants in estuarine or coastal waters. The marine species, about 3000 presently known, are widely distributed in the more temperate seas. Most red algae are multicellular and exhibit considerable variety of form and coloration. One group, the genus *Gelidium*, is important as a major commercial source of agar.

The red algae pigments include chlorophyll *a* and chlorophyll *d*, xanthophyll, carotene, phycocyanin, and phycoerythrin, the last two being chemically different from pigments of the same names in cyanophytes. Food reserve in the reds is in the form of floridean starch, which does not react with iodine in the standard starch test. The presence of the pigment phycoerythrin apparently accounts for the ability of these algae to exist at depths greater than most other plants. This pigment efficiently absorbs light in the spectrum's blue range which, as we have seen in Chapter 7, is usually transmitted more deeply than others in water. Another interesting feature of this group is the presence of nonmotile gametes, the meeting of sex cells being dependent upon water movements despite the almost universal rule that sexual reproduction in aquatic organisms involves gametes capable of self-propulsion.

Division Mycophyta (Fungi)

The fungi are plants lacking chlorophyll. The group includes, generally, such well-known forms as mushrooms, yeasts, and molds. Representatives of this group are essentially saprophytic or parasitic and are found everywhere that living matter exists. Out of a considerable array of kinds only relatively few are found in the aquatic environment. These, however, engage in the same processes as the nonaquatic fungi, *i.e.,* aiding in the decay of organic substance and, in some instances, causing disease.

A majority of the aquatic fungi are classified in the class Phycomycetes, the tubular fungi. The fundamental structural unit of these fungi is the hypha, a slender filament of cytoplasm containing many nuclei and lacking cross walls. A mass of hyphae constitutes a mycelium, which may be loosely organized, as in bread mold, or compactly bound, as in a toadstool. Asexual reproduction is by flagellated spores borne in a sporangium. Sexual reproduction involving gametes also occurs. The fungus *Saprolegnia,* called "water mold," is a typical example of an aquatic phycomycete. It commonly causes infection of small fishes and appears as a whitish, fuzzy mass on the animal. Other molds, very similar in appearance, attack fish eggs during development, and in the natural habitat or in hatcheries may be of serious consequence.

A small number of sac fungi (Ascomycetes) occur in natural waters. These usually resemble miniature mushrooms (which they are not) on decaying plants.

Many sac fungi cause diseases, and others, such as the famous *Penicillium*, produce antibiotic substances. The importance of these antibiotic materials in the natural environment is not known. As it relates to human welfare, the field of exploration for new compounds from aquatic fungi is as yet relatively untouched.

Division Bryophyta (Liverworts and Mosses)

The bryophytes are relatively small plants lacking the flowers and specialized water-conducting tissues characteristic of "higher" plants. Water may be retained in the tissues of these plants for considerable periods of time, however, due to presence of numerous empty cells scattered throughout the plant body. The life cycle of bryophytes typically consists of two phases: the leafy, green gametophyte that produces motile gametes, and the usually brownish, nonleafy sporophyte generation that produces spores.

As pointed out previously, bryophytes do not occur in marine waters. About 45 genera of these plants occur in or near fresh waters of North America. Of these, about 12 are liverworts (class Hepaticae), which are often small, flattened green plants. Some of these lack distinct development of stems and leaves; others possess these structures. *Riccia* has a slender, branched thallus usually floating individually just beneath the surface of ponds, canals, and ditches. Often common in the same environment, *Ricciocarpus* is a notched, semicircular form about 1 cm in diameter, with black rhizoids on the undersurface. *Jungermannia* is a leafy liverwort found in streams and slow waters.

The class Musci comprises the true mosses, those bryophytes with distinct stems and leaves. Peat moss, *Sphagnum*, is probably the best-known of the mosses. It is widespread and under proper conditions forms extensive bogs. *Fontinalis* and a number of other true mosses often occur in large waving masses, particularly in spring runs and mountain streams. These are typically long, sinuous plants with thick-set leaves arising in various fashions along the stem. Interesting assemblages of small plants and animals usually inhabit masses of these mosses.

Division Tracheophyta (Vascular Plants)

As the name implies, these plants are characterized by conducting tissues (xylem and phloem) for the transport of materials throughout the plant body. Most of the tracheophytes exhibit structural specialization into true roots, stems, and leaves. The development of these organs is associated ecologically with life in a terrestrial habitat. Aquatic vascular plants comprise only a minority of the group.

The lower tracheophytes are not tolerant of salt concentrations normally found in sea water. Their distribution and importance in estuaries is, therefore, limited. Even in fresh waters the number of species of these lower vascular plants is relatively small. Water horse-tail *(Equisetum)* and the grass-like quillwort *(Isoëtes)* are frequently common along the shores of fresh waters and sometimes grow submerged in a lake or stream. Among the ferns, the water shamrock *(Marsilea)* inhabits shallow zones; its leaves, consisting of four broad leaflets, arise from rhizomes growing in the substrate. Water fern *(Azolla),* shown in Figure 14.5 is a small, floating fern with overlapping leaves and fine roots hanging from the undersurface. The leaves, usually no more than 5 mm long, are green during summer, but turn reddish in the fall. Where abundant, these plants may completely obscure the water surface, the shading effect thereby inhibiting production in the water below.

Class Gymnospermae The most prominent gymnosperm species are the conifers, the group that includes such familiar species as pines, spruces, hemlocks, and cedars, characterized usually by needle- or scale-like foilage. Very few gymnospern species occur in water, and none are marine. Among the few aquatic gymnosperms are the bald cypress *Taxodium,* which forms dense swamps in the southeastern United States, often pendant with Spanish moss. In the northern bog mats and forests, one finds Tamarack *(Larix)* and white cedar *(Chamecyparis)*.

Class Angiospermae (flowering seed plants) The angiosperms are among the plants that have, through the course of evolution, become quite successfully adapted to life on the land. The characteristics which distinguish these forms and also make possible the terrestrial habit include (1) the presence of specialized parts that enclose the seeds; (2) the seed itself, which protects and nourishes the plant embryo; (3) the presence of flowers and fruits; and (4) the development of an elaborate system of tissues that function in conduction of

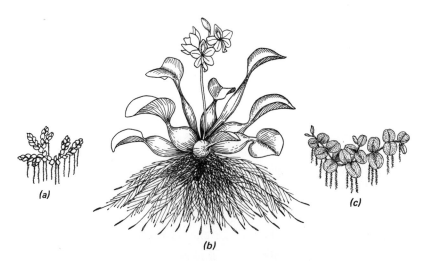

(a)

(b)

(c)

Figure 14.5 Floating plants of freshwater environments: *(a)* the fernlike *Azolla; (b)* the flowering water hyacinth, *Eichornia crassipes* (Mart.) Solms.; *(c) Salvinia.*

food and water and in support. The continuous xylem vessels of angiosperms represent the ultimate in conductive structures of tracheophytes generally.

Of some 200,000 known species of flowering plants, relatively few grow as "true" aquatics in fresh waters, and only about 30 species inhabit salt water. This distribution, however, presents an interesting illustration of the part played by natural waters in the evolution of organisms and their invasions of "new" habitats. The angiosperms first appeared in the early Mesozoic. Some time during subsequent years some of these land-adapted plants invaded fresh waters, carrying with them the features necessary for terrestrial existence. In the water, adaptive radiation has resulted in many species occupying a variety of habitats and exhibiting considerable morphological diversity. From two widely adapted families of freshwater angiosperms, the Najadaceae, and the Hydrocharitaceae, arose most of the present-day marine angiosperms.[1] Their evolution was probably one of slow adaptation to increased salt concentration accompanied by migration down freshwater streams through the estuary and into the sea.

The recognition of a category to be labeled "true aquatic plants" demands a certain amount of latitude and qualification, for within the world of plants we find a very broad spectrum of adaptation to water. The adaptations relate to water requirement and water tolerance, and range from strongly xerophytic, through mesophytic, to hydrophytic forms. We may define "aquatic plants" as those whose seeds germinate in either the water phase or the substrate of a body of water and which must spend part of their life cycle in water. This ecological grouping includes plants which grow completely submerged (except when flowering) as well as a variety of emergent types. This general idea is shown in Figure 14.6. In the United States about 50 families of flowering nonwoody plants may be considered primarily aquatic.

The family Najadaceae is said to be the largest of the families of aquatic plants. In the United States this group includes about 13 genera representing nearly 60 species, some of which are estuarine and marine. The predominantly marine forms are *Zostera, Phyllospadix, Cymodocea, Thalassia,* and *Halodule. Zostera,* or eelgrass, is the most widely distributed marine angiosperm, occurring in coastal waters and estuaries throughout a vast portion of the cooler regions of the northern hemisphere. It grows commonly in shallow zones on sandy mud bottom. During 1931-1932, great mortality of eelgrass was experienced over extensive areas, particularly on the Atlantic coast of North America. This was serious because the plant was an important food of some waterfowl. Since about 1940, however, there has been considerable return of the plant. The cause of the mortality has never been fully explained. *Phyllospadix* inhabits coastal embayments and rocky shores of the Pacific coast of North America.

1. The meaning and validity of this generalization depend upon the point of view of systematics. Some authorities relegate several of the marine angiosperms to separate and dictinct families. Nevertheless, the basic idea of evolution of the marine forms from freshwater ancestors remains sound.

Cymodecea and *Halodule* are restricted to the warm coastal waters of southern and tropical zones. *Thalassia* is common in these regions. Broad, shallow "flats" of sandy mud and the mouths of small estuaries on the northern Gulf Coast of Florida support abundant growths of both of these plants, together with another genus, widgeon grass *(Ruppia),* the latter being also common in alkaline waters throughout most of the United States. Two genera of pondweeds, *Potamogeton* and *Najas,* are essentially freshwater inhabitants, although a few species do extend into brackish waters. In the United States, these two genera include over 80 percent of the species of Najadaceae. One or another of the species occurs in nearly all types of fresh waters. These plants are mostly submersed, some with floating leaves, exhibiting a variety of leaf form ranging from linear to broadly ovate. Ecologically, the pondweeds are of great importance in the cycles of nutrients and respiratory gases, and in often providing very dense habitats which supply food and shelter to numerous small organisms living on and among the plants. Many of the potamogetons serve as a major item of food for ducks and geese.

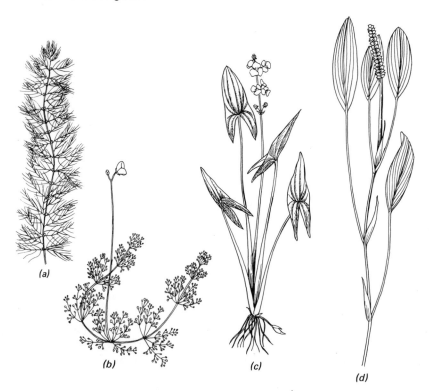

Figure 14.6   Flowering plants of fresh waters: *Myriophyllum (a)* and *Utricularia (b)* are essentially submersed plants. The various species of *Sagittaria (c)* and *Potamogeton (d)* exhibit diverse adaptations, some bearing floating leaves, some being emergent, and others being completely submerged.

The water plantain family (Alismaceae) contains about 50 species of emergent marsh plants and submersed aquatics. In most of these the leaves possess long slender petioles with flattened blades. Two of the genera, *Alisma* and *Sagittaria,* are widely distributed in a variety of waters, some species of the latter entering the upper regions of estuaries. The conspicuous flowers of *Sagittaria* often add color to marsh communities. Also noteworthy here is the family Hydrocharitaceae, mentioned previously. This group contains the well-known genus *Anacharis* (*Elodea*) so popular with aquarists. One species was introduced into the United States from South America; another species, native to North America, was introduced into Europe. Both became serious pests, clogging streams and canals. Hydrocharitaceae also includes two marine genera, *Halophila* and *Thalassia,* of warm waters, and *Vallisneria,* a freshwater form with long ribbon-shaped leaves, widely consumed by waterfowl.

Many grasses (Gramineae), sedges (Cyperaceae), and rushes (Juncaceae) typically inhabit marshes and shore zones; these include manna grass, cut grass, reed (*Phragmites*), cord grass, wild rice, sedge, saw grass of the Florida Everglades, spike rush, true rush, and others. These plants normally grow in the very shallow waters, but usually in profusion, and upon death and decay contribute to the richness of the body of water.

A number of angiosperms have become adapted to floating on the water surface. Their role in community activities is interesting in that the roots of the plants, dangling as they do in the water beneath the plant, extract nutrients from the water phase in competition with phytoplankton. The leaves, borne above the water, contribute little to the cycle of photosynthetic and respiratory gases in the water. As the plants die, nutrients are returned to the water. Where abundant, these floating forms often shade the water from sunlight, thereby inhibiting production. *Pistia,* the growth form of which resembles garden lettuce, is common in the Gulf Coast states. The family Lemnaceae includes the duckweeds, which are minute plants of flattened or spherical body form. *Lemna* and *Spirodela* are duckweeds widespread in the United States, often in association with liverwort, *Ricciocarpus,* and the fern, *Azolla,* described previously. Another floating species, but of the family of pickerelweeds, the water hyacinth (*Piaropus,* or *Eichornia*), shown in Figure 14.5, was introduced into the United States from South America during the latter part of the 19th century. Because of its very showy flower, it has been introduced widely. In favorable areas, however, the plant quickly becomes a pest—navigationally, economically, and ecologically.

Various members of the water lily family are familiar to almost everyone. To this family belong the white water lily (*Nymphaea*), the yellow pond lily (*Nuphar*), and the lotus (*Nelumbo*). In many lakes and streams patches of water lilies are favorite nesting areas for certain sunfishes. The water shield *Brasenia,* a small member of the water lily family, is found chiefly in acid waters in the eastern half of the United States; the stem and undersurface of the leaf of *Brasenia* are thickly coated with a gelatin-like substance.

Of the various habits and adaptations exhibited by aquatic plants, one more is worthy of mention. This is found in *Utricularia* (Figure 14.6); most species of this plant are free-floating forms lacking roots. They may float near the surface of the water or hang near the bottom. In some, the leaves may function as roots in anchoring or absorption. Most of the species possess small bladders, developed from leaf segments, which have trap doors that can be sprung to capture minute aquatic organisms.

GENERAL REFERENCES TO AQUATIC PLANTS OF THE UNITED STATES

Fassett, N. C. 1957. *A manual of aquatic plants,* revised edition. University of Wisconsin Press, Madison.

Muenscher, W. C. 1944, 1973. *Aquatic plants of the United States.* Comstock Press, Ithaca, New York. (Reprinted by Cornell University Press, 1973.)

Smith, G. M. 1950. *The freshwater algae of the United States.* McGraw-Hill, New York.

Tiffany, L. H., and M. E. Britton. 1952. *The algae of Illinois.* University of Chicago Press, Chicago.

Animal Species of Aquatic Habitats

15

As we saw in Chapter 14, plants are grouped into approximately ten generally recognized divisions. In the other major kingdom of organisms—Kingdom Animalia—there is considerably greater diversity, and the number of major taxonomic groups (phyla) is generally put at between 30 and 33. Representatives of every phylum are represented in aquatic habitats, either by free-living forms, by parasites inhabiting other aquatic species, or by both. An enormous number of animal species spend all or part of their life cycles in natural waters. At least 30,000 insect species are known to be aquatic in some way, and, with few exceptions, these are restricted to fresh water. The number of echinoderm species is estimated at 4800; all of these are marine. In the United States alone, some 8500 invertebrate species (apart from protozoans) are known to occur in fresh waters. These numbers may seem at first to be rather large, but in fact these estimates are probably rather conservative, since they are limited to "known" species. It should be remembered that in many groups only a fraction of the existing species have been "described" scientifically; the rest still await accurate study.

Of the phyla considered here, the first three are sometimes grouped separately in a single phylum, Protozoa. Some authors add to this phylum the Mastigophora, or at least the nonphotosynthetic forms. In still another scheme, all of the unicellular organisms are treated as Protista. For reasons stated previously, the classification adopted herein places the flagellated unicells with protists, and on the basis of strong differences in locomotor adaptations of the nonflagellated forms recognizes four phyla of single-celled animals.

PHYLUM SARCODINA (PSEUDOPOD-BEARING UNICELLS)

In addition to the well-known *Ameba,* a free-living sarcodine, and *Entomoeba,* a pathogenic parasite, the phylum Sarcodina includes a great number and variety of forms inhabiting freshwater or estuarine environments. Pseudopods, the distinguishing feature of the group, are used for feeding and locomotion. Many sarcodines are naked species; others possess shells composed of various substances.

Some of the materials of which sarcodine shells are composed contain ions that are important in biological and chemical processes; thus these animals enter prominently into biogeochemical cycles. In the sea, and to a small extent in fresh water, sarcodines of the order Foraminifera secrete coiled and spiralled, snail-like shells composed predominantly of calcium carbonate. Deposition of these shells in ocean sediments essentially "locks up" considerable quantities of carbon and calcium. The magnitude of such deposition can be appreciated from the knowledge that the famous White Cliffs of England are of marine origin and are composed mainly of "foram" shells. Some of the common marine forams are *Globigerina, Globigerinoides, Globorotalia,* and *Hastigerina.* The sarcodine order Radiolaria is composed entirely of marine forms. These animals secrete variously shaped shells in which silica is abundant. In fresh waters, *Difflugia* (order Lobosa) builds urn-shaped coverings of sand grains cemented by a chitinous substance. *Arcella* secretes a hemispherical shell with a single opening, or **foramen**, in the center of the flattened surface; the pseudopods extend through the foramen. Some of the sarcodine shells contain a bubble of gas, apparently given off by the animal, which is probably an adaptation for floating. The composition of the coverings of the shelled sarcodines varies considerably from mucilaginous materials to the above-mentioned chitinous substance.

The Heliozoa are sarcodines usually enclosed in gelatinous sheaths, and characterized by the possession of numerous long, stiff radiating projections of cytoplasm. The animals, among which *Actinophrys* and *Actinosphaerium* are typical examples, resemble sunbursts. Helioza are primarily freshwater forms.

In the activities of the community, sarcodines feed upon organic detritus, bacteria, small algae, and even other unicells. This they do by taking the food item into cytoplasmic vacuoles.

Hydramoeba hydroxena is a large sarcodine that preys upon hydra, a coelenterate animal. The ameba, shown in Figure 15.1, attacks the hydra, loosens and devours the cells, and eventually kills the prey. Sarcodines, in turn, serve as food for larger organisms.

PHYLUM CILIOPHORA (CILIATED UNICELLS)

This group is characterized by the possession of numerous hair-like projections, or **cilia**, which in some forms serve almost wholly for locomotion, and in others for creating currents which deliver food to the ciliate. These animals are the largest of the unicells and also exhibit the greatest complexity of structure. A variety of morphological types are known. The "slipper-shaped," actively swimming *Paramecium* is generally the best known. *Stentor* is a horn-shaped form that frequently attaches to some object in the water; when swimming it is somewhat globular. *Lacrymaria* has a serpent-like form when extended and moving. A number of genera, such as *Epistylis* and *Carchesium,* are sessile, colonial forms in which the cell bodies are borne on stalks. The family Tintinnidae is a marine group of vase-shaped ciliates bearing feathery projections about the oral region. Among marine ciliates, there is a large number of tintinnids which secrete a chitinous or pseudochitinous **lorica**, or case, in which they live. Some 300 kinds are known. They are common in the plankton, and the funnel-shaped *Tintinnopsis* is common in the Northeast.

Although the body form of ciliates exhibits considerable diversity, the internal structures are highly developed as permanent organelles. The ciliates

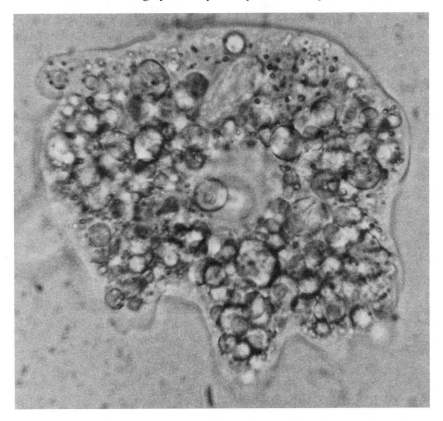

Figure 15.1 *Hydramoeba hydroxena* (Entz), a sarcodine predator on *Hydra.* Just to the right of the clear center region can be seen a freshly ingested cnidoblast of a hydra. (Photograph by Helen Forrest.)

typically possess a mouth-like opening which leads into a cytopharynx, contractile vacuoles, neurofibrils which coordinate, contractile fibers for movement, and locomotor structures. A form of sexual reproduction involving exchange of nuclear substance as two animals conjoin is characteristic of ciliates. In all, the structure and organization of ciliates represent a considerable evolutionary "advance" over the much simpler structure and organization of the protists and the sarcodines.

PHYLUM SPOROZOA

This phylum includes unicellular nonmotile organisms all of which are parasitic, many being pathogenic. In our province of interrelationships of free-living inhabitants of natural waters, sporozoans are of little interest.

PHYLUM CNIDOSPORIDIA

A special phylum, Cnidosporidia, has been erected for those sporozoan-like organisms which produce stalk-forming spores. Once again, these are of little interest to the aquatic ecologist.

PHYLUM PORIFERA (SPONGES)

The bodies of the poriferans consist of loosely organized masses of more or less independent cells. Some of the cells are specialized for the production of **spicules**, which are variously shaped structures that serve primarily as skeletal elements. The chemical composition of spicules is used as a basis for classification of sponges, some being made up primarily of calcium carbonate ("chalk sponges"), some of silica ("glass sponges," including the freshwater forms), and others of protein compounds, with or without inorganic spicules ("horny sponges"). The basic structure of sponges involves a somewhat porous or sac-like body, with numerous pores or canals ramifying throughout the tissues. Many variations in form and color exist, especially among marine species. Food is obtained from water circulated through the pores or canal system. Although the larvae may be motile, the adults of all sponges are sessile, and most are attached to the bottom or to some object in the water.

Nearly all sponges are marine. They are found in all seas, from shallow waters down to depths greater than 400 m. Most species are rather narrowly restricted by temperature and salinity. Marine species vary in size from small crustose masses to large vase-like or globular forms over 2 m high. Not many animals eat sponges, but the poriferan does frequently harbor an assemblage of algae, worms, insects, and crustacea, and fishes often nibble at the sponge in seeking the more tasty items.

One family of sponges (Spongillidae) occurs in unpolluted fresh waters. These are characterized by the formation of specialized reproductive structures, known as gemmules, which sometimes contain unique spicules. Although highly variable in form and size, the freshwater sponges do not resemble their more elaborately fashioned marine relatives. Most freshwater species are inconspicuous, green or dull-colored flat masses encrusting objects in lakes or streams. Among the 150 or so species, considerable variation in adaptation to environmental conditions is shown. Some sponges inhabit soft, or acid, situations; others are found in hard waters. In spite of the requirement of silica for spicule construction, many species exist in waters of remarkably low silica content. Over 375 species of sponges have been observed in marine waters. They are particularly abundant in tropical waters, but occur throughout all ocean regions.

PHYLUM COELENTERATA

The coelenterates are a varied group of simply constructed animals, the greatest number and most conspicuous of which are marine and estuarine. Included in the group, typically under three classes, are sea anemones and corals (Anthozoa), jellyfishes (Scyphozoa), and freshwater and marine hydroids (Hydrozoa). These animals are constructed of two relatively well organized layers of tissue in the form of a vase. The single opening is usually bordered by tentacles, which aid in feeding. The cavity of the animal is the digestive chamber. A unique aspect of coelenterates is seen in the presence of stinging cells, or **nematocysts**, variously modified to eject filaments which ensnare or penetrate and poison prey, or protect the animal itself. The larvae of coelenterates are ciliated and motile; the adult stage may be as a polyp attached to a substrate, or as a free-swimming (or drifting) **medusa**. Alternation of these forms is typical in the life cycle of hydrozoans such as the marine and estuarine *Obelia,* and the freshwater *Craspedacusta.*

Only a very few hydrozoan coelenterates have become adapted to the freshwater environment. These have evolved from marine stocks using the estuary as a route to freshwater habitats. No scyphozoans or anthozoans have apparently invaded fresh water, even though the phylum is a very ancient one and has, in the marine world, become highly diversified. In fresh waters, the hydras, well-known to almost anyone who has had general biology, are probably the most widespread. Of these, however, only about a dozen species are known in the United States. Two species, *Hydra oligactis,* the brown hydra, which ranges from several western states to New England, and *Hydra americana,* of the eastern United States, are shown in Figures 15.2 and 15.3. *Chlorohydra viridissima,* rather widely distributed in North America, and *Chlorohydra hadleyi,* of the northeastern United States, are of ecological interest, for the tissues of these forms normally contain large populations of an alga. This unicell, usually *Chlorella,* imparts a rich green color to the coelenterate. As pointed out in our consideration of algae, this relationship may not be an obligate one, for the algal

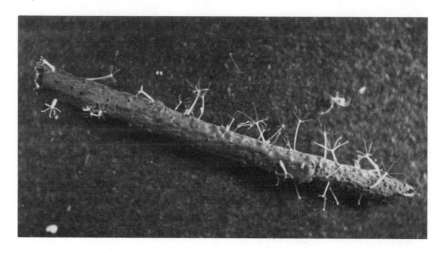

Figure 15.2   A small bit of submerged twig, colonized by *Hydra americana* Hyman. (Photograph by Helen Forrest.)

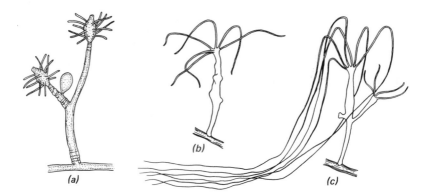

Figure 15.3   Freshwater coelenterates. *(a) Cordylophora lacustris* Allman; *(b) Hydra americana* Hyman; *(c) Hydra obligactis* (Pallas), the brown hydra.

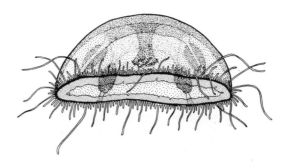

Figure 15.4   The medusa of *Craspedacusta sowerbii* Lankester, the freshwater jellyfish.

cells can be removed from the hydra by glycerin treatment and the alga continues to exist. Similarly, the hydra can survive without the presence of the alga. The other hydrozoans of fresh water include a single species of "jelly fish" (*Craspedacusta sowerbii*), which appears to have been introduced into North America from China or other regions of the Far East. Its distribution in North America is not fully known, for it is sporadic in occurrence, sometimes not being seen for years in a locality where it is known to have once been present. The medusa of *Craspedacusta* (Figure 15.4) is generally about 15 mm in diameter and bears numerous tentacles. The hydroid is small and often difficult to find, because it is usually covered with silt or other debris. One species of a colonial polyp, *Cordylophora lacustris* (Figure 15.3), is widely distributed in fresh waters and in low-salinity regions of estuaries. Two types of polyps occur in the colony, a globular reproductive member which produces sperm or eggs, and a feeding polyp bearing tentacles. *Protohydra leuckarti* is a small hydra-like hydroid no larger than about 5 mm, and lacking tentacles. It typically inhabits estuaries, being known at present from localities along the northern Atlantic coasts of North America and Europe.

Nearly 10,000 species of coelenterates are known from marine waters. Many of these fall within our estuarine province to varying extents, being limited by currents and salinity. These include a number of colonial hydroids, *Cordylophora* for example, and some anthozoans, particularly the soft-bodied sea anemones. Medusoid jellyfishes, such as the sea nettle *(Dactylometra quinquecirrha)* and the white sea jelly *(Aurelia aurita),* invade estuaries, sometimes being observed in salinity as low as 16‰. The anemone *Aiptasia* and coral *(Astrangia)* occur in the more saline zones of some estuaries, in Texas, for example. The upstream distribution of sessile, attached hydroids and anthozoans (and probably other animals as well) in estuaries is generally limited more by water level and salinity fluctuations during tidal cycles than by actual salinity. Given a comparatively "quiet" estuary, we should expect further penetration of the more strictly marine species.

Marine coelenterates are many and varied. The so-called "air fern" is a common marine hydroid, *Sertularia*. Sea anemones include the large green flowerlike *Anthopleura* of the Pacific coast and the smaller *Metridium* along the northeastern coast. *Cyanea artica* is the largest known jellyfish, with tentacles reaching a length of 30 m; another jellyfish, *Physalia* (the Portuguese man-of-war) is smaller, but generally regarded as one of the most deadly. *Physalia* and *Velella* float abundantly in warm seas with tentacles hanging, the "sails" protruding a bit into the air.

Ecologically, the marine corals (Figure 15.5) are divided into two functional categories, the **hermatypic** and the **ahermatypic** corals. The hermatypic corals, mostly reef species, have endodermal tissue with symbiotic chlorophyllous alga, the **zooxanthellae**. The ahermatypic, or "deep-sea", corals lack zooxanthellae. The latter have a far wider range of distribution (Wells, 1957). The great concentration of corals is, of course, in the coral reefs and atolls, where a single atoll may contain more than 150 different species. Corals are colonial forms and include the stony corals (Madreporaria) such as staghorn coral (Madrepora) as well as such soft corals as dead-man's fingers (Alcyonium), common on coral reefs. The gorgonians, or horny corals, include sea fans *(Gorgonia)* and ostrich-plume coral *(Aglaophenia).*

In the over-all economy of natural waters, coelenterates are, in the main, active predators. Many of the anemones are highly carnivorous, feeding upon fishes that venture too close to the coelenterate's tentacles. In the laboratory, hydras are regularly fed planktonic crustacea; under natural conditions, however, such organisms are seldom found in the digestive cavity.

PHYLUM CTENOPHORA (COMB JELLIES)

Of the nearly 100 known species of this group, all are primarily marine. Some, however, do penetrate estuaries. In certain Texas bays, *Beroe* occurs abundantly during summer months in a salinity of 10 to 12‰. Ctenophores are generally small, nearly transparent, ovoid animals containing a large amount of gelatinous material in the body layers. They possess a pair of tenacles, which lack stinging cells, and eight rows of "comb plates," extending from pole to pole, which are used in locomotion. A number of species exhibit biolumines-cence. Ctenophores may occur in great numbers in coastal ponds and embay-ments, where they nearly deplete the phytoplankton stock. *Pleurobrachia* and *Idyia* are especially common in coastal waters of the northeastern United States.

PHYLUM PLATYHELMINTHES (FLATWORMS)

This phylum includes a large number of free-living and parasitic species characterized by dorso-ventrally flattened bodies showing greater differentiation and specialization of parts than do the preceding groups. The basic plan of con-struction is similar to that of the coelenterates in that a single opening leads to a digestive cavity. Three classes are usually recognized, and at least certain

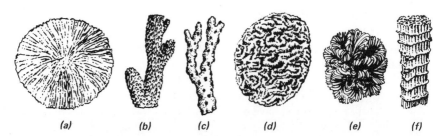

(a)   (b)   (c)   (d)   (e)   (f)

Figure 15.5 Representative coral types: *(a)* solitary coral, order Madreporaria; *(b through e)* colonial stony corals, order Madreporaria; *(f)* Horny coral, order Gorgonacea.

representatives of each are familiar to students of introductory biology. These groups include the free-living flatworms of the class Turbellaria, and the flukes and tapeworms, the last two being parasitic. Although parasitism is of great interest and importance for its own sake as well as in the health of a community, our interest here is in the animals which range freely in aquatic situations. Therefore, emphasis will be placed mainly on the turbellarians.

Turbellarians are flattened or cylindrically shaped forms found widely in marine and freshwater habitats. These animals "creep" over bottoms and submerged objects by means of cilia and subtle body undulations. Some marine forms, such as *Leptodoplana*, may reach a size of 40 mm; others are microscopic. Dorsal coloration of turbellarians is highly varied, but usually dark and often changeable to blend with the habitat. The ventral surface is typically creamy or grayish. Some species are green, due to the presence of zoochlorellae. Turbellarians are characteristically carnivorous, feeding on small organisms by means of a highly extensible pharynx, which is extruded through a ventrally located mouth.

Five orders have usually been recognized in the class Turbellaria, although in recent years there has been a tendency to promote some of the subordinate groups to higher rank. The Acoela are marine, usually inhabiting intertidal zones. These flatworms lack a hollow gut and muscular pharynx. Most acoels are distinctly flattened, and some, such as *Convoluta,* contain algal cells. The Polycladida are marine forms with thin, leaf-like bodies. The gastrovascular cavity is highly branched and a sucker is present on some species. One group of polyclads is of especial biological interest in that a planktonic, ciliated larval stage is included in the life history of the species. This would seemingly be of adaptive value in aiding in dispersal of these types. The three remaining orders, Tricladida, Alloeocoela, and Rhabdocoela, include both freshwater and marine forms, and some which are **euryhaline,** or widely salt-tolerant. In the triclads, the gastrovascular cavity is hollow and three-branched, one branch passing anteriorly and two directed posterolaterally. This group includes the more familiar "planarians," among which *Dugesia tigrina* is probably the most widespread in the United States. Triclads, generally, are stream-adapted forms. Some, however, such as *Procotyla fluviatilis,* are found in a variety of habitats ranging from streams and ponds to brackish waters. The Alloeocoela and Rhabdocoela posses a nonbranched intestine, the two groups being distinguished on the basis of pharynx construction. These forms are typically more common in standing waters than in streams. One rhabdocoel species, *Gyratrix hermaphroditus,* is known from freshwater, estuarine, and marine habitats. Another rhabdocoel, *Microstomum,* feeds on hydra and in the process obtains stinging cells which the worm uses to its own advantage. The bodies of *Gyratrix hermaphroditus* and other worms such as *Stenostomum* and *Catenula* are composed of chains of segment-like zooids associated with asexual reproduction. Sexual reproduction occurs in many turbellarians, often followed by the formation of a cocoon around the eggs. The cocoon is stalked and attached to a submerged object in the habitat.

PHYLUM NEMERTEA (PROBOSCIS WORMS)

Proboscis worms are unsegmented, slender, often highly colored animals of fresh and salt waters. Nemertea bear a general resemblance to flatworms. Unlike the flatworms, however, these animals possess a true digestive tract, with mouth and anus. The most distinctive feature of the group is a protrusible proboscis, which is extended to grab prey. Some species are equipped with a hypodermic-like **stylet** and glands which secrete a toxin into the puncture made by the stylet. The Nemertea of fresh waters are relatively small, seldom attaining a length greater than 20 mm, and are greatly contractile. At present, only one species is known for the United States. This is *Prostoma rubrum,* which inhabits plant surfaces and bottom debris from coast to coast, but which is seldom collected.

About 500 species of nemerteans are known from marine communities. Many are common inhabitants of shells, rocks, and bottom muds of estuaries. Others are free-swimming components of plankton. The marine and estuarine species of proboscis worms exhibit greater size variations than their freshwater kin. *Cerebratulus lacteus* is probably the largest of the American shallow-water forms, often reaching a width of 25 mm and a length of over 6 m. A European species may be over 30 m in length.

As suggested by the food-capturing adaptations, nemerteans are carnivorous. They feed upon a variety of organisms and, in turn, enter into the diet of various worms, fishes, and other larger predators. Special habitats such as oyster reefs, mussel assemblages, and algal mats characteristically support proboscis worms as conspicuous members of the local communities.

PHYLUM NEMATODA (ROUNDWORMS)

Nematodes are small, cylindrical worms with tapered ends, a chitinous cuticle, and straight digestive tract from mouth to anus. Locomotion is by vigorous, whip-like thrashing of the body. Although a great majority of these animals is parasitic, the free-living forms occur wherever organic substances are present, and usually in tremendous numbers. A single spadeful of humus soil or cupful of lake, stream, or estuary bottom will contain thousands of individuals. In spite of their great abundance, aquatic nematodes have been studied but little.

These worms, in general, play an important role in the turnover of organic matter in aquatic communities. Many cause disease and ultimately death of plants and animals. The diversified feeding adaptations of the free-living forms include detritus utilization, predation upon other living animals, and plant consumption. With respect to habitat restriction, only a few genera, such as *Chronogaster* and *Tripyla,* are found exclusively in fresh waters. The genus *Haplectus* has representatives in fresh, brackish, and marine habitats. Of ecological and evolutionary interest is the fact that a peculiarly estuarine, or brackish-water fauna of nematodes apparently has not developed.

PHYLUM NEMATOMORPHA (HORSEHAIR WORMS)

As indicated by their common name, the Nematomorpha are long (up to 1 m) and slender, thereby resembling horsehair. In fresh waters, the adults are usually most common in summer in pools or slow-moving brooks. The marine forms typically swim in a rapid writhing fashion near the surface. The developmental stages are spent as parasites of certain arthropods or mollusks. The adults do not feed.

PHYLUM GASTROTRICHA

The gastrotrichs are microscopic animals ranging in length to somewhat over 500μ. All are more or less ciliated ventrally, and some possess conspicuous spinous processes; others are more or less naked, with but a few spines. The internal morphology of some has been likened to that of rotifers (see below), and others have been compared with nematodes. Locomotion is produced by the beating of cilia and the action of spines.

Of some 400 known species, most are freshwater forms. However, little work has been done on this group and it would not, therefore, be proper to say that gastrotrichs are predominantly of one environment or another. The genus *Dasyatis* (Figure 15.6) is one of a half dozen genera restricted to fresh water. A few genera, including *Chaetonotus* (Figure 15.6), have representatives in both fresh and marine habitats. Within their environment, gastrotrichs are usually closely associated with debris or submerged objects. Some are occasionally taken with plankton. Quiet pools, sphagnum bogs, and shallow zones of lakes are their

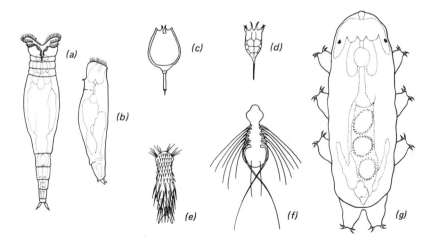

Figure 15.6   Rotifers: *(a) Philodina; (b) Proales; (c) Monostyla;* and *(d) Keratella.* Gastrotrichs: *(e) Chaetonotus* and *(f) Dasydytes.* Tardigrade: *Macrobiotus (g).*

most common habitats. Few gastrotrichs have been found in rapidly flowing streams. Class Kinorhyncha (Echinodera) comprises some 30 species of small marine sediment-dwelling worms in shallow to deep seas. They are up to 5 mm in length, with 13 or 14 rings (Storer, 1951). Class Priapulida (Priapulids) includes cylindrical yellow or brown wormlike marine species feeding on algae and detritus in shallow-water sediments. The food of gastrotrichs consists mainly of smaller organisms such as bacteria, algae, and protozoans.

Reproduction and sexual development among gastrotrichs is of more than usual interest. A number of marine species are hermaphroditic, the opening of the female structures being posterior to that of the male, both being situated on the ventral surface of the animal. Males are unknown in freshwater forms, the females being produced parthenogenetically, *i.e.*, without fertilization.

PHYLUM ROTIFERA (ROTIFERS)

The Rotifera are generally microscopic animals possessing conspicuous organ-systems and characterized by an anterior corona of cilia surrounding a mouth. The cilia aid in feeding and swimming. Posteriorly, most rotifers possess toe-like appendages and glands which secrete a cementing substance, these adaptations serving to anchor the animal or aid in a creeping locomotion. In many species the outer body covering, or cuticle, is stiffened to form an armor-like lorica. The extent of development and the form of the lorica vary greatly and are useful taxonomic features.

Sexuality and reproduction in rotifers are similar to those of gastrotrichs, this similarity suggesting close phylogenetic relationships of the two groups. In certain of the rotifers, males are unknown, the populations consisting entirely of parthenogenetic females. In other groups, certain types of females produce eggs which develop into degenerate males whose sole purpose appears to be that of fertilizing ''special'' eggs. These eggs are highly resistant to adverse environmental conditions and thus represent an evolutionary adaptation of considerable survival value.

The distribution and ecology of rotifers also have interesting evolutionary implications. Of the more than 1500 species, over 90 per cent are freshwater inhabitants. This suggests a freshwater origin of the group. Rotifers have undergone considerable evolution and adaptive radiation, resulting not only in morphological differentiation (Figure 15.6), but also in their occupancy of a great variety of habitats and their filling of a number of roles in the economy of fresh waters. Many species are cosmopolitan, their occurrence being nearly worldwide in suitable habitats. Within a body of water, various species are found in the sediments of shallow shore-zones and in the bottom deposits of deep water. Many are typically planktonic; others inhabit masses of vegetation, certain species showing close affinities for particular plants. A few rotifers are ectoparasitic, and some have entered into other symbiotic relationships. Feeding habits are similarly varied; some species are carnivorous, others utilize the sap of algal cells, and many are detritus feeders or wholly omnivorous.

PHYLUM BRYOZOA (MOSS ANIMALS)

Bryozoans are found in both fresh and salt water, but predominantly in the latter. The common name is derived from the fact that these forms typically occur as colonies encrusting objects in the water. In fresh waters, *Pectinatella* grows as a large gelatinous mass, while colonies of *Plumatella* and *Fredricella* (Figure 15.7) are loosely developed and mat-like. *Membranipora,* a cosmopolitan estuarine form, grows as calcareous encrusting mats on submersed structures. Many marine bryozoans, *Bugula* for example, resemble large ferns; others may be coral-like, possessing a rigid framework of carbonates or chitinous substance. The basic unit of the colony is the individual animal, or zooid. The mouth of a zooid is bordered by ciliated tentacles which create currents and bring in food. The digestive tract of these animals is a tube bent upon itself, with the result that the mouth and anus lie close together. Sexual reproduction by hermaphroditic individuals occurs for a short time each year. Asexual reproduction by budding contributes to the increase in size of the colonies. In addition to budding certain freshwater species produce asexually formed bodies (statoblasts) containing germinative tissue. These bodies are constructed of a tough chitin-like material and are very resistant to desiccation. One type, called a spinoblast, is produced by *Pectinatella* and is shown in Figure 15.7. Other statoblasts may lack spines or hooks and are termed floatoblasts. Both spinoblasts and floatoblasts are free-floating and often taken in plankton collections. A third type, such as that produced by *Fredricella* and *Plumatella*, is attached to the colony; these are sessoblasts.

Freshwater bryozoans occur in a variety of habitats, but seldom in polluted waters. Highly turbid and acid waters also inhibit colony development. Summer is the favorable period for growth of these animals. During this season, bryozoans commonly develop on submersed objects in shallow shore zones of lakes and ponds, and in both slow and rapid waters in streams. Estuarine forms are

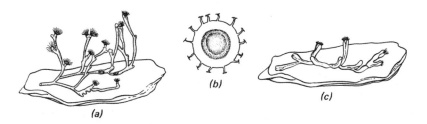

(a) *(b)* *(c)*

Figure 15.7   Common bryozoans of fresh waters: *Fredricella sultana* (Blumenback) *(a)* the statoblast of *Pectinatella magnificata* Leidy *(b)*; and *Plumatella repens* L. *(c)*.

particularly attracted to pilings, mollusk shells, and rocks. Many species, especially those forming statoblasts, are found worldwide in favorable habitats. Such widespread distribution is due in part to transport of the statoblasts by many agents, including waterfowl. Marine bryozoans include the common genera *Alcyonidium* and *Bugula*.

At times, some of the bryozoans enter prominently into human economy. Large colonies of *Pectinatella* may grow on intake pipes and screens of water-supply plants and in time clog the pipes, demanding manual removal. In estuaries and marine communities, bryozoans contribute to fouling of boat hulls and pier pilings; efforts to control this condition also involve labor and expense.

PHYLUM CHAETOGNATHA (ARROWWORMS)

In terms of numbers of species, this is a relatively small phylum; about 30 species, all marine, are known. At certain times, however, the abundance of individuals places the group high in importance in energy relationships in marine waters, for seasonally at given localities these animals form a major item in the diet of fishes. Chaetognaths are slender, transparent animals with lateral and caudal fins, usually no longer than about 40 mm. The head bears several curved spines which serve to seize prey consisting of plankton, small fishes, and, frequently, other chaetognaths. Being restricted to generally high salinity, these forms probably inhabit positive estuaries no farther than the lowermost reaches. The genera *Sagitta* and *Krohnitta* are cosmpolitan in distribution. They are small (about 25 mm long) and are common in all seas, where they are both planktonic and benthic.

PHYLUM ANNELIDA (SEGMENTED WORMS)

Annelids are typically cylindrical, elongate animals with serially segmented bodies, bearing a mouth at one end and an anus at the other. Although the terrestrial earthworm is the most familiar annelid, the greatest variety and abundance of species are found in estuaries and the marine environment. Of four recognized classes, only three need receive our attention in this synopsis; these are: Polychaeta, the sandworms; Oligochaeta, the earthworms; and Hirudinea, the leeches.

CLASS POLYCHAETA

This class consists mostly of marine animals characterized by the presence of serially arranged pairs of fleshy appendages (parapodia) usually bearing numerous stiff bristles (setae). The surface of the parapodia serves as a gaseous exchange region in "breathing"; the beating of the appendages moves water across the surfaces and also aids in locomotion. Although mostly small, some species of the genus *Nereis* may attain 50 cm in length. Many polychaetes inhabit burrows in the substrate, others build tubes in which the animal lives permanently, and some species are free-swimming. In the estuary and the sea, a number of species typically inhabit shells of living mollusks. Sexual reproduction predominates in this group, and the sexes are separate. Breeding takes place seasonally, and in many species is marked by precisely timed periodicity. The life history of polychaetes involves an unsegmented larva (the trochophore) which metamorphoses into an adult.

Of the more than 3000 known species of polychaetes, less than 20 are typically freshwater forms. A number of families are represented in fresh waters, suggesting their separate evolution and invasion of a different environment. In North America, less than a dozen species, mostly of the family Nereidae, inhabit fresh waters. Most of these are euryhaline species distributed in estuaries and in their freshwater tributaries. *Neanthes succinea* is found in estuarine and nearly fresh environments on both coasts of temperate North America. The freshwater genus *Manayunka* (family Sabellidae) has representatives in streams in Pennsylvania and New Jersey, and in Lake Erie and Lake Superior. Notorious as marine fouling organisms, the family Serpulidae is represented in fresh waters by a species found in a California lake and in streams entering the western Gulf of Mexico. Of special interest is the occurrence of a sabellid species in Lake Baikal, Russia. This is a very ancient lake and is over a thousand miles from salt water.

Polychaetes are characteristically more abundant in marine and estuarine environments than in freshwater communities. In the zone of intermediate salinity in the Elbe Estuary polychaetes were found at a density of 25 per square meter. This is to be compared with zero for the freshwater zone and $19,000/m^2$ in the seawater region of the same stream. Polychaetes contribute to the metabolism of the estuarine community in several ways. Being active feeders, in large numbers they consume considerable quantities of organic substance. Many of them are integral items in the diet of larger organisms; certain species of fishes, for example, consume great stores of polychaetes. The continual workings of the burrowing forms turn over much material of the sediments. These and other animals join with various plants to produce dense local associations contributing to productivity generally.

CLASS OLIGOCHAETA

The characteristics of the class Oligochaeta, as typified by the earthworm, are familiar to all and need not be considered here. The structure of aquatic oligochaetes deviates little from that of earthworms. In contrast to polychaete reproduction, however, some of the aquatic oligochaetes generally reproduce by asexual budding; this is true of the Naididae and Aeolosomatidae. The other oligochaetes engage in sexual reproduction involving cross-fertilization between two hermaphroditic individuals. Direct development from eggs and embryos contained in cocoons attached to submerged objects follows.

Also in contrast to polychaete adaptations, nearly all of the aquatic oligochaetes are inhabitants of fresh waters. Freshwater species occur widely in all types of aquatic conditions, but the greatest abundance is usually associated with organically rich substrates. The family Tubificidae is distinguished by a number of species adapted to existence in estuaries and in near-anaerobic sediments in deep lakes. *Tubifex tubifex,* a red worm, found world-wide, reaches greatest population densities in highly polluted waters. A number of tubificid species are tolerant of the low oxygen concentrations present during winter stagnation in lakes. Many oligochaetes, such as *Tubifex* and the naid genera *Dero* and *Nais,* construct mud tubes for habitation. The construction of tubes and the burrowing by other forms contribute to the overturn of sediments, indicating that one ecological role of aquatic worms is similar to that of their terrestrial counterparts.

CLASS HIRUDINEA

This class includes the leeches, found primarily in fresh waters, although a few marine and terrestrial species are known. These annelids are flattened, and usually possess sucking devices which aid in locomotion and in attachment while eating. Some members of this group are ectoparasites, others feed on detritus, and still others are carnivorous. Leeches reach maximum abundance in standing waters containing large amounts of debris.

The phylum Annelida presents an interesting picture of evolution and ecological and physiological adapation with respect to the freshwater–estuary—marine series. As we have seen, the density per unit area of individuals of marine species of polychaetes is much higher than that of estuarine forms. Conversely, the density of oligochaetes decreases from fresh water through the estuary to the marine. It would appear that physiological adaptations of each group to environmental salinity have been slow to change. This could be the entire answer. On the other hand, ecological adaptations may be important. Even if large numbers of species should migrate, they might be unable to successfully colonize because their particular roles and habitats might be already filled. The answer to the problem is not fully known. Certainly laboratory experiments have shown that many present-day forms can not transgress the salinity barrier, but the tests tell us little with respect to the possibility of change over the many millions of years that these forms have been in existence.

PHYLUM TARDIGRADA (WATER BEARS)

This is a relatively small phylum of animals of ''uncertain systematic position.'' Some authors include the forms with the arthropods, others consider the water bears to be more closely related to annelids, and still others simply place the animals in a separate phylum. Tardigrades are small, typically less than 500μ in length. As shown in Figure 15.6, these are "well-formed" organisms possessing a distinct head, usually with eyes, four pairs of legs with claws, and a segmented body. Internally, a brain and ventral nerve cords coordinate locomotion and other functions. The digestive and reproductive systems appear well formed, but respiratory and circulatory systems are absent. Water bears develop from eggs laid by a female and fertilized beneath a shedding cuticle by males. Cuticles are shed several times during the life of the animal. In many species, individuals are capable of contracting into a small mass inside the cuticle and surviving many years of adverse conditions of dryness and low temperature.

About 350 species are known to inhabit aquatic and semiaquatic habitats; the latter includes damp masses of algae, mosses, and sand. Only a few species are marine.

PHYLUM ARTHROPODA (JOINT-LEGGED ANIMALS)

This phylum is said to be the most successful (in terms of evolution and adaptation to numerous environments) and the largest (in respect to numbers of species). Representatives are found everywhere that life can be supported, and rather more than a million species are known. The classes with aquatic representatives are: Crustacea, Insecta, and Arachnida. Our survey of the arthropods here must be brief and quite general, with attention being directed toward the less familiar but ecologically important types. For a deeper appreciation for the great variety of these animals, consult the several references suggested at the end of this chapter.

CLASS CRUSTACEA

The larger members of this class, such as the crabs, lobsters, shrimp, crayfish, and barnacles, are quite familiar. Less commonplace, however, by virtue of their size, are the myriads of crustacean species which occur in the plankton of fresh and marine waters. The tremendous importance of these animals rests upon their indispensable position in the food web and energy relationships in lakes, streams, estuaries, and the sea. An important concept to bear in mind is that for practically every major type found in fresh waters there is an ecological counterpart, often similar in appearance, inhabiting the sea; many of these meet in the estuary.

Four of the crustacean orders are sometimes called the branchiopods. These include superficially different groups of small animals. Three of these, the fairy shrimps, tadpole shrimps, and clam shrimps, are entirely freshwater forms typically found in small ponds or temporary puddles and pools; in summer even a water-filled cow-track may harbor a population of one or more of these. The eggs are quite resistant and are disseminated by a number of agents, but primarily by wind. The fairy shrimps (Figure 15.8) generally are elongate forms, usually about 20 to 30 mm in length, which swim slowly and gracefully on their backs. One species of fairy shrimp, *Artemia salina,* the brine shrimp, is found throughout the world; in the United States it occurs in abundance in Great Salt Lake. The clam shrimps give the appearance of being fairy shrimps bent and enclosed between bivalve shells. Some of these attain a length of about 16 mm, although most are smaller. The tadpole shrimp is characterized by an arched carapace over much of the body; these branchiopods have not been found east of the Mississippi River.

The fourth group of the branchiopods includes the water fleas of the order Cladocera (Figure 15.8). These are small (up to 3 mm in length) crustaceans found in all types of fresh waters; a few species occur in the estuary and the sea. Over 130 species are known in the United States. The animal is contained within a folded bivalve carapace. In a number of species the posterior of the carapace is produced as a long spine. The surface of the carapace may exhibit fine etchings,

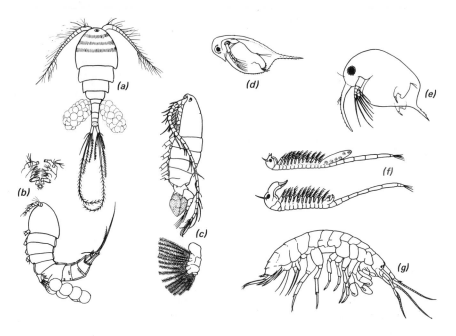

Figure 15.8   Some representative freshwater crustaceans: *(a)* Cyclopoid copepod, *Macrocyclops ater* (Herrick) with two egg sacs; *(b)* Harpacticoid copepod, *Moraria virginiana* Carter, with single egg sac and antenna approximately the same length as first body segment; *(c)* Calanoid copepod, *Diaptomus birgei* Marsh, with one egg sac and antenna about same length as body; *(d)* and *(e)* Cladocerans, *Daphnia* and *Bosmina; (f)* fairy shrimp, *Branchinecta*—female above, male below; *(g)* Amphipod, *Gammarus.* Also shown, scraping mouthparts *(1)* of harpacticoid copepod and filtering mouthparts *(2)* of calanoid copepod.

or *striae*. A large compound eye is conspicuous on the head. Also in the head region is a pair of antennae used in locomotion. Under favorable environmental conditions cladoceran populations consist mostly of females produced parthenogenetically from eggs carried in the upper part of the carapace. As unfavorable conditions arise, some of the eggs develop into males and, simultaneously, a different type of female is also produced. When she is fertilized, her carapace thickens and forms a protective **ephippium,** or extra (outer) shell, about one or two eggs. With the following molt, the ephippium is shed and becomes sealed. Thus, the embryo is able to withstand the unfavorable times, and upon return of proper environment the developing animal(s) are released. This adaptation serves to ensure repopulation of waters following periods of drought or other unfavorable circumstances.

A most interesting and noteworthy aspect of cladoceran biology involves seasonal changes in the body form of several species (Figure 12.5) in some geographical areas. During the warm months, *Bosmina coregoni* becomes noticeably humpbacked and the rostrum greatly elongates; *Daphnia cucullata, D. longispina,* and *D. retrocurva* are marked by the development of a conspicuous "helmet" in summer. Termed **cyclomorphosis,** this phenomenon has not been fully explained. It was early thought that such enlargements served to increase surface area as an aid in flotation during times of lowered viscosity of water. Nutrition, temperature during development, and various physiological conditions have also been suggested to account for cyclomorphosis. It has recently been found that circulation of water might influence the development of helmets in *Daphnia cucullata,* the suggestion being that the produced helmet is an adaptation to resistance of water current.

Within a given lake, certain groups of cladocerans exhibit affinities for a particular set of environmental conditions. Nearly all species of the family Chydoridae are found in stable embayments, or "backwaters." Such genera as *Bosmina, Diaphanosoma,* and *Daphnia* are typically found in the limnetic, or open-water, region of the lake. Others are associated with the mud zone, and a number of forms inhabit various intermediate zones.

Cladocerans are worthy of much attention because of their tremendous importance in food and energy relationships, especially in lakes and ponds. All carnivorous fishes pass through an early growth stage in which zooplankton is the major source of food. From one third to more than one half of the zooplankton in the vegetated littoral zone, a favorite haunt of sunfishes and perch, consists of cladocerans. Upon reaching larger size and becoming piscivorous, large fishes are more or less dependent upon the plankters, for the small fishes taken as food usually feed regularly upon the microscopic forms.

The order Ostracoda includes small pea-shaped, bivalved crustaceans generally about 1 mm in length. Of the 2000 known species of ostracods, most are marine and estuarine. The valves of many species are often colored and patterned. Some of both the freshwater and marine species are cosmopolitan and found in a great variety of habitats. In both major environments ostracods are usually most abundant near the surface of bottom muds or among dense vegetation. The development of ostracods following hatching includes a larva known as a **nauplius**, which is often abundant in plankton. Although generally of less importance than other microcrustaceans as food for fishes, ostracods are consumed by the gizzard shad (*Dorosoma cepedianum*), an important forage species in waters of the southern United States.

The crustacean order Copepoda embraces an ecologically important group of small animals widely distributed in freshwater, estuarine, and marine environments. Copepods are mostly less than about 2 mm in length, and, of the 6000 or more species, a majority are marine. A few forms are parasitic. The free-living species fall into three suborders showing relatively distinct body forms, living habits, locomotion, and ecological roles. These groups are: **Calanoida, Cyclopoida,** and **Harpacticoida** (Figure 15.8).

Calanoid copepods are generally characterized by long antennae of 23 to 25 segments extending the length of the animal, and by an elongate **metasome** (anterior body) of generally similar segments. Eggs are carried in one or two closely appended sacs. All of the calanoid families except one have representatives in fresh and marine waters. The genus *Eurytemora* includes several euryhaline species, some of which are the dominant calanoids of estuaries of certain regions. *Acartia* is primarily a marine genus, but some species occur abundantly in estuaries. The family Diaptomidae includes the most common calanoids of standing fresh waters. Calanoids are typically planktonic in the limnetic regions of lakes, *Diaptomus* being especially abundant under certain conditions. Calanoids are also distinguished by their feeding habits in that they are primarily "filter-feeders," consuming small organisms and detritus delivered by currents set up by the antennae. Locomotion in these copepods is typically by swimming.

Cyclopoid copepods possess short antennae (of 6 to 17 segments) seldom more than about one half the length of the metasome; the metasome is generally compressed longitudinally and tapered posteriorly. Eggs are characteristically carried in two sacs attached laterally. Species of *Halicyclops* inhabit mesohaline waters. *Cyclops* is distributed widely in fresh waters. Within a given body of water, cyclopoid copepods are mainly inhabitants of the littoral, although one or two species may be abundant in the limnetic. These forms possess mouth parts adapted for seizing and biting small organisms. Locomotion in cyclopoids is essentially of a leaping nature involving both antennae and legs.

In harpacticoid copepods the antennae are quite short, consisting of no more than nine segments, and seldom longer than the cephalothorax. The body is nearly cylindrical, with little differentiation between anterior and posterior regions. Nearly all genera of North American harpacticoids are represented in fresh, estuarine, and marine waters. In fresh waters, *Canthocamptus* and *Bryocamptus* are probably the most common, being found frequently in bottom debris in both shallow and deep zones. Several genera, including *Tachidus,*

Nitocra, and *Mesochra,* are notably estuarine, some forms being widely distributed. *Nitocra spinipes,* for example, is known from estuaries of northern Europe and North America, being reported from such widely distant localities as Hudson's Bay, Mexico, and Alaska. In all bodies of water, harpacticoids show extremely close affinities for the littoral zone and its cover, being especially common in masses of vegetation, even above water and on beaches. Interstitial waters of beach sands may often harbor these copepods.

Emphasis must be given to the tremendously important place occupied by copepods, generally, in the economy of natural waters. These crustaceans occur in great numbers and consume great quantities of phytoplankton and detritus. Under certain conditions the density of limnetic copepods in lakes may exceed 1000 organisms per liter. As much as 85 to 90 percent of estuarine zooplankton may consist of cyclopoids. In turn, plankton-feeding fishes ingest copepods; a single gizzard shad may contain hundreds of the crustaceans. The great fishing areas of the world are usually in plankton-rich regions, and the zooplankton is largely composed of copepods.

During the course of evolution a few copepods have become ectoparasitic, mainly upon fishes. These include the families Lernaeidae and Argulidae, which are found in both fresh and salt waters; the genus *Argulus* is known from estuaries and fresh waters.

The crustacean order Cirripedia includes the barnacles, all of which are marine or estuarine species. All barnacles are placed in two groups, the true barnacles (Thoracica) and the acorn barnacles (Balanomorpha). Among the former are the stalked barnacles, such as the gooseneck barnacle *Lepas.* The acorn barnacles include common intertidal species such as *Balanus improvisus,* widespread in mesohaline waters, and the rock barnacle *B. balanoides,* which is extremely common in the northeast Atlantic. *Chthalamus* is a high tide barnacle common along warmer coasts. The young of the barnacles are motile, but as development proceeds they become attached to objects in the water, including other animals.

The crustacean subclass Malacostraca includes some twelve orders characterized by having abdominal appendages, eight segments in the thorax, six segments in the abdomen, and by being generally large in size. Many of the malacostracans are bottom dwellers possessing highly muscular pincers and strong, biting mouthparts; these include crayfish, crabs, lobsters, and the like. Others are small, but have similarly developed, diminutive appendages; many of these animals, such as shrimps, are swimming forms, and a few species are conspicuous components of marine plankton. Only about one third of the dozen orders is represented in fresh waters; the remainder are marine, with estuarine species.

The order Mysidacea (opossum shrimps) includes about 300 mostly marine species of small, shrimp-like crustaceans usually found in cool waters. Only one wholly freshwater species is known for the United States; this is *Mysis oculata,* found in cold lakes. Another species, *Neomysis mercedis,* inhabits fresh waters and estuaries of the Pacific slopes from California northward. *Neomysis vulgaris* commonly occurs in estuaries of northern Europe. Where abundant, mysids are taken in considerable numbers by fishes, especially those that feed near or from the bottom where the mysids normally live.

Sowbugs (in the order Isopoda) are flattened malacostracans found in damp terrestrial habitats and in practically all types of aquatic communities. Of the more than 3000 aquatic species only a few are adapted to fresh water. The freshwater species, such as those of the genus *Asellus,* are often very common in certain situations, such as near the littered shores and bottoms of small streams. *Cyathura carinata* and species of the genus *Idotea* occur in estuaries of northern Europe and the Atlantic coast of North America, *C. carinata* being found as far south as Virginia. Isopods are generally nocturnal in their activity. Jetties, piers, and shores are often overrun with the animals at night.

The order Amphipoda includes about 3000 species of laterally compressed crustaceans, commonly called "scuds" or "sandhoppers," ranging to about 20 mm in length (see Figure 15.8). Most of these are marine and estuarine. A number of amphipods, including *Gammarus locusta* and *Corophium volutator,* are known from estuaries of northern Europe and the Atlantic coast of North America. The genus *Gammarus* is represented in freshwater, estuarine, and marine communities, several species being found in a variety of habitats. In thickly vegetated spring runs *Gammarus* may occur in densities of several thousand per square meter (see Figure 16.7). A common freshwater amphipod is *Hyallela azteca;* in southern lakes and streams, the dense hair-like roots of a single water hyacinth (*Eichornia crassipes*) may harbor hundreds of *Hyallela.* Amphipods are eaten in great numbers by many fishes. Some amphipods serve as intermediate hosts in the life cycle of many parasites. *Hyallela,* for example, harbors a stage of the acanthocephalan, *Leptorhyncoides thecatus,* which has its adult stage in a vertebrate animal.

The order Euphausaicea includes the "krill" *Euphausia,* a type of shrimp which reaches a length of about 10 cm and can be very abundant in cooler oceans at intermediate depths. Krill forms a favored food of several important species of whale and seal.

Familiar to all are the shrimps, crabs, crayfishes, and other crustaceans of the order Decapoda. Of some 8000 known species, most are marine and estuarine. The freshwater representatives include only the crayfishes and some freshwater shrimps.

Various habitat affinities are exhibited by the crayfishes. *Orconectes propinquus* and *Cambarus longulus* are typically stream inhabitants. *Cambarus diogenes* is a burrowing form found from New Jersey to Texas and the Great Lakes. *Procambrus blandingi* occurs widely in various habitats from New England to Mexico. Underground waters and caves often contain crayfishes; *Orconectes pellucidus* inhabits such situations from Indiana to Alabama. There appear to be no species particularly characteristic of lakes.

The freshwater shrimps are represented by the family *Palaemonidae,* which also has marine members, and by the genus *Macrobrachium,* the large river shrimps. These are characteristically associated with vegetation of debris. In certain lakes of Florida, *Palaemonetes paludosus* occurs abundantly among the roots and tangled plant fragments in the underside of floating islands. In estuaries, members of the Palaemonidae commonly live in patches of grasses growing in shallow water. The river shrimps *(Macrobrachium),* some reaching nearly 200 mm in length, are erratically distributed from Virginia to Texas, and in the Ohio-Mississippi River drainage.

Many species of crabs are found in estuaries, and in southern regions the brown shrimp *(Peneus setiferus)* is especially common. Mesohaline waters are, in fact, important "nursery grounds" for the shrimp, indicating that an early phase of the life cycle must be spent in such zones. Marine forms well known for their commercial value are the shrimp *Crago,* prawn *Peneus,* lobster *Homarus,* spider crab *Libinia,* edible rock crab *Cancer,* and blue or soft-shelled crab *Callinectes.* Of special ecological interest are the hermit crab *Pagurus,* the fiddler crab *Uca,* and the little beach-dwelling sand crab *Emerita.*

CLASS ARACHNOIDEA

The most familiar members of this class are the terrestrial spiders and ticks. The only truly aquatic representatives, however, are the mites of fresh and marine waters, and the king, or horseshoe crabs, and sea spiders (Pycnogonida) of the marine and estuarine environments. The water mites (Hydracarina) are common in a wide variety of freshwater habitats. Hot springs, lakes, ponds, stagnant pools, and rapid streams normally support hydracarinid populations. Some species occur in salt waters. In many instances the populations are rather narrowly restricted to their particular habitats, with the result that each type of community usually has a distinctive fauna. Many species are vigorous, active swimmers; others are retiring and crawl about in masses of debris, or algae, or simply over the substrate. Some of the mites are parasitic. Mostly, however, mites are carnivorous, feeding upon small crustaceans, worms, and other animals. The mites, in turn, are consumed by a number of larger organisms, including fishes. Many hydracarinids are brilliantly colored, often with distinctive patterns. The widespread stream mite, *Limnesia,* is red with black designs on the dorsum. *Diplodontus despiciens,* a pond form found in North America and Europe, is frequently a brilliant red. The marine mites are usually dark colored.

The king crab, or horseshoe crab *(Limulus),* is a bizarre arachnoid. The appendages and body are contained beneath a broadly domed, leathery carapace. *Limulus* is a bottom inhabitant, often burrowing and pushing along through the sediments, uncovering small animals, which it eats. Pycnogonids, sometimes placed in a separate class, are found along the costs in vegetation, and in the deep sea. In the latter habitat the animals may have a spread of several feet.

CLASS INSECTA

The number of North American insects known to spend part, or all, of their lives in fresh waters exceed 5000 species. Included in this figure is a great array of forms exhibiting a remarkable variety of adaptations to all types of fresh waters. Although the number of strongly marine forms is inconsequential, many species, derived from fresh waters, inhabit estuaries, the nature of the fauna depending largely upon salinity. In our synopsis here we can give only cursory consideration to the orders represented in aquatic communities.

On the basis of development from egg to adult, insects can be classified into essentially two categories: one group in which metamorphosis is gradual or incomplete, and another in which metamorphosis is complete (egg→larva→pupa→adult). In the first category the nymph resembles the adult, at least superficially, and changes in form are gradual. This group includes the stoneflies, dragonflies, mayflies, and true bugs. In the second category are the forms which pass through abrupt stages, including a worm-like and a pupate form, before reaching adulthood. Insects in this group include the alderflies, dobsonflies, spongeflies, caddisflies, moths and butterflies, beetles, and the two-winged flies. These developmental and physiological adaptations are of ecological interest because they pertain to seasonal occurrence and abundance of insect forms in the food and energy relationships within a given community.

Stonefly nymphs (Plecoptera) are common inhabitants of swift, cool streams and the shores of temperate lakes. They are rather sluggish creatures, usually associated with stones or aquatic vegetation. The series of nymphal stages may persist for several years, followed by mass emergence, usually in early summer. Stonefly nymphs superficially resemble those of mayflies (to be considered subsequently), but differ from the latter in characteristically possessing only two tail appendages and lacking tracheal gills on the abdomen. Some of

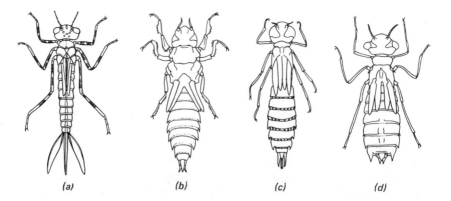

(a)　　　　(b)　　　　(c)　　　　(d)

Figure 15.9　Some Odonata nymphs of North America fresh waters: *(a) Argia emma* Kennedy, a damselfly. All others are dragonflies: *Progomphus obscuris* Rambur; *(c) Anax junius* Drury; and *(d) Pachydiplax longipennis* Burmeister.

the stoneflies (Perlidae for example) are carnivorous; others are plant and detritus feeders. Larger organisms, including stream fishes, feed on stoneflies.

Damselflies and dragonflies of the order Odonata are represented in fresh waters by nymphal stages. These forms are recognized by the high degree of development of the labium (lower lip) into a mask-like seizing device. Odonate nymphs are quite predaceous, feeding on a variety of animals, often of considerable size. These insects are widely distributed in most types of fresh waters. A number of species, such as *Anax junius* and *Pachydiplax longipennis* (Figure 15.9), are typically inhabitants of ponds and sluggish stream waters. *Erpetogomphus designatus* is primarily a stream inhabitant. In the estuary of Weekiwachee Springs, Florida, a damselfly, *Enallagma durum,* reaches maximum population density in a zone in which chloride concentrations approach 5.5‰. In many ponds and streams odonate nymphs constitute a standard item in the diet of some fishes.

The nymphs of mayflies (Ephemeroptera) are common inhabitants of most streams and standing waters containing sufficient oxygen. At least one species inhabits brackish waters. These animals are readily recognized by the presence of tracheal gills attached to the abdomen and two or three caudal filaments (Figure 15.10). Although the series of nymphal stages may last for several years, the

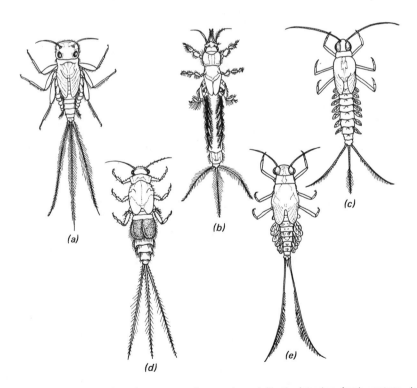

Figure 15.10   Nymphs of some mayfly species of North America fresh waters: *(a) Stenonema exiguum* Traver; *(b) Hexagenia munda marilandica* Traver; *(c) Baetis spiethi* Berner; *(d) Caenis diminuta* Walker; *(e) Pseudocloeon alachua* Berner.

adults live but a few days at the most, that time being spent in reproduction. Mayfly nymphs are variously adapted for living in lakes or streams. In streams, some species make their way freely over the bottom, others inhabit debris and masses of plant material, and still others are strongly associated with submerged objects in swift water. Lakes contain burrowing types and bottom dwellers. *Hexagenia* (Figure 15.10) and *Ephemera* are burrowing forms widely distributed in North America. *Stenonema* is usually found wherever there is some movement of water, often along lake shores. *Baetis* is a widespread stream inhabitant. *Caenis* is a common bottom form in lakes and ponds. *Pseudocloeon* occurs widely in moving water. Mayfly nymphs are primarily herbivorous in their feeding habits, and thus form a short "chain" to many fish species which feed upon the insects.

The true bugs (Hemiptera) are characterized by the modification of the anterior wings into tough horny sheaths and the mouthparts into a rostrum specialized for sucking. Some species are wingless. The Hemiptera exhibit various modifications directed toward inhabiting the shore, the surface film, the bottom sediments, and actively swimming in the water of ponds and streams. Some forms, such as *Trichocorixa, Rheumatobates,* and *Halobates,* occur in (or on) marine waters. A number of species, including the "light bugs" (Belostomatidae) and the water boatmen (Corixidae) are capable of sustained flight and frequently leave the water. Some representative types are show in Figure 15.11.

The larvae of alderflies, dobsonflies, and spongeflies (Neuroptera) are found widely, though usually not abundantly, in lakes, ponds, and streams throughout much of North America. The most familiar, at least to fishermen, is

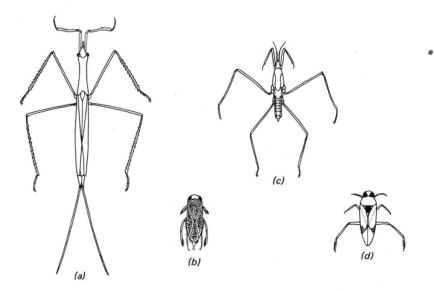

Figure 15.11   Some representative aquatic hemipterans: *(a) Ranatra; (b) Corixa; (c) Gerris; (d) Notenecta.*

the large, stream-dwelling larva of the dobsonfly, *Corydalis*, known as a **hellgrammite**; it may grow to over 80 mm in length. The larvae of the alderfly, *Sialis*, are found in a variety of habitats, including plant debris on lake or stream bottoms. Neuropteran larvae are lively predators, possessing strong mouthparts. Caddisfly (Trichoptera) larvae, or at least their cases, are well known to those who have probed into streams and standing waters, and the adults are familiar to trout fishermen. Many of the larvae build a variety of cases to protect themselves and the pupae of the developing caddisfly (Figure 15.12). These cases are usually in the form of elongate cones or cylinders constructed of cemented sand grains, sticks, leaves, and other materials, the type of case sometimes being characteristic of a particular family. In flowing waters, a number of species build nets between stones, the nets, directed upstream, serving to catch food carried in the current.

The beetle order (Coleoptera) is represented in fresh waters by a variety of forms which have aquatic larvae, terrestrial pupae, and aquatic adult stages. Many aquatic beetles are typically carnivorous, both as larvae and as adults; others, including the Haliplidae, feed on plant material. Many of the species, such as the diving beetle *(Dytiscus)* and some of the scavenger beetles (Hydrophilidae), leave the water on occasion and fly about. An impressive array of feeding, breathing, locomotion, and reproduction adaptations are found in this order. These are too numerous to describe here; however, a concise synopsis of the families is found in the appropriate section in Edmondson (1959).

The order Diptera includes a vast number of insects such as the midges, houseflies, mosquitoes, blackflies, and craneflies. All of the adults of these are aerial or terrestrial. As in the Coleoptera, the larvae of the flies display a wide assortment of morphological and ecological adaptations. Most of the species are

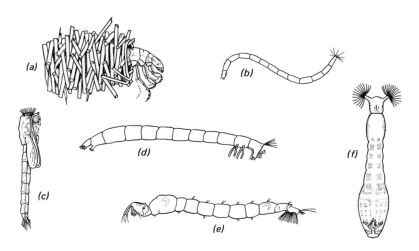

Figure 15.12   Some insects of inland waters: *(a)* the caddisfly *Oecetis; (b)* a dipteran, *Ceratopogon; (c)* the pupa of a dipteran, *Tendipes; (d)* the larva of *Tendipes; (e)* the larva of the dipteran *Chaoborus; (f)* the larva of the blackfly *Simulium.*

found in fresh waters. Some, however, such as the mosquitoes of the family Culicidae, occur abundantly in salt marshes and pools. Some representative dipterans are shown in Figure 15.12.

Of particular interest to us from the general ecological point of view of food relationships are the true midges of the Tendipedidae, the biting midges (Heleidae), and the mosquitoes. Tendipedid larvae are found widely in all types of fresh waters, and many species develop dense populations. Such large populations are possible because the larvae feed primarily on detritus and algae. Tendipedid larvae are small, ranging up to about 25 mm in length, and slender, some possessing a deep red color ("bloodworms") due to a particular blood pigment. Due probably to their size and abundance, these forms are consumed in great quantities by many fishes. *Tendipes (=Chironomus)* and *Calopsectra (=Tanytarsus)* are frequently reported as major items in fish diets. The transparent culicid larva *Chaoborus* is a predaceous form found on organically rich lake and pond bottoms. It is unique in that it can exist without surfacing for air. It does, however, migrate to the upper waters, and during this time it is captured by fishes. Although we have emphasized the importance of dipterous larvae in the food of fishes, we should point out that many pupae are also consumed. In a study of approximately 1000 specimens of the black crappie *(Pomoxis nigro-maculatus)* in Florida, it was found that over 25 per cent of the fishes had eaten diptera pupae, and over one third of the stomachs contained larvae; in the fall, an average of five pupae per fish were recorded. It is of interest to note that of the two truly aquatic insects, one is a marine chironomid of Samoan tide pools and the other is a hemipteran. Of the many other marine insects, none is entirely aquatic (Usinger, 1957).

PHYLUM MOLLUSCA

This phylum comprises over 70,000 species of externally diversified animals. Internally the animals exhibit common features, including a muscular foot, a mantle covering a visceral mass, and a calcium carbonate shell secreted by the mantle. Although a fairly large number of species inhabit fresh waters, the group has been most successful in the marine environment, for there the greatest variety of adaptations is found. Three of the five recognized orders are not found in fresh waters, although a few species enter estuaries. One order includes the small, slightly arched, gliding animals of the bottom, the chitons. Another order is that of the tusk shells, and the third marine order includes such animals as the octopus and squid. A small squid, *Loliguncula brevis,* commonly inhabits mesohaline waters from Chesapeake Bay southward and along the Gulf coast.

CLASS AMPHINEURA

The Amphineura include the chitons, which are animals elongate elliptical in shape with eight calcareous overlapping plates. They creep slowly by a muscular foot and curl up like a pill bug when dislodged from the substrate. They occur mostly on shallow or intertidal rocks, but some are found at depths as great as 4700 m. The chitons feed upon seaweeds and microorganisms scraped from rocks (Storer, 1951).

CLASS GASTROPODA

This class includes the familiar snails and other mollusks possessing a single coiled shell. Gastropods, in general, are widely distributed in fresh, estuarine, and marine waters. More specifically, however, many families, genera, and species exhibit restricted distributional patterns which may reflect the physical and chemical characteristics of the waters of the region. For example, two groups of freshwater snails are recognized on the basis of breathing mechanism: the **pulmonates,** which possess a lung and utilize atmospheric gases, and the **prosobranchs,** which have gills and make use of dissolved gases. Although the pulmonates are essentially inhabitants of fresh waters and the land, a few species are found in the marine environment; *Melamphus coffeus,* for example, occurs in estuaries of the Gulf coast. Similarly, in fresh waters of North America the large family Viviparidae is common east of the Mississippi drainage but is absent to the west. The occurrence and abundance of gastropods are greatly influenced by pH, carbonate concentration, and dissolved oxygen. In addition to these factors, food, of course, is important. Most of the species feed on plant material, particularly algal deposits; others are omnivorous and scavengers. In the food web, snails are consumed in large numbers by certain fishes such as some of the catfishes, sunfishes, and suckers. In estuarine regions of the Florida Gulf coast, the batfish (*Ogcocephalus*) ingest considerable quantities of gastropods, including *Nassarius, Urosalpinx, Bittium,* and *Mitrella.*

CLASS PELECYPODA

The Pelecypoda are the mollusks typically enclosed between two calcareous shells. In our province this includes the clams, mussels, and oysters, most of which are marine. In all situations, pelecypods appear to be most common in regions of relatively stable substrate, free from pollution, and with low silt turbidity. These factors relate, of course, to the fact that many of the animals characteristically lie buried, or partially so, with their siphons protruding into the water above. Through a ventral siphon water is taken in and food extracted; therefore shifting sediments and heavy silt load would be detrimental. In fresh waters the greatest abundance and number of species are typically found in large streams, the most numerous group being the family Unionidae. The Sphaeriidae, particularly the genus *Pisidium,* are often locally abundant. Of particular interest with respect to the estuary are the oyster reefs. The American oyster (*Crassostrea virginica*) builds extensive reefs which become inhabited by a most interesting assemblage of euryhaline plants and animals. Similarly, massive clumps of mussels, such as *Mytilus* or *Brachidontes,* come to harbor more or less characteristic associations, the nature of the associations being largely dependent upon local ecological factors and type of regional fauna.

In fresh waters, a number of kinds of pelecypods are eaten by various fishes and other animals, including muskrats and waterfowl. Man also takes large quantities of certain species for the pearl-button industry. In the estuary, small clams and oysters are preyed upon by fishes, and in many areas the large forms are eaten by starfishes and large crustaceans.

CLASS SCAPHOPODA

The Scaphopoda (tooth shells or tusk shells) are slenderly elongated marine species with tubular shells open at both ends. They live partly buried in sediment at depths ranging from shallow to 4600 m.

CLASS CEPHALOPODA

The cephalopods are the most highly evolved mollusks, all marine species, such as nautilus, squid, and octopus. They have large heads, with two conspicuous eyes, a central mouth, and a number of sucker-bearing arms. Locomotion is by walking and by "jet" action using a siphon. A common squid is *Loligo;* *Nautilus* is the pearly nautilus of the Indo-Pacific Ocean. Some squids are quite long, with museum specimens measuring over 18 m long. A large octopus measures up to 8.5 m across (Storer, 1951).

PHYLUM ECHINODERMATA (SEA CUCUMBERS, STARFISHES, SEA URCHINS)

Echinoderms are radially symmetrical animals possessing a spiny skeleton composed of calcareous plates. Nearly 5000 species are known, all of which are primarily marine. Many species, however, are characteristic of estuaries, the distribution of the forms being determined mainly by salinity. The sea cucumbers are soft-bodied, creeping forms, several of which, such as *Thyone* on the Atlantic coast, and *Thyonacta* on the Gulf coast, enter estuaries. A number of species of starfishes occur in estuaries; some, such as *Asterias,* are commonly associated with oyster or mussel reefs where, as noted above, the echinoderm constitutes a major menace to the mollusks. The disk-shaped sand dollars and the fragile brittle-stars also inhabit mesohaline waters and are fed upon by bottom-dwelling fishes such as flounders. Sea urchins (*Arbacia*) are widely distributed along our coasts and are often conspicuous components of the estuarine fauna.

PHYLUM CHORDATA

The Chordata phylum includes the familiar vertebrates and the less well known "invertebrate chordates," or protochordates. The protochordates possess a nonsegmented "backbone" at some stage of their life, and are essentially marine, with many species found in estuaries but none in fresh waters. Of these primitive forms, the tunicates (Urochordata) are often abundantly represented in estuaries. Ascidians, or "sea squirts," are commonly associated with oyster reefs and other submerged or intertidal objects. Along the Atlantic coast *Molgula* and *Botryllus* occur as members of encrusting communities on buoys, pilings, stones and shells, and other such structures. Cephalochordates, commonly called "Amphioxus," are found on sandy beaches below the high-tide mark. The genus *Branchiostoma* is found on both coasts of the United States, and a southern species, *B. caribaeum*, enters estuaries, at least along the west coast of Florida.

SUBPHYLUM VERTEBRATA

This subphylum covers the vertebrates, which are distinguished, among other things, by the possession of vertebrae. The classes of this great group include the jawless vertebrates, the cartilaginous fishes, the bony fishes, the amphibians, the reptiles, the birds, and the mammals. All of the classes are represented, to various extents, in fresh and marine environments and in the intermediate communities. It is this group that usually comes to mind as that of the highly predaceous animals, representing, as it were, the culmination of the aquatic food chain. Although this is generally true, the over-all ecological picture is complicated by a wide range of size and feeding adaptations among the vertebrates.

Class Agnatha.

This class is represented by the jawless vertebrates, including the hagfishes and lampreys. The hagfishes are wholly marine, usually found in offshore waters. Lampreys occur in both freshwater and marine environments, but the marine forms must enter fresh water to spawn. The sea lamprey *(Petromyzon marinus)* is parasitic as an adult, and has become permanently established in some bodies of fresh water, including the Great Lakes. Some species inhabit inland streams and are not parasitic.

Class Chondrichthyes

The members of this class, the sharks, skates, and rays, are cartilaginous vertebrates which are almost wholly marine. A few species enter estuaries, and one shark, *Carcharhinus nicaraguensis,* is apparently adapted to fresh water in Central America. These are mainly carnivorous animals. In estuaries skates and rays devour mollusks, worms, crustaceans, and other animals.

Class Osteichthyes

The bony fishes, which make up this class, are widely distributed in practically all natural waters. Many are narrowly restricted to fresh waters, others to marine waters, and some from both elements freely inhabit estuaries or move from one extreme to the other. The salmon are anadromous and move from the sea into fresh waters to spawn; eels are catadromous, migrating from inland waters to the sea for reproduction. In inland waters, many species are clearly adapted to standing waters, others to running water. Within a body of water, fishes are generally associated with particular zones such as open water, bottom, or shallow vegetated regions. In the food web, fish species are variously adapted, some obtaining nutrients from organic detritus, others from plankton, and many from predation upon a variety of larger invertebrate animals, and, indeed, other fishes.

Class Amphibia

None of the frogs, toads, and salamanders, members of this class, is truly marine, although the larvae of a few have been found in brackish pools, and adult toads (*Bufo*) and frogs (*Rana*) have been reported in estuaries. Adult frogs and seasonally abundant tadpoles are common to nearly all fresh waters. The tailed amphibians, or salamanders, such as the newts (*Triturus*), are less conspicuous due to their secretive habits. They are, nevertheless, often common in ponds and streams among litter or stones. The larvae of amphibians are essentially scavengers; the adults feed upon insects, worms, and other small organisms.

Class Reptilia

As a class, the reptiles are represented in marine and fresh waters and the intermediate estuary. The Galapagos iguana is apparently the only marine lizard. Snakes, particularly *Natrix,* are common in estuaries within the geographical distribution of the genus. A few snakes are fully marine, being found in the open seas. Crocodilians inhabit fresh waters and estuaries. Turtles, of course, occur in both marine and freshwater environments, but there is apparently little mixing of the two faunas in the estuary. Along the Atlantic and Gulf coasts of North America, the most typical estuarine species is probably the diamond-back terrapin (*Malaclemys terrapin*). The aquatic reptiles are primarily carnivorous, although some freshwater turtles feed upon plants to a certain extent.

Classes Aves and Mammalia

The birds and mammals are associated with inland waters and estuaries in a host of ways. Various waterfowl use the plants and animals of lakes, streams, and estuaries for food, and the marshes often associated with these bodies of water provide nesting sites. Wading birds and shore birds are characteristic of shallow zones and shores of bodies of water, and these animals also take food produced in or near the water. In inland waters beavers, muskrats, otters, and other fur-bearing mammals enter into the ecology of various communities, the dam-building trait of beavers often changing the nature of streams. Some of these mammals may be found in and around estuaries. Of the large "sea-going" mammals, some, such as the manatee and the dolphin (porpoise) enter estuaries. A considerable portion of the Amazon River of South America is inhabited by the freshwater dolphins *Inia geoffrensis* and *Sotalia* (Layne, 1958).

GENERAL REFERENCES THAT INCLUDE AQUATIC ANIMALS OF THE UNITED STATES

Blair, W.F., N.P. Blair, P. Brodkorb, F.R. Cagle, and G.A. Moore. 1957. *Vertebrates of the United States*. McGraw-Hill, New York.

Edmondson, W. T., editor. 1959. *Fresh-water biology*. John Wiley & Sons, New York.

Hyman, L.H. 1940-1959. *The invertebrates*. McGraw-Hill, New York.

Pennak, R. W. 1953. *Fresh-water invertebrates of the United States*. Ronald Press, New York.

16 Aquatic Communities

In Chapter 1 we learned something of the origin of the term, and the concept, **ecology.** The idea of the ecological **community,** like the idea of ecology itself, was first set forth in scientific terms late in the nineteenth century. In an 1883 translation of an earlier paper, the German zoologist Karl Möbius discussed a relatively simple marine ecosystem in the following terms:

> Every oyster-bed is thus, to a certain degree, a community of living beings, a collection of species and a massing of individuals, which find here everything necessary for their growth and continuance, such as suitable soil, sufficient food, the requisite percentage of salt, and a temperature favorable to their development.

Möbius coined a new term, **biocoenosis,** to refer to such a collection of species inhabiting a well-defined habitat with a given set of environmental factors. This concept stresses the idea of pronounced interrelationships—the interrelationships that link plant and animal species to one another, and those that link all species to the physicochemical features of the medium in which they live. As we saw in Chapter 12, the various species of any environment are linked by harmful and beneficial interactions, as well as by those that involve only the tolerance of one species for another.

An oyster reef is an easily identified community, composed (under any given set of environmental conditions) of characteristic species populations. The same is true of a riffle community in an upland stream, a community of the marginal zone of a lake, or that of a wharf piling in an estuary. These, and a host of others, are all relatively distinct aggregations of organisms. Each of them has evolved into a generally distinct assemblage through adaptation to conditions determined by the fitness of the environment. In the process, mutual adjustment and harmonious interactions among the component organisms have reached a high level of organization. One result of this is that the aggregation, or community, given adequate nutrients and living space, will maintain its distinctiveness, through the processes described in Chapter 12. From this we see that a community typically possesses a certain degree of structural unity, so far as its component species populations are concerned. In studying communities, our first concern will therefore be with the **species composition** of each.

Composition alone, however, is only one of the two basic aspects of the natural community concept. The second aspect relates to **organization** of communities. A given community is more than simply an assemblage of species living together without pattern. As we saw in Chapters 12 and 13, there is organization within the community with respect to food and energy relationships, or **metabolism.** The arrangement, or location, of populations of a community is generally organized and manifested in **stratification** of the inhabitants. The activities of the components of a community are strongly time-related, lending to the community temporal organization as seen in **periodism.** Change within a community is, under normal conditions, orderly, and often with predictable direction; this aspect or organization discussed in earlier pages is called **succession.**

On the basis of these ideas, we may now broadly define a community as a local assemblage of species populations maintained in an area delimited by environmental features. Note that, in addition to "community," two other key ideas, "maintenance" and "environment," are inherent in the definition. **Community** refers to the living components; **environment** embodies the total of the physical and chemical factors which exert an effect upon the living assemblage; the dynamic interaction and coaction of environment and community, involving nutrient cycles and energy flow (**maintenance**), combine in the concept of **ecosystem.**

On the basis of our definition, a community may consist of only a few ecologically related individuals, or it may be composed of a vast assemblage. A small puddle of water with its plants and animals, temporary though it may be, constitutes a community; at the other extreme the total biota of the ocean, with its shore-zone inhabitants, its bottom assemblages, and its plankton of the open waters, is also a community. We may, therefore, recognize major communities, such as those of the sea, an estuary, a lake, or a pond. As we have just seen, major communities are essentially self-sufficient and generally independent of other major communities. Within a major community are found varying numbers of minor communities. The undersurface of a lily pad, for example, forms the substrate for a unique community, the **aufwuchs.** The bottom muds from which the lily pad arises contain a bottom community, the **benthos.** The water phase between the lily pad and the bottom harbors the **nekton** and **plankton.** In some instances the change in species composition between two communities is marked and conspicuous. For example, the gradation from stream type to pond type, encountered where a fast brook enters a pond, is often abrupt. Similarly, the intertidal community of a wharf piling in an estuary is usually decidedly different from that below the low-tide mark, the latter usually containing a greater number of species even though some forms are common to both zones on the piling. On the other hand, some communities grade insensibly, and recognition of boundaries and components is often difficult. This condition is frequently encountered in the transition from freshwater to marine communities in relatively stable estuaries of considerable length. In estuaries, the community composition grades quantitatively, with respect to number of species (Figure 16.1) as well as qualitatively in terms of kinds of organisms. There is also evidence that the

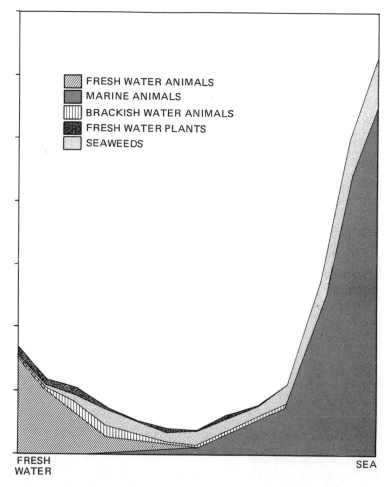

Figure 16.1 Species composition profile of the Tees Estuary, England. Groups of organisms distinguished are freshwater, mesohaline, and marine. (After Macan and Worthington, 1951.)

number of individuals is lowest in the estuarine transition between freshwater and marine communities.

Unlike species populations, which are delimited by inherited biological traits, communities are bounded, generally, by local environmental factors or geomorphological features of the land or water. The **type** of community developed in a certain locality depends, to a great extent, upon the relationships between environmental factors, species characteristics, and species functions. For example, an oyster reef community and a stream riffle community harbor predictable populations of ecological kinds of organisms performing particular roles, and therefore may be thought of as types of communities; many other cases exist in nature. Observe that we refer to **ecological kinds** of species. The species composition of a given community may vary taxonomically from one geographic locality to another. The types, or ecologically adapted kinds, however, are quite characteristic wherever similar environmental conditions prevail. These kinds may, indeed, belong to the same genera. The macrofauna of a stream riffle in California is ecologically similar to that in a riffle community in a New Jersey stream, in that insects are conspicuous and dominant organisms. Many genera of mayflies, beetles, and dipterans are shared by the two widely separated communities.

Regardless of the type of community, however, general principles discussed earlier, pertaining to community composition, to the activities of the component populations, to the ecological roles of groups of populations, to energy relationships, and to evolutionary processes, are common to all. In this chapter we wish to deal with selected principles pertaining to some of these aspects of the community concept. For more comprehensive treatment of these and other community principles, the student is invited to consult Allee *et al.* (1949), Dice (1952), and Odum (1971).

COMPOSITION OF SOME REPRESENTATIVE COMMUNITIES

In our preceding chapters we have given a great deal of attention to geological, physical, chemical, and biological features of natural waters. We have repeatedly emphasized that one of the major points of view in the consideration of such features derives from our interest in the "fitness of the environment" with respect to habitation by plant and animal populations. In the opening pages of this chapter we suggested that communities are predictable assemblages of organisms inhabiting a given environment or segment thereof. Let us now view some of the characteristics of a few selected aquatic communities.

STANDING WATERS (LENTIC HABITATS)

The category of standing waters includes, obviously, pools, ponds, bogs, and marshes. Lakes too are usually considered to be "standing" waters, although large lakes may have considerable circulation and wind-driven water movement. The oceans show these same dynamic properties, but on a vastly larger scale, and they are considered to be quite distinct; they are therefore treated separately in a later section. The communities found in different types of standing-water habitats are generally similar in their structure and biological processes. Therefore we will consider first the broad ecological patterns of the various lentic habitats. The conspicuous regions of a lake as shown in Figure 16.2 are normally inhabited by typical (but by no means rigidly delimited) associations of plant and animal species. The populations of these species, in their typical associations, may constitute recognizable **minor communities.** Within the minor communities are to be found even more restricted assemblages. For the purpose of the aquatic ecologist, a number of distinct regions are generally recognized in bodies of standing water: the lake surface; the **limnetic** zone, comprising the **trophogenic** and the **tropholytic strata;** the **littoral** zone; the **profundal** zone; and various submerged structures, which support a number of ''aufwuchs'' communities.

The Lake Surface

The lake surface, as the interface between air and water, serves as the substrate upon which an interesting and diverse community has developed. The organisms associated with the surface film are collectively referred to as **neuston.** Some snails and flatworms and, seasonally, the larvae and pupae of dipterous insects (*e.g.* mosquitoes) occupy the undersurface of the water film; these are the **hyponeuston.** On the surface, freely floating plants such as *Lemna* and *Azolla,* and the floating leaves of such rooted species as *Nymphaea,* combine in a distinctive plant community. Add to this a number of species of Hemiptera (*Ranatra, Hydrometra,* and others) which tread upon the surface film, swimming ''whirligig'' beetles (Gyrinidae), and a number of other insects lighting upon the plants, and a colorful community, the **epineuston,** results. As living, mature individuals, at least some of these organisms might conceivably be considered as belonging to something other than the aquatic realm. The hemipterans named above, for example, contribute little, while living, to the food web of the water phase of a lake or stream.

The Limnetic Zone

The **limnetic** zone of a lake is that region of open water, the horizontal extent of which is bounded peripherally by the zone of emergent vegetation. Shallow bodies of water in which vegetation extends across the basin, of course, lack a limnetic zone. Vertically, the limnetic zone extends from the surface to the bottom and includes both the **trophogenic** and **tropholytic** zones, if the latter is present (see below and Chapter 10). Recall from earlier considerations that these zones are usually marked by quite different regimes of physical and chemical conditions. We should expect, therefore, that the animal and plant communities would also differ.

The upper, or trophogenic, zone normally corresponds to the lighted zone and serves as environment for a community of microscopic photosynthetic plants and a plant-based animal community. The lower, or tropholytic, zone is characterized by an abundance of bacteria and a paucity of organisms generally. The animals which inhabit the tropholytic zone are typically those adapted to low-oxygen conditions, many of the species obtaining food from the rain of organic particles from the trophogenic zone above. In shallow lakes and ponds, or in relatively clear waters, the tropholytic zone may be poorly defined or missing. On the basis of locomotion, the inhabitants of the limnetic zone are classified as **nekton** and **plankton**. The former includes the large free-swimming animals capable of sustained, directed mobility. The nekton (from the Greek *nektos,* "swimming") includes primarily the fishes and certain insects. The second grouping, the plankton (Greek *planktos,* "wandering"), comprises an abundant and varied assemblage of essentially microscopic to submicroscopic plants and animals.

Although numerous references have been made previously to reactions between physical and chemical factors and plankton, and to coactions between

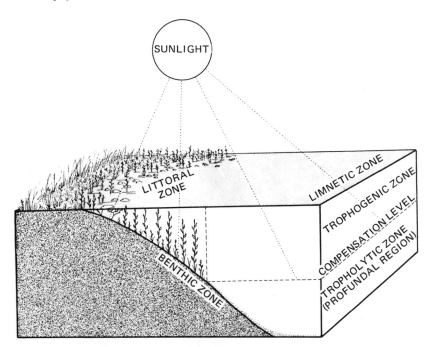

Figure 16.2   The major horizontal and vertical life zones of a lake.

plankton and other plants and animals, it seems appropriate to consider, briefly, some general features of the composition of a limnetic plankton community. Both zooplankton and phytoplankton range in size from the smallest protozoans and bacteria to algae and crustacea easily seen without magnification. On the basis of size, **nannoplankton** includes the organisms that pass through a collecting net of bolting silk. The usual material (No. 25 silk) has a mesh opening of 30 to 40μ. Nannoplankton is collected by centrifuging water samples (see Welch, 1948, for details). **Net plankton** is that held in the collecting net.

The phytoplankton community of lakes and ponds is composed mainly of diatoms, blue-green algae, green algae, and photosynthetic flagellates. It must be recognized that the quantitative and qualitative composition of phytoplankton communities vary greatly from one body of water to another, and that no truly "typical" community exists.

In a small Minnesota pond, *Fragilaria* and *Navicula* were found to be the most prominent diatoms, the population density increasing during colder months, and reaching a maximum of nearly 4000 cells per liter in December (Dineen, 1953). In a Colorado lake (Gaynor Lake) some 12 genera of diatoms were reported (Pennak, 1949), and on two occasions, in April and in May, the number of cells in a mixed population reached nearly 12 million and 20 million per liter, respectively. Seasonal "pulses" in diatom populations have been reported for many lakes. These pulses may occur as a single vernal increase in numbers, or as two conspicuous blooms, one in early spring and another in the autumn. Sometimes three or more pulses per year are common. The precise mechanisms involved in such population fluctuations are not fully understood. They appear, however, to involve complex interactions between environmental factors, such as temperature and nutrient supply, and the physiology and reproductive potential of the plants. It does seem that diatoms are characteristically the most abundant algae of larger lakes.

The blue-green algae are often abundant and represented by a notable variety of kinds in ponds and lakes. In the Minnesota pond, this group was predominant. *Microcystis, Anabaena,* and *Oscillatoria* were common, and a bloom of *Aphanizomenon flos-aquae* in the early summer produced over 1.5 million cells per liter. In large lakes, blue-green algae typically exhibit a late summer pulse, and sometimes an additional spring pulse. In Boulder Lake, Colorado, a spring pulse, mainly of *Coelosphaerium,* produced over 13 million cells per liter. During an autumn pulse, chiefly of *Schizothrix,* the population density reached nearly 41 million cells per liter. In another lake of the same region, *Chroococcus* was the predominant blue-green in a pulse that reached a peak of over 62 million cells per liter in summer.

A considerable variety of kinds of green algae may ccur in ponds and lakes. *Pediastrum, Cosmarium, Scenedesmus,* and *Staurastrum* are commonly found in both small and large bodies of water. Small-scale blooms of green algae have been reported, some occurring simultaneously with pulses of blue-green algae and diatoms.

Aquatic flagellates constitute a conspicuous segment of pond and lake plankton, at least seasonally. Among this group, *Synura* often becomes abundant in small lakes and ponds. *Trachelomonas* and the common lake plankter, *Ceratium hirundinella,* are characteristic of most large and small lakes of the temperate zone.

The zooplankton of ponds and lakes is composed predominantly of rotifers and microcrustaceans (Figure 16.3). In the Minnesota pond, 29 species of rotifers were collected. These reached maximum abundance in late summer; during one year a population density of over 4000 individuals per liter were recorded. Two common, cosmopolitan species of rotifers, *Keratella cochlearis* and *K. quadrata,* were predominant during winter, the maximum density being attained during summer. The rotifer genus *Brachionus* was represented in the summer community by six species. Several other forms occurred commonly. Fifteen species of rotifers were found in the Colorado lakes, the most common forms being *Keratella cochlearis* and *Polyarthra trigla.* In neither the pond nor the lakes did rotifer populations follow any predictable pattern with respect to seasonal population density and occurence of species. Although a considerable literature suggests cyclical trends in seasonal abundance of particular species, intensive studies such as those of the above-mentioned pond and lakes indicate that irregular phenomena of rotifer activity are probably the more usual.

Among the plankton crustaceans of the limnetic regions of ponds and lakes, Cladocera and Copepoda constitute the conspicuous elements. Although cladocerans are characteristically more abundant in ponds than in lakes, certain species, such as *Bosmina longirostris, Daphnia longispina,* and *D. pulex*, are commonly found in both types of waters. In the Minnesota pond, *B. longirostris* was the predominant cladoceran throughout the year, the greatest density occurring in late summer. Eight other species inhabited the pond, some being primarily summer inhabitants. The Colorado lakes were inhabited by five limnetic species, only three of which also occurred in the Minnesota pond. In the open

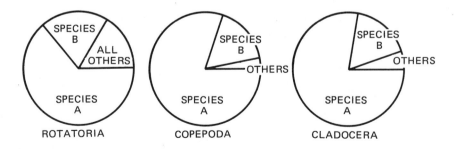

Figure 16.3   Momentary species composition of a typical limnetic zooplankton community. The relative proportions of the sectors indicate percentages of both dense and scanty zooplankton populations. (After Pennak, 1957.)

waters of large lakes, copepods are frequently more abundant than cladocerans, although the number of species of the former may be less. *Cyclops bicuspidatus* is a common inhabitant of both ponds and lakes in temperate regions of North America. In one of the Colorado lakes, this species reached a population density of over 2000 individuals per liter. Seasonal maxima in both ponds and lakes may be erratic with respect to the number of pulses; in some lakes two, one or no pulses may occur. The limnetic plankton of the Colorado lakes contained only one other species of copepod; this was *Diaptomus siciloides*. In addition to *C. bicuspidatus*, the Minnesota pond supported *Diaptomus eiseni* and *Canthocamptus staphylinoides*, a bottom littoral form, the latter population arising rapidly in spring and existing for only one month. Although the densities of the various species populations in the limnetic zooplankton of a given lake often exhibit unpredictable and considerable variation, the species composition of the open-water zone remains relatively constant in time. A considerable amount of research has revealed that the number of dominant microcrustacean species in the limnetic plankton is typically small (frequently less than a half dozen). Furthermore, the component populations of any lake, under normal circumstances, apparently persist year after year. Studies of the plankton of Lake Michigan, for

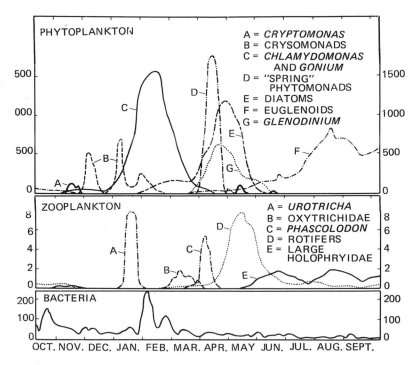

Figure 16.4　Seasonal variation and succession of dominant organisms in a small artificial pond. The curves are smoothed by fifteen-point moving averages. (After Bamforth, 1958.)

example, have shown that little change in the animal and plant composition occurred over a 40-year period. Of the species abundant in 1887-1888, only one, the copepod *Epischura lacustries,* was absent from the 1926 investigations. Associated with ecological succession (see below), the gradual change in community structure, however, is a natural order of replacement of species. Depending upon a number of biogeochemical factors, this process may be exceedingly slow or quite rapid.

The preceding descriptive consideration of the limnetic plankton of lakes and ponds serves to illustrate some of the general principles pertaining to plankton dynamics. We have seen, for example, that the composition of plankton varies both quantitatively and qualitatively from lake to lake, and seasonally within a given body of water (Figure 16.4). These variations are related to a number of environmental factors, including light, temperature, and dissolved substances (particularly nutrients necessary for phytoplankton production), and to biological features such as competition and population growth attributes. In general, fertile ponds and small lakes contain a richer plankton than do large lakes. The volume of nannoplankton is greater than that of net plankton, in some bodies of water the ratio reaching three, or more, to one. In ponds and lakes of temperate North America, phytoplankton volume usually exceeds that of zooplankton by two to six times. The relationships between phytoplankton and zooplankton pulses are not clear. Although it has been suggested often that sudden, conspicuous increases in population density of zooplankton follow phytoplankton blooms, a number of studies have shown a lack of significant correlations between the two events. Furthermore, in some lakes, at least, food (phytoplankton) does not appear to constitute an important limiting factor with respect to the overall zooplankton population.

The composition of phytoplankton communities can often be correlated with trophic stages of lakes. Oligotrophic lakes are generally characterized by a relatively low quantity of total plankton, and population pulses are uncommon. Eutrophic lakes, on the other hand, typically support a large quantity of phytoplankton composed of few species; pulses are common and frequent. Table 16.1 lists algal species which contribute a large percentage of the phytoplankton community in lakes of Western Canada during summer.

In the temperate regions of the world two more or less characteristic assemblages of phytoplankton may be recognized. One community, in which diatoms predominate, appears to be prevalent in cold-water lakes of the more northern latitudes. The other community, characterized by dominance of blue-green algae, is more commonly found in warmer climes in which the summer surface temperature of lake waters exceeds about 19°C. Temperature also relates to annual cycles of production of plankton. As a rule, plankton abundance is greater in summer than in winter. Some species, however, exhibit winter pulses, often producing large numbers underneath ice cover. In some North American lakes the population density of plankton may increase four-fold or more during a short time in spring.

The very integrity of the plankton community and, indeed, much of the organic maintenance of standing waters are ultimately dependent upon flotation of the organisms in the medium. In order to carry on anabolic activities, autotrophic species must remain within the lighted zone. It follows, therefore, that a great proportion of the herbivorous organisms must be suspended in somewhat the same region. Since the specific gravity of protoplasm and the various body coverings secreted by plants and animals is greater than that of water, most organisms tend to sink. The rate at which a body sinks is determined by the relationship between weight beyond the specific gravity of water and frictional resistance between the surface of the organism and the environment (see Stokes's Law, Chapter 6). Increased surface area increases friction. Thus the small size of plankton organisms is of distinct advantage in flotation, for the ratio of area to volume is thereby increased. In addition to inherent small size, various morphological and physiological adaptations contribute to suspension of plankton in water. The development of spinous projections, such as those of *Ceratium* and *Scenedesmus* (Figure 14.2), increase surface area and decelerate sinking rate. Needle-shaped *(Synedra,* Figure 14.4) and curved *(Closterium,* Figure 14.2) body forms tend to reduce sinking. The formation of radial and chain-like colonies of cells such as *Asterionella* and *Tabellaria* (Figure 14.4) also contributes positively toward flotation. Organelles such as flagella (see *Synura* and *Eudorina* in Figure 14.2) are aids in maintaining position. Many flagellates, diatoms, rotifers, and plankton crustaceans produce oil globules which serve to decrease the specific gravity of the organisms. The zooplankton typically possesses some swimming ability. This, coupled with produced filaments and spines, oil droplets, and air bladders (in the larva of the dipteran *Chaoborus,* for example) found variously among the animal plankton enable these forms to inhabit the limnetic zone. Of further importance in maintaining position in open

TABLE 16.1   Limnetic algae: approximate trophic distribution in lakes of western Canada[a]

Oligotrophic	Mesotrophic	Eutrophic
Asterionella formosa	Fragilaria crotonensis	Microcystis flos-aquae
Melosira islandica	Ceratium hirundinella	
Tabellaria fenestrata	Pediastrum boryanum	
Tabellaria flocculosa	Pediastrum duplex	
Dinobryon divergens	Coelosphaerium naegelianum	
Fragilaria capucina	Anabaena spp.	
Stephanodiscus niagarae	Alphanizomenon flos-aquae	
Staurastrum spp.	Microcystis aeruginosa	
Melosira granulata		

[a] From Rawson, 1956.

water is the movement of water itself. We have seen previously that turbulence, development of eddy viscosity, and other water motions are characteristically a part of the dynamics of standing waters. Depending upon the intensity of these factors in particular lakes and ponds, the plankton components may be benefited with respect to maintaining position in the medium.

The plankton of standing waters is rarely distributed uniformly throughout the vertical extent of the water. We have already considered the fact that lakes typically contain two regions with respect to biological activity. The upper, lighted stratum, in which photosynthesis exceeds total plankton respiration, we termed the trophogenic zone; the lower, unlighted region, where decomposition predominates, was called the tropholytic zone. Between these two lies a level at which there is a balance between algal photosynthesis and total plankton respiration; this is the **compensation level** (Figure 16.2). In highly turbid or strongly colored lakes this level may lie within a meter or so of the surface, or it may occur at considerable depth in clear lakes. With regard to vertical distribution of plankton generally, the compensation level marks the lower limit of occurrence of populations. Phytoplankton is restricted by light requirements to the zone above the compensation level. Since the zooplankton is directly or indirectly dependent upon the plants, the animals are, likewise, found in greatest abundance in the trophogenic zone. Because of diurnal migration patterns, some zooplankton is often more abundant in the trophogenic zones during the night and in the tropholytic during the day.

It is now apparent that the entire water mass of a lake is not uniformly inhabited by plankton. Vertical stratification, determined by light, temperature, water movement, and nutrient supply, is characteristic of the limnetic zone. Even within the euphotic zone, the vertical distribution of plankton is typically marked by stratification of various populations of the community, this zonation being determined by a number of environmental factors. Many plankton species populations undergo vertical movements, or migrations, thereby changing the community structure at various levels. We have previously seen that excessive sunlight can limit photosynthesis; thus, the zone of maximum density of phytoplankton is normally at some depth below the lake surface rather than in the uppermost waters.[1] The level at which the greatest abundance of phytoplankton is found is usually determined by transparency. In highly turbid lakes, for example, phytoplankton may be narrowly restricted to near-surface regions; in deep, clear lakes the photosynthetic zone may extend to considerable depths. In summer, thermal stratification may serve to limit the vertical distribution of both animal and plant components of the plankton, an important factor in this instance being the lack of oxygen in the hypolimnion. In winter, with more uniform

---

1 An exception to this distribution is seen in some blue-green algae. Recall that these plants typically possess gas-filled vacuoles and tend to float at the surface. Some of these, such as *Anabaena* and *Microcystis,* often develop dense floating masses at the surface of lakes and ponds.

conditions throughout the water column, various planktonic forms may be mo\. uniformly distributed to greater depths. Nocturnal-diurnal migrations of many plankton species change the nature of the limnetic community regularly. Figure 16.5 illustrates vertical movements of one such migratory species.

The Littoral Zone

The **littoral zone** is that portion of a body of water extending from the shoreline lakeward to the limit of occupancy by rooted plants. In many shallow lakes and ponds this zone extends completely across the basin, particularly during the growing season. In the deeper bodies of water the development of a littoral zone is influenced by factors which are limiting to plant growth. These may include depth of water, vertical extent of effective light transmission, movement of water (particularly wave action), and fluctuations in water level.

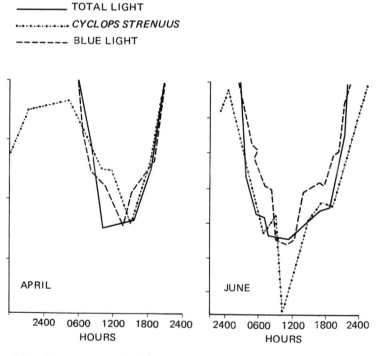

—————— TOTAL LIGHT

••••••••••• *CYCLOPS STRENUUS*

— — — — — BLUE LIGHT

APRIL

JUNE

2400 0600 1200 1800 2400
HOURS

2400 0600 1200 1800 2400
HOURS

Figure 16.5 Diurnal movements of the copepod *Cyclops strenuus,* relative to light penetration in Lake Windemere, England. The graph of total light shows (at hourly intervals) the depth to which light of an intensity of 32,800 erg/cm²/sec in April, and 108,000 erg/cm²/sec in June, penetrated; the broken line shows blue light treated in the same manner, recorded at intensities of 9600 erg/cm²/sec in April and 305 erg/cm²/sec in June. The third line shows the depth at which most individuals were caught, at hourly intervals. (After Macan and Worthington, 1951.)

To these may be added such obvious features as nutrient supply and texture of the substrate. Of the limiting factors, depth of the water is important to rooted emergent species and those with floating leaves. Given sufficient light, submersed plants may exist at any depth. The amount of wave action is, of course, important, for we are all aware of the lack of plant growth in the shallow, wave-washed regions of lakes, as compared with the great profusion often found in quiet, protected areas. Fluctuating water levels are usually very damaging, and reservoirs and lakes subject to such fluctuations generally have little rooted aquatic vegetation.

The littoral zone usually is predominantly vegetated by "higher" plants. These plants normally occur in relatively distinct associations and form conspicuous physical and ecological "subcommunities" within the littoral zone. Without reference to any specific example, let us view the general features of plant zonation and the animal associations in the shallow regions of a lake. It is in the littoral that we find the greatest variety of species of any of the lake regions, for here occurs abundant cover for protection, substrate for travel and attachment, food, and dissolved substances.

The zone extending lakeward, from the shoreline to a depth determined primarily by the ability of rooted plants to reach the surface of the water, is referred to as the zone of emergent vegetation. Throughout most of the world, this border is typically dominated by grasses, rushes, and sedges possessing long leaves and stems. In the temperate zone, bur-reed *(Sparganium)*, spike rush *(Eleocharis)*, bulrush *(Scirpus)*, and cat-tail *(Typha)* combine in a fairly typical and commonly found assemblage. These may be joined by herbaceous types such as arrowhead *(Sagittaria)*, water plantain *(Alisma)*, and, in eastern North America, arrow arum *(Peltandra)*. The plants of the emergent zone comprise a somewhat "in between" stage in energy relationships. The photosynthetic organs are typically above water, and there utilize atmospheric oxygen and carbon dioxide; gases given off by photosynthesis and respiration are contributed to the atmosphere. Mineral nutrients and water, however, are taken from the lake substrate. Upon death, the emergent structures fall into the water where, as litter, they form a habitat for numerous plants and animals and provide bits of matter used by numerous detritus feeders. Decay of the fallen plants contributes to nutrient supply and cycles of the lake. Some of these emergents are used as food by waterfowl and mammals, and as nesting sites and cover by certain birds.

Lakeward from the zone of emergent vegetation, an association of plants bearing long stems or petioles and floating leaves typically parallels it. This second community is termed the **zone of floating-leaf plants.** Sometimes the zones are abruptly and clearly marked; in other instances a subtle transition from one to the other is seen. The assemblage in this zone characteristically includes rooted plants with floating leaves, such as the familiar (at least in central and eastern North America) white water lily *(Nymphaea)*, the more widely distributed yellow water lily *(Nuphar)*, water shield *(Brasenia)*, and several species of pondweed *(Potamogeton)*. Frequently this zone includes a number of free-floating plants; for example: water fern *(Azolla)*, duckweed *(Lemna* and *Spirodela)*, and in some of the southern states, water lettuce *(Pistia)* and water hyacinth *(Eichornia)*. The free-floating hydrophytes and rooted plants with floating leaves influence energy relationships and community composition in various ways. Both groups may contribute to a surface massing of such extent as to shield the underlying water from sunlight, thereby inhibiting organic production. In warm regions the water underneath such mats has frequently been found to be nearly oxygenless. The roots of free-floating plants extract dissolved nutrients from the water, and, together with the undersurfaces of floating leaves, often harbor rich communities of organisms (aufwuchs) and serve as the substrate for egg deposition of a number of animals. Certain snails *(Physa* and *Planorbis,* for example), beetles *(Donacia* and *Gyrinus),* hemipterans, and at least one mayfly deposit their eggs on the undersurfaces of lilypads. This zone is commonly used as a breeding and nesting area by sunfishes and various other fishes.

The **zone of submersed vegetation** typically forms the innermost belt in the lake border of plants. The plants in this zone are characterized by long, sinuous leaves, or by a bushy growth-form with fine, highly branched leaves. These plants may be considered as wholly aquatic, for they derive gases and nutrients for photosynthesis and respiration from the water and, in turn, most of their substance returns to the water through decomposition. The genus *Potamogeton* includes over 40 species of typically aquatic plants, many of which occur widely throughout North America. Hornwort *(Ceratophyllum),* naid *(Najas),* milfoil *(Myriophyllum),* waterweed *(Anacharis),* and certain species of arrowhead *(Sagittaria)* are commonly found in the zone under consideration. In certain lakes, stonewort *(Chara)* may be prominent in the submersed vegetation zone; we have previously considered some of the conditions conducive to stonewort development (see Chapter 10 on chemical factors).

The plankton of the littoral zone is typically rich in number of species, and sometimes in number of individuals. A considerable amount of this plankton may, however, consist of organisms which have been displaced from an aufwuchs community. These elements, termed **tychoplankton,** include many solitary, unicellular, and filamentous algae as well as larger animals normally found on some substrate. Diatom species are abundant in the littoral plankton, often more so than in the limnetic. These may be supplemented, particularly in stagnant waters of embayments, by filamentous blue-greens, such as *Oscillatoria* and *Anabaena,* and form dense blooms. Green algae, *Spirogyra, Ulothrix,* and *Oedogonium,* for example, frequently form dense mats in these waters. Desmids such as *Closterium, Micrasterias,* and *Cosmarium* occur commonly in the still waters of the littoral of lakes and ponds. The zooplankton of the littoral zone includes many forms which do not occur commonly in the limnetic region. Mites (Hydracarina) and ostracods, not typically found in the open waters of lakes, are often common in the littoral zone. Cyclopoid and harpacticoid copepods are more characteristically littoral than limnetic, and frequently abound in the shore zone. Certain cladocerans are more abundant in the littoral plankton. The nekton of the littoral zone includes many fishes, beetles, hemipterans, and dipterans not normally found in the limnetic region. Larger animals, such as water snakes, turtles, newts, and frogs, are conspicuous members of the littoral community.

Bottom Communities of Lakes

The association of species populations of plants and animals that live in or on the bottom of a body of water is called the **benthos**. Probably no other community within a lake exhibits greater variety of kinds and numbers than does the benthos. Grading, as it does, from shore to considerable depths, and often consisting of a number of different substrates, the bottom frequently presents an impressive variety of environmental complexes. These complexes, in turn, are normally inhabited by more restricted communities which are quantitatively and qualitatively rather varied. The composition of benthic communities is controlled by a number of factors, some of which appear to operate directly to limit certain species. In other instances, the influences are more subtle, suggesting interrelationships of several factors. Among the more conspicuous and better understood of those influencing the kinds of organisms and their distribution in lake bottoms are, first, physicochemical features of the water such as temperature, transparency, dissolved oxygen content, and water currents (these are typically related to lake-basin morphology and the nature of the drainage basin and local climate), and, second, biological factors such as food, protection, and competition.

The lake bottom is usually considered as including three zonal subdivisions: the **littoral** bottom, which extends from shore to the lakeward limits of the zone inhabited by rooted hydrophytes; the **sublittoral** bottom, or the zone between the littoral and deep (profundal) region; and the **profundal** bottom, or the area of the lake bottom contiguous with the hypolimnion of the limnetic region. Depending upon the area and slope of shoal, these zones and their communities exhibit extreme diversity, not only within the limits of a single lake, but from lake to lake. In shallow, clear bodies of water, the sublittoral or profundal zones may be quite reduced or even entirely absent. As an alternative, the benthos is often described and delimited by depth alone; in such instances the three zones given above are designated, somewhat arbitrarily, by depth, or are disregarded. In any case, the biotic differentiation of littoral and profundal is usually more distinct than is that of the sublittoral, the last constituting an **ecotone**, or "buffer" zone, between two communities. Of the major importance is the fact that each of these zones, if present in a lake, typically supports a relatively distinct community. Furthermore, each of these communities usually contains a number of assemblages based primarily on the depth and nature of the bottom sediments.

The profundal bottom of eutrophic lakes is characterized by absence of light, low oxygen content, and high carbon dioxide concentration. The sediments are typically soft, flocculent, or ooze-like materials rich in organic substance.

Fungi, bacteria, and certain protozoans are usually abundant near the mud surface. Green plants are normally absent. The macrofauna consists predominantly of tubificid oligochaetes, bivalved mollusks of the family Sphaeriidae, and dipteran insect larvae of the genera *Chaoborus* and *Tendipes* (*=Chironomus*). Some species of *Chaoborus* occasionally take the form of plankton in that they inhabit the bottom during the day and migrate to the lake surface at night; the migration is often rapid, being about 20 m per hr. As indicated, the number of kinds of animals adapted to bottom dwelling in the deeper profundal region is relatively small. The population densities of these species may, however, be high. Table 16.2 gives the average number per square meter of the major groups of animals at various depths in Lake Simcoe, Ontario. Note, especially, that at depths greater than 20 m the benthos is dominated by dipterans (*Chaoborus* and several species of the family Tendipedidae) and by oligochaetes (*Tubifex*). Although both gastropod and pelecypod mollusks inhabit the profundal bottom, the latter (represented by Sphaeriidae) is much the more abundant. The miscellaneous animals include nematodes, leeches, hydrachnids, and odonates and other insects. These data illustrate clearly a characteristic feature of deep eutrophic lakes of the temperate regions, that is, a typically rich community (in terms of numbers or mass) maintained by a small number of species. As shown in Table 16.2, there does not appear to be a great difference between the average mass of organisms produced in the littoral and that in the profundal. In some lakes, however, this differential is more pronounced. In Lake Texoma, on the Oklahoma-Texas border, the annual mean mass of bottom fauna in the profundal (Sublette, 1957) was found to be 102.9 kg per hectare (10.29 g/m²/yr); in the littoral the annual production amounted to only 21.75 kg/ha (2.175 g/m²/yr).

Analysis and description of the benthos of the littoral zone in most lakes are complicated by kinds and abundance of both plants and animals present, and by variations in the nature of the bottom sediments. The presence of plant growths offers opportunities for some animals to desert the benthic environment and to climb, thus adding an essential vertical stratification to the over-all littoral pattern. Variations in vegetation density in the shallow zone also make for irregularities in animal distribution and abundance. Water movement, especially wave action, is a major factor influencing the nature of the sediments and, in turn, the composition of communities.

Within a given lake there is usually a high correlation between the nature of the substrate and the number of species and population density. Furthermore, each substrate type typically supports a relatively distinct community of organisms. In Lake Texoma, for example, four minor communities based upon bottom sediments in the littoral zone have been recognized (Sublette, 1957). A gravel shoal, or bar, kept clean of fine sediments by wave action, contained a community of nine species of macroscopic animals, including a sponge, a gastropod, and insects. Three of the species populations were not found in any other situation in the lake; these were a mayfly, a damselfly, and the sponge. A pure sand substrate was inhabited by only one species, a caddisfly (*Oecetis*

TABLE 16.2 Major groups of bottom organisms: average number of organisms per square meter at different depths in Lake Simcoe, Ontario, 1926–11928[a]

	Tendip-edid	Ephem-erid nymphs	Gastro-poda	Pele-cypoda	Amphi-poda	Oligo-chaeta	Chao-borus	Trichop-tera	Miscel-laneous	Avg no. of all organisms per sq m	Avg dry wt of all organisms[b] in mgm per sq m
Shore zone, 0 to 1 m	152	48	54	22	28	16	0	1	84	405	1028
0 to 5 m	300	90	124	82	112	36	0	14	30	788	1280
5 to 10 m	450	54	130	78	138	30	2	10	34	926	1480
10 to 15 m	240	52	54	88	94	24	9	4	9	574	654
15 to 20 m	540	28	82	118	10	26	15	8	17	844	1170
20 to 25 m	620	0	9	114	2	80	30	0	92	947	1420
25 to 30 m	780	0	8	102	0	106	62	0	10	1068	1340
30 to 35 m	860	0	7	84	0	98	74	0	11	1134	1220
35 to 40 m	760	0	5	52	0	118	70	0	7	1012	950
40 to 45 m	740	0	6	34	-0	120	72	0	6	978	852

[a] Data from Rawson, 1930.
[b] Weight of mollusk shells not included.

inconspicua) that builds sand cases. Where the substrate was composed of a mixture of sand, clay, silt, and organic detritus the community contained nearly 60 taxonomic groups; of these, 25 were restricted to the particular substrate. A clay substrate was inhabited by only one species, a large blood-red dipteran larva *(Xenochironomus festivus)*. Beds of the spermatophyte, *Potamogeton,* were inhabited by 28 species of animals, of which only six were restrictrd to the habitat. The depth distribution of some species was found to overlap substrates, thereby demonstrating the position of an ecotone between the habitats (Figure 16.6).

Communities on Submersed Structures: The Aufwuchs

The communities of the **aufwuchs** include all the organisms that are attached to, or move upon, a submerged substrate, but which do not penetrate into it.[1] The benthos is, by definition, excluded from this community. The physical (and probably the ecological) base of the aufwuchs is composed, typically, of an assortment of unicellular and filamentous algae (such a substrate was shown in Figure 12.11). Located among the plants may be various attached protozoans, bryozoans, and rotifers. The free-living members of the aufwuchs may include representatives of most of the animal phyla found in water. Free-living protozoans, roundworms, rotifers, annelid worms, crustaceans, and insects are variously found in the aufwuchs, the kinds and numbers of these animals differing with substrate, water movement, depth, and chemical composition of the water.

Although we are considering the aufwuchs under the discussion of lakes, the concept and principles are applicable to stream and estuarine conditions also. No fixed object, living or nonliving, submersed in natural waters is devoid of aufwuchs. The nature of the object (substrate) frequently determines the composition of the community, however. Bryozoans seldom inhabit the alga *Chara,* due probably to the sulfur content of the plant. The bryozoan *Plumatella* typically colonizes on broad, flat plant surfaces to a greater extent than on other surfaces; *Fredricella,* on the other hand, occurs more commonly on finely branched leaves. The hydra, *Pelmatohydra,* is seldom found in a dense aufwuchs dominated by algae. Clumps of algae often serve as habitat for free-swimming protozoans, the community relationships becoming complex. Continued population increase of the protozoans leads to destruction of the algae, ending in breakdown of the community. Various species of rotifers exhibit different

---

1. In much of the English literature the term "periphyton" has been used. This term, however, has usually been applied to the total assemblage of sessile or attached organisms on any substrate. Under this definition, the creeping and crawling forms commonly found in association with the attached organisms are not included. The connotation of the German term *Aufwuchs* is broader and includes both attached and free-living plants and animals. It is adopted herein and somewhat Anglicized by dropping the capitalization of the first letter.

affinities for certain substrates. Generally, plants possessing finely divided leaves support a greater number of rotifer species than do broadleaved plants. Of particular interest with respect to substrate occupancy is the aufwuchs developed on the carapaces of a number of species of turtles. Some half-dozen species of filamentous algae are known to occur on such a substrate. Two species, comprising the genus *Basicladia*, are found only on turtle carapaces; these algal growths may be sufficiently dense to support a number of other organisms. Certain crustaceans (copepods and crayfish) sometimes serve as substrate for a community of sessile protozoa and protists.

The great importance of the nature of the substrate in the establishment and maintenance of aufwuchs is seen by comparison of a stone and a leaf, for example, in an upland brook of moderate current in summer. A portion of the stone is apt to be covered by a dense growth of filamentous algae and diatoms. Mayfly nymphs and blackfly larvae are common on the exposed surfaces of the stone. Other animals, such as caddisflies, flatworms, and snails, may also be present. The aufwuchs of the leaf, on the other hand, may consist of a sparse association of algae, stalked protozoa, and, perhaps, hydra. The aufwuchs of the stone gives, in general, an appearance of "permanence" and complex interrelationships born of the more lasting substrate. The community on the leaf is

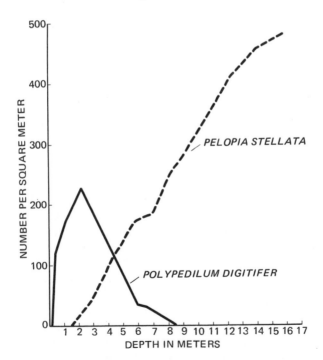

Figure 16.6 The depth distribution of two species of tendipedids (summer populations) in Lake Texoma, Oklahoma and Texas, showing zones of population overlap. This area of overlap is interpreted as an ecotone. (After Sublette, 1957.)

typically less dense, giving an impression of temporariness and reflecting, actually, the impermanence of the substrate.

The development, maintenance, and composition of aufwuchs are also influenced to a great extent by movement of water and by fluctuation in water level. In streams, maximum growth and density of the community are usually found in zones of clear water and moderate current velocity. In pools where rate of flow is reduced, silt deposition inhibits community development. In zones of rapid water, the scouring effect serves to prevent extensive colonization of bottom structures. Indeed, freshets in streams may essentially remove the established aufwuchs. In the littoral zones of lakes, objects in shallow regions of considerable wave action normally support a meager aufwuchs, the more productive zones being those of extensive growths of larger plants. The aufwuchs of submerged plants, or plant parts, in lakes and ponds is especially rich and varied.

The chemical characteristics of the environment, operating under the principle of limiting factors, act upon the biotic structure of aufwuchs, especially as the various members of the community exhibit different reactions to dissolved substances in the water. The precise relationships between aufwuchs animals and the chemistry of the medium are not completely known. This is due, in part, to the often complex interactions of dissolved substances, such as we have studied in earlier chapters, and the interrelationships among organism, substrate, and chemistry. There exists a considerable literature on the tolerance of a great number of aufwuchs organisms to various chemical factors, especially with respect to pH. It has been shown that certain species exhibit great affinities for acid waters, others for waters of high pH. Since, as we have seen, pH involves a number of processes and forms of carbon dioxide, it is difficult to recognize particular factors involved. There is good evidence, however, that certain sessile rotifers, *Ptygura melicerta* and *Collotheca algicola* for example, are absent from aufwuchs of certain lakes due to the high bicarbonate content of the water. Dissolved oxygen and chemical pollution limit the distribution of a great number of aufwuchs animals.

Studies of the aufwuchs algae of four types of Florida springs have revealed that each spring type supports a relatively characteristic aufwuchs community (Ferguson *et al.*, 1947). The composition of these communities appears to be determined to a great extent by the chemical nature of the water. As is a frequent practice, a community may be named for its dominant components. With respect to kinds of algae and water chemistry, the following relationships have been shown for the Florida springs: (1) a *Cocconeis-Stigeoclonium* aufwuchs of hard, freshwater springs of constant temperature and flow (this community is illustrated in Figure 12.11); (2) a *Cladophora-Cocconeis-Enteromorpha* community of "oligohaline" springs, these springs being similar to the hard freshwater kinds except for greater content of chlorides (100 to 1000 ppm); (3) an *Entermorpha-Lyngbya-Licomphora*-dominated aufwuchs of "mesohaline" springs (100 to 10,000 ppm chlorides); (4) a *Phormidium-Lyngbya* community found in sulfide springs, these springs having constant temperature, an absence of oxygen, and high concentrations of sulfates and hydrogen sulfide.

STREAM COMMUNITIES (LOTIC HABITATS)

The communities of streams are under the influence of several major environmental features not encountered by lake inhabitants. One of the most important of these is current, including, as it were, the numerous hydrological processes related to stream flow. Moving water poses problems in maintenance of position for most organisms, for animals caught up in the current are apt to be carried to unsuitable environments. The corrosive and erosive action of flowing water modifies the chemical and physical characteristics of given habitats. Variable discharge may deposit quantities of silt at one moment and scour a zone at another time. On the other hand, the considerable exchange between land and stream often serves to enrich the nutrient supply. Current-created turbulence tends to maintain a relatively uniform set of physicochemical conditions (at least within a given segment) normally free from the stratification found in lakes. Sorting of bottom materials resulting from variable stream velocity produces a great variety of substrates for colonization and the development of communities. In this last-mentioned aspect is found the keynote of stream communities: astounding variety. From the slow-moving, often lake-like lower stream course to the rapidly flowing upper reaches, an impressive array of communities exists throughout the length of a stream. These include pool, run, riffle, and various shore and bottom variations.

Stream Plankton

There seems to be general agreement among aquatic biologists that no "true," or distinctive, plankton community exists in streams. This does not mean, however, that streams do not possess plankton, for many of them do. The plankton may be derived from headwater lakes or ponds, or from quiet backwaters of the stream. Plankton developed in quiet waters and introduced into a stream is often lost rather rapidly. As we have already learned, streams frequently contain an abundant tychoplankton consisting of organisms dislodged from the bottom or from submerged objects. Among the more common plants of stream plankton (potamoplankton) of some temperate regions are: *Asterionella formosa, Fragilaria capucina, Synedra ulna, Tabellaria fenestrata, Melosira granulata,* and species of *Pediastrum, Scenedesmus, Ulothrix,* and *Navicula.* Zooplankton in upper stream courses is quite sparse and, indeed, may be nonexistent in rushing brooks. Rotifera and Cladocera, common planktonic animals of standing waters, seldom occur in swift streams. In pools and slower waters, Copepoda and Cladocera may become prominent.

The Shore Zone of Streams

In the middle and lower courses of many streams shallow-water communities may develop. Often these communities do not differ greatly from those of lakes and ponds in the same drainage area. The growth and maintenance of the near-shore communities are acutely dependent upon favorable substrate, its slope and sedimentary character, and upon stability of discharge. Where present under optimum conditions, the community contains a zone of emergent vegetation near shore, a zone of floating-leaved plants lying parallel and streamward to the emergent zone, and an inner zone of submersed vegetation (Figure 16.7); where the shore slopes steeply the emergent zone may be absent. Swift current, depth, and silt content limit the extent of submersed vegetation.

In an extensive shallow-zone community such as described above, a "true" plankton may be meager or absent, but a rich tychoplankton, derived from normally abundant benthos and aufwuchs, is usually present. The nekton among the plants contains many insects and a number of fishes; turtles and water snakes are often common. Current-delivered nutrients and continual flow of water across the community contribute to an often luxurious aufwuchs. This latter

Figure 16.7   Zonation of floating and emergent plants along a small tidal tributary of the St. Johns River, Florida. In the background, at the shoreline, a zone of sawgrass (*Mariscus jamaicensis*) predominates. A zone of dark-leaved pickerel week (*Pontederia lanceolata*) lies in front of the sawgrass. Next, and mostly left of center, is a zone of smartweed (*Persicaria portoricensis*). The zone bordering open water comprises water hyacinth (*Eichhornia crassipes*) and parrot's feather (*Myriophyllum proserpinacoides*). The pronounced zonation shown here is probably a result of the relatively stable water level. (Courtesy of A. M. Laessle.)

factor is an important one, and deserves greater study with respect to microclimate. In standing waters, for example, the rate of delivery of nutrients and respiratory and photosynthetic gases and the rate of removal of metabolically produced substances are relatively slow; maintenance of sessile members of the aufwuchs is probably limited thereby. In streams with current sufficient to overcome the "drag" at the surface of each organism, the environment in contact with each individual is constantly moved, thus favoring maximum community growth.

The Benthos of Streams

We have previously emphasized the overwhelming variety in numbers and kinds of organisms inhabiting stream floors from the upper reaches to the lower stream course. This over-all change (or **longitudinal succession**) results, for the most part, from manifestations of the dynamics of flowing water described in Chapter 4. Downstream differences in velocity, discharge, and load contribute to the formation of numerous habitats differing with respect to water movement, composition of substrate, and chemical characteristics. The occurrence of plants and animals along a stream course can be limited by any one of a number of factors encompassed by the foregoing categories, or by interactions of several factors. For example, the oxygen available in an aquatic habitat affects the distribution of, among others, the endemic fish species (Figure 16.8). In this case

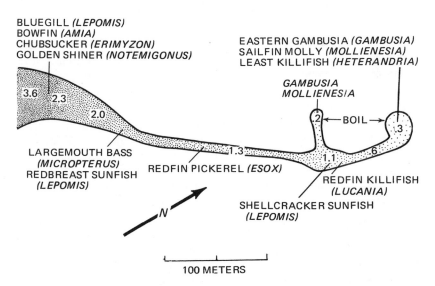

Figure 16.8   Distribution of twelve species of fish in Beecher Springs, near Welaka, Florida. The locations shown for each species indicates the point closest to the head of the spring at which fish were observed on a sunny day in June, 1953. Dissolved oxygen values are indicated in parts per million. (After Odum and Caldwell, 1955.)

we can observe the "law" of the minimum operating to limit the upstream penetration of several species. The water issuing from the "boil" is oxygenless, but with downstream flow atmospheric oxygen soon enters the water, with consequent changes in the fish species composition of the stream.

On a smaller scale we can see the effects of velocity and bottom composition on the distribution of animal species in three short segments of a spring run in the southwestern United States. Lander Springbrook, New Mexico (Figure 16.9), is a small stream, only 77 m long. It is relatively chemostatic and thermostatic, and is inhabited by a generally depauperate fauna, ill-nourished and undergrown. It serves well to illustrate relationships between community structure and various features of the environment.

Along the course of a large stream we find environment–community interrelationships magnified and multiplied many times over those of Lander Springbrook. The nature of bottom sediments along the course of a stream has been described more fully in Chapter 4. We can, however, review this as we consider, briefly, some general aspects of benthic communities in running water.

The substrate of lower stream courses is characteristically of fine sand, silt, mud, or mixtures of these. The water is usually turbid. Animals of the benthos are not unlike those of the same community in lakes. Tendipedid larvae, inhabiting self-made tubes, burrowing mayflies, odonate nymphs, and the larvae of the alderfly are common insects of this region. The annelid *Tubifex* may be abundant. Mollusks, such as the small clams *Sphaerium* and *Pisidium* and the larger forms *Unio* and *Anodonta,* occur commonly in or on the substrate. Under favorable conditions, the population density of *Sphaerium* becomes exceedingly high, on the order of 5000 or more per square meter. In regions of the Mississippi River over 30 species of pelecypods may be found (Riley, 1937). The number of molluscan species decreases toward the upper reaches of streams. Several species of mud-grubbing or detritus-feeding fishes commonly frequent the bottom community of lower stream courses. These include the carp, suckers, catfishes, and, in estuarine regions mullets *(Mugil)* and young croakers (Sciaenidae). All of these forms feed primarily on detritus.

The upper stream course is characterized by pools, stretches of fast water, and riffles; each of these harbors rather distinct communities. In pool communities a supraneuston commonly occurs, and the nekton consists of fishes, such as the sunfishes (Centrarchidae) and minnows (Cyprinidae). The riffle sections are inhabited predominantly by insects, the community frequently containing a large number of species. A single riffle area in a California stream was found to be inhabited by nearly 40 kinds of insects (Reimers *et al.,* 1955).

The benthic communities of rapids and mountain brooks are composed mainly of organisms adapted for living in the swift current. The plants of the community are mostly filamentous algae which are firmly attached to the substrate, or diatoms which grow as crustose masses usually covered by slippery mucous secretions. The moss *Fontinalis* occurs in this region. The benthic animals exhibit a variety of adaptations which fit them to the rapid waters.

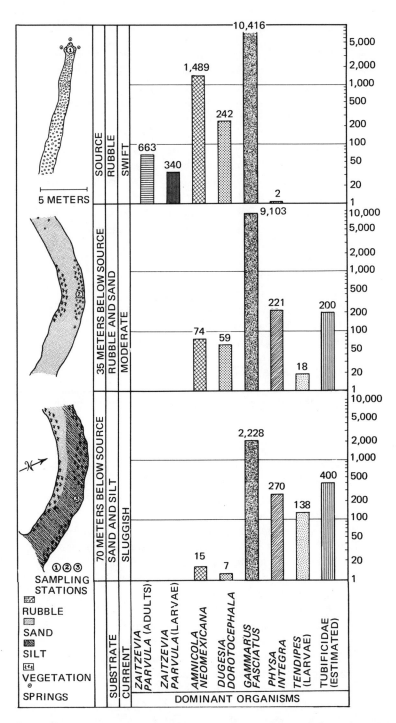

Figure 16.9 Habitat characteristics and mean annual populations of some animals in Lander Springbook, New Mexico, 1950-1951. (Data from Noel, 1954.)

Streamlined or flattened bodies reduce friction with the current or permit the animal to inhabit cracks and interstitial spaces among the rocks; rapids-dwelling mayflies, stoneflies, and dragonflies exhibit these adaptations. The larvae of the blackfly *(Simulium),* a common inhabitant of swift water, attaches to upper surfaces of stones by suckers on the posterior end of the body (Figure 15.12), or by means of strands secreted by the animal. Some caddisfly members of this community add small stones to their pupal cases, presumably for additional weight; the larvae of some species may not build cases but cling nakedly to objects in the stream. Many insects possess strong claws for holding and for locomotion against the current. Snails and flatworms are adapted by means of adherent surfaces and the secretion of mucilaginous material. Some animals, such as crayfishes and some insects, inhabit regions of fast water by more or less avoiding the major factor of current. These forms live behind stones or on the downstream surfaces of stones, thereby gaining protection from the current. Fishes of the benthos are predominantly darters of the family Percidae. These have streamlined bodies and strong pectoral fins used for bracing against the current. Sucking mouthparts of amphibian tadpoles and the armored catfishes of South America are further adaptations to life in mountain streams.

Pollution and Communities of Pollution

In many parts of the world there remain but few primitive streams unaffected by the ways of man. Throughout much of North America, most streams have, to varying extents, been modified from their original state. In many instances, entirely too many, these man-made changes are the results of pollution, *i.e.,* the introduction of substances which make the waters abnormal in comparison with natural, undisturbed waters. Pollution, and the subsequent change in community composition, comes about in essentially three ways: (1) by the introduction of erosional products, such as silt and clay, through improper control of soil in mining, timbering, and agricultural practices; (2) through inflow of materials from industrial operations which directly poison the water or otherwise make the environment uninhabitable; and (3) by the dumping of domestic sewage or industrial substances which enter into biological processes, resulting in a lowering of the oxygen concentration below the limits of tolerance of the original inhabitants.

Many agencies concerned with water use have been prone to disregard or minimize the roles of turbidity and sedimentation in pollution. In the sense that these factors can exert adverse effects upon the original aquatic communities, as well as upon the proper use of natural waters by humans, suspended and deposited substances should be considered as pollutants. We have already seen in Chapter 7 that turbidity is a major influence in determining development of phytoplankton and rate of photosynthesis which, in turn, influence overall productivity in bodies of water. In earlier sections of this chapter, we learned that communities based upon submersed plants contain a great number of species and are highly productive; deposition of silt on these communities can eradicate the entire assemblage. Examples of this effect are common throughout the land; one has only to look at downstream sections of streams located near mining operations, housing developments, road-building operations, farming, or lumbering activities.

Limnological studies concerned with detection and control of sedimentation as a pollution process would involve considerable application of principles of stream hydrology and limiting factors. In practice, the establishment of rigid standards of turbidity and sedimentation may be difficult, due to the complexities of stream flow and the biology of communities encountered from one area or drainage basin to another.

Communities found in turbid and sedimented stream regions are generally those described earlier for lower stream courses; these typically include a small number of species, such as annelids, dipterans, mollusks, and fishes. The development of such communities is, however, highly dependent upon rate of sedimentation and degree of turbidity. Under extreme conditions of either of these factors, the community may be severely reduced.

Polluting effluents from industrial plants are highly varied and may affect aquatic communities in many ways. One category of substances might include those which impart a disagreeable odor or taste to the receiving water; the effects of these might affect not only the aquatic communities, but nearby human assemblages as well. A second group of materials includes chemicals such as lead, phenolic compounds, sulfite, acids, and numerous others which could be directly toxic to all organisms or to certain ones which may be important in food relationships in the community. In either case, there are apt to be deleterious effects on the animal and plant groups. Chemical effluents may also act to make the environment untenable by changing the density or chemistry of the water. Brine from oil fields causes streams to become highly saline, and, more recently, it has been found that wastes from mining of uranium ore, if unrecovered, may greatly lower the pH of streams in the vicinity. Radioactive wastes from research and industrial installations would fall into this general category; the effects of these materials are poorly known at present but are certainly worthy of considerable study. A third category of industrial substances encompasses organic compounds which through rapid decomposition utilize great quantities of oxygen, or through slow biochemical digestion form flocculent masses which increase turbidity and suffocate organisms. Included here are fats and coal-tar derivatives common to a great number of manufacturing processes, and cellulose carbohydrates of paper-making. A fourth category is concerned with heating of stream waters brought about by the use of the water in cooling processes in large industries. Water returned to streams after circulating through cooling systems may often be of considerably higher temperatures than when taken in.

The dumping of domestic sewage and organic substances often exerts great effects on stream communities, primarily through uptake of oxygen beyond the "normal balance" of natural photosynthesis and respiration processes in streams. Depending upon local conditions, the inflow of oxidizable matter may be of sufficient magnitude as to bring about complete depletion of dissolved oxygen, in which case aerobic bacteria fail to function in decomposition of the material; this latter process is continued, however, by anaerobic forms. As a result of such conditions, gases (such as hydrogen sulfide and methane) are formed, and a species-depauperate community exists in the highly polluted region.

It is well established that the effects of pollution become less pronounced with increasing distance downstream, assuming of course, that no additional pollution occurs. This process is reflected in a form of longitudinal succession of stream communities, the composition of each being determined by the degree of pollution throughout a stream segment. A number of schemes devoted to classification of pollution zones have been devised. One such system, applied to the River Trent in England, is shown in Figure 16.10. This figure also serves to illustrate longitudinal succession associated with certain chemicals introduced into the stream. Because the natures of pollution and of the streams themselves differ greatly, it has been difficult to establish a standard scheme for describing zones of pollution. It is generally agreed, however, that the following zones are found in heavily polluted streams:

Zone of recent pollution   Much organic material present. Early stages of decomposition. Dissolved oxygen content usually high. Green plants present. Fish may be abundant.

Zone of active decomposition–septic zone, polysaprobic zone   Great amount of oxidation occurs here. Zone may be nearly depleted of dissolved oxygen. Much carbon dixoide and hydrogen sulfide. Bacteria abundant. Green plants mainly absent (blue-green algae may be present). Many kinds of proto-zoans live in this zone. The rat-tailed maggot, *Tubifera (=Eristalis)*, abounds. If a small amount of oxygen is present, tubificid worms may be found, often in great abundance. Through decomposition of organic matter, this zone grades into the

Strongly polluted zone–alpha-mesaprobic zone   Green algae present, although in reduced numbers. Bacteria continue abundant. Dissolved oxygen content low, particularly at night. Fauna more varied and a greater number of species present than in the preceding zone.

Mildly polluted zone–beta-mesosaprobic zone   Green algae and "higher" plants common in this zone. Dissolved oxygen usually above about 5 ppm. Diminishing oxidation of organic matter. Conditions within range of tolerance of many animals. Fishes such as eels, carp, and minnows may inhabit this zone.

Zone of cleaner water–oligosaprobic zone   This section is essentially free of pollution, although decomposition occurs. Dissolved oxygen content gener- ally above 5 ppm even at night. Green plants are abundant. Animal populations are generally those of typical "healthy" streams of the region.

The nomenclature used here to refer to the different zones has been chosen from several proposed classifications. For a more complete treatment of the subject, see Fjerdinastad (1964). The determination that a stream is polluted, and measurement of the extent of pollution, may be carried out in any of several ways. The most important are:

(1) direct chemical analyses of water;

(2) bioassay, in which test organisms are placed in water samples and their reaction compared with controls;

(3) observation of "indicator" organisms, *i.e.,* certain plants and animals which experience has shown to be significantly characteristic of kinds and degrees of pollution. It has also been suggested that the absence of organisms known to be highly intolerant of pollution might serve as an indication of unnatural conditions. At present, the concept of indicator organisms is receiving much reconsideration and appraisal.

Another closely related ecological approach to recognition of pollution involves the effect of the condition upon community composition. This approach relates the limiting factors principle to species number and population density of a community. In other words, when a factor, oxygen for example, becomes limiting, a number of narrowly tolerant species are lost, thus decreasing community-wide competition for the factor. The remaining, more tolerant, species are then able to undergo population growth. Comparison of community structure, when well known, provides, therefore, an index to pollution;

(4) measurement of the amount of oxygen required to stabilize the demands for aerobic biochemical action in the decomposition of organic matter. This is not the amount required to completely oxidize all organic matter, but rather the volume necessary to restore balance between oxidation and bacterial activity. This measure, known as the **biochemical oxygen demand**, or **BOD**, is widely used in sanitary engineering work. For details of this and other methods of water analysis see A.P.H.A. (1971); for further information on various aspects of pollution consult publications from the U.S. Department of Health, Education, and Welfare, especially those of the Robert A. Taft Sanitary Engineering Center, Cincinnati, Ohio.

ESTUARIES

As we have seen, estuaries are characterized by tidal movement in the lower basin, and consequent mixing of salt water with fresh. There is often a continuous changing of salinity at any one point in an estuary and a moving of the whole salinity gradient pattern up and down stream. In order for organisms to

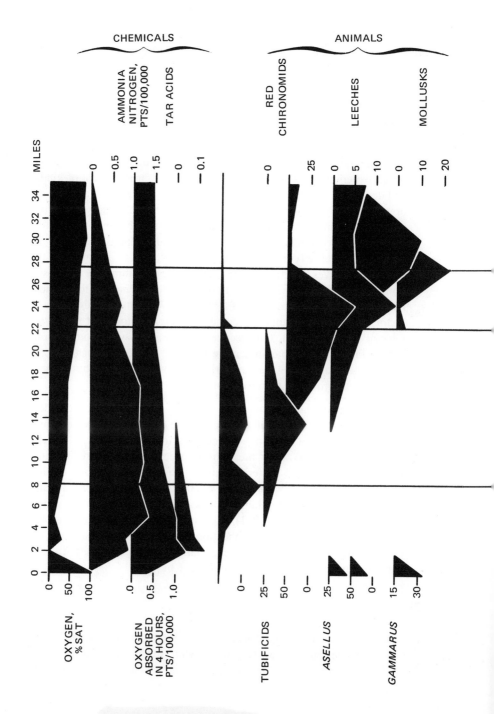

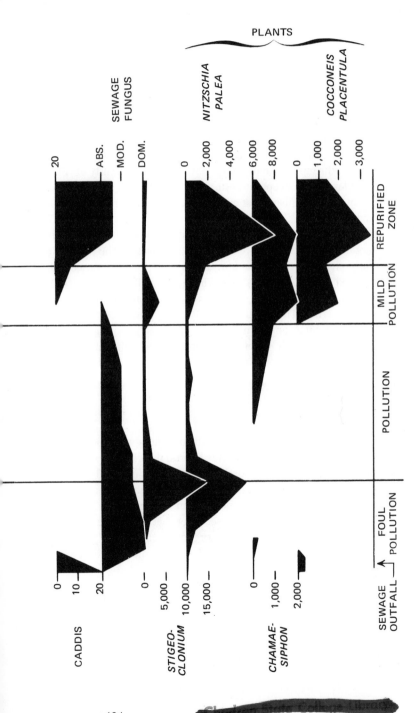

Figure 16.10 (Overleaf) Longitudinal succession associated with pollution in the River Trent, England. Beginning at the area of sewage outfall, four zones are recognized: (1) The zone of foul pollution, where sewage fungus flourishes; *Stigeoclonium* and tubificids also inhabit the zone. (2) The zone of pollution, where tubificids are abundant but other animal life, except for chironomids, is rare; fungus is common and a few algal species become more common. (3) The zone of mild pollution, characterized by great numbers of the isopod *Asellus* and of algae; sewage fungus declines in quantity. (4) The zone of repurification, in which, other factors being similar, the biota resembles that living above the outfall. (After Macan and Worthington, 1951.)

survive in estuarine conditions, attached forms must be able to tolerate frequent changes in salinity and accompanying osmotic pressure. Plankton and nekton, in contrast, can move with the water, remaining in the concentration which they can endure. In addition to variations in salinity, other problems faced by estuarine organisms include waves, increased water temperature, and high runoff (from land and from municipal waste), which increases the concentration of biogenic salts.

The value of estuaries is linked to the sea. The outflow from estuaries adds nutrients to the ocean and has a major effect on the abundance of life in adjacent coastal waters. Estuaries are also important because they provide protection for a wide range of ocean organisms which spawn or spend a portion of their lives in estuaries. For instance, shrimp spawn offshore, but much of their early development occurs after the larvae are swept into estuaries. It is this crucial function as a nursing ground of marine organisms that makes the estuary so very important to marine ecology and economy.

Estuarine species occur in a consistent pattern, the number varying considerably from the mouth at the ocean upstream to the fresh water. The number of individuals is usually greatest near the mouth, declining to a minimum near the headwater, in the region of mixing and extreme changes in salinity (Figure 16.1). Here only a few marine species persist while a few freshwater forms are also found; it is an area of overlap. The structure and dynamic processes of estuaries result in several types of habitats and communities. The most important areas are the headwater region where fresh water enters and mixes with seawater; the upper coves or basins; the lower bay; the tidal salt marsh; and tidal and mud flats.

Headwater Communities

The headwater of an estuary is often near a falls or a dam, at the point where fresh water first mixes with tidal salt water. This intertidal region is exceedingly hostile to living things because of the sudden great shift in salinity. The severity of this stress will be greatest in those estuaries where marine waters reach the headwater at each tide; the stress will be much less in a long estuary, or in one whose mouth is restricted, and in which the seawater that enters the estuary is diluted before it reaches the headwater. Aquatic organisms are virtually absent in the more severe conditions of headwaters. Little but blue-green algae occurs in some headwater habitats.

Communities of Upper Basins

In some estuaries, upper basins, or bays connected to the upper basin, may have the inflow of tidal waters restricted sufficiently that the seawater is always

dilute, with salinity values of 0.5 to 15 parts per thousand. These brackish-water habitats support a luxuriant community of specialized organisms. Narrow-leaved cattail *(Typha angustifolia)* is the predominant emergent aquatic, while water-weeds such as sago pondweed *(Potamogeton pectinalis)* and Dutch grass *(Ruppia maritima)*, often mixed with one or more species of stonewort *(Chara)*, form dense, attached submerged stands. Brackish-water animal life includes numerous fish, snails, crabs, and oysters. Caspers (1958) counted in one square meter, 6500 amphipods, 500 tubificids, and 25 polychaetes in brackish-water of the Elbe estuary. He measured the total biomass at 37 g/m². Bacterial action is high, and sediments usually are black and reducing and smell of H_2S.

Lower Bay

The major portion of the estuary, from the mouth upstream, might be considered as the lower bay. Tidal action is maximal in this region and the salinity approaches that of seawater.

The plankton in the lower bay consists primarily of marine species, such as the diatoms *Coscinodiscus* and *Chaetoceros,* the dinoflagellate *Ceratium longipes,* and the silicoflagellate *Distephanus.* Freshwater forms which are swept in from upstream include *Asterionella formosa* and *Fragilaria capucina.* More typically brackish-water genera include the flagellates *Exuviaella* and *Prorocentrum.* Zooplankters prominent in estuaries include *Mesodinium, Codonella,* and *Tintinnopsis. Keratella* usually a freshwater rotifer, is found occasionally. In Narragansett Bay, there are generally two main seasonal pulses, one in early summer (June) and another in late fall (September to December). During a pulse, the numbers of organisms can exceed 1,000,000 cells per liter.

"Red tides" are blooms of a single kind of phytoplankton, commonly *Gonyaulax* or *Gymnodinium,* which appear to build up in an estuary and then to spread out along the coastal waters. This "tide" produces powerful toxins which can kill fish and even affect man when they are spread ashore with the spume from breaking waves. What it is that triggers these blooms, and what causes otherwise innocuous blooms to produce dangerous toxins, are still not known.

The benthos of sedimented lower-bay regions is often characterized by a dense vegetation of eelgrass *(Zostera)*; in estuaries of the southern United States, this is replaced by turtle grass *(Thalassia)*. In European waters, Redke (1933) reported the benthic fauna of one estuary to include the coelenterate *Cordylophora,* the annelid *Nereis,* 7 copepods *(e.g., Eurytemora* and *Acartia),* 5 amphipods *(e. g., Gammarus),* the mysidian *Neomys,* one decapod, one mollusk, and two bryozoans. Caspers (1958) measured the biomass of the benthic fauna of the Elbe estuary, Germany. The lowest value recorded (less than 40 g/m², mostly amphipods) occurred upstream; the highest (6068 g/m², mostly mollusks) occurred at the mouth.

Tidal Salt Marshes

The shores of protected portions of estuaries are often occupied by tidal salt marshes. These broad, flat plains extend from the water's edge to the uplands, at the extreme limit of tidal influence. The vegetation of a Connecticut marsh (Miller and Egler, 1950) consists of three or four zones. The outermost zone, next to the bay or channel, is dominated by tall cord grass, *Spartina alterniflora*. Inshore there is a broad pasture of short, often somewhat matted salt-meadow grass, *S. patens*. Still further inshore, one finds the black grass *Juncus gerardi*. The limit of tidal influence at the highest strandline is marked by the occurrence of the switchgrass *Panicum virgatum*. On the surface of the tidal marsh itself, various algae, such as the green alga *Enteromorpha*, form weblike tangles amid the grass culms. This algal layer has been found to be very productive. Crabs may abound in such marshes, making burrows that extend throughout the marsh. Little is known about the life in these burrows, but their openings are known to play an important part in aerating the roots of the grasses.

Gross primary productivity of salt marshes is generally reported to be about 20,000 kcal/m²/yr. However, in a Georgia salt marsh, productivity values for *Spartina* have been estimated at 34,580 kcal/m²/yr (Smalley, 1959), while values

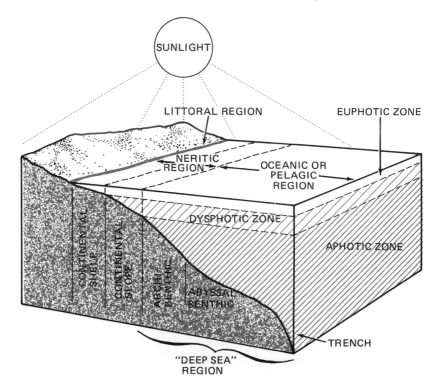

Figure 16.11   The major horizontal and vertical life zones of the sea. Note the extreme thinness of the euphotic zone.

for algae have been estimated at 1800 kcal/m²/yr (Pomeroy, 1959). These are among the highest natural productivity rates recorded.

Tidal Mud Flats

Extensive estuarine mud flats are often exposed at low tide, particularly along river banks and just beyond the limit of tidal salt marshes. Hard-shell or quahaug clams abound in such flats. When covered by the tide, these filter-feeders project their siphons up from their holes and pump water through their systems, extracting the food. Extensions of tidal flats out into the mouths of estuaries support the oyster reefs typical of southern waters, while to the north such reefs may become equally populated with mussels. Mud flats appear at first to be devoid of life, but they are often teeming with bacteria, protozoa, worms, ghost shrimps, and clams.

MARINE COMMUNITIES

The distinction between freshwater and marine life is controlled by two main environmental factors—salinity and history. If not for salinity, the environmental factors affecting a submerged rock at a depth of several meters would be largely the same in either fresh or salt water. However, the salt content of seawater makes it so different osmotically that the organisms of the two rocks would in fact be totally different. Most organisms are adapted to either a fresh- or a salt-water environment, and only a very few species are able to survive in both. The historical factor is mentioned to suggest that the striking differences observed in organisms of inland and marine waters is a result of long evolution which has occurred independently in the two environments. Angiosperms, amphibians, and insects have become major elements in the life of inland waters, but are relatively unimportant in the seas. Macroscopic seaweeds, coelenterates, mollusks, and echinoderms, on the other hand, have evolved into the major life forms of many marine environments.

The ocean is usually thought of as being naturally divided into regions and zones corresponding to those distinguished in lakes; these are shown in Figure 16.11. The major regions, each supporting more or less characteristic communities, are the **oceanic** region (waters lying beyond the continental shelf); the **neritic** region (over the continental shelf); and the **littoral** region (the intertidal zone). The coral reef is also distinguished as a separate region, because of its unique geological and biological structure.

The Oceanic Region

The oceanic region, by far the greatest part of the entire ocean, is the area that lies beyond the limit of the continental shelf. It is the region of blue water and great depths. The bottom and coastal regions are so remote that living things are continuously suspended. Dead organisms may take months or years to sink to the bottom, and organisms swept along by currents may take weeks or months to reach shore. Organisms that sink gradually are repeatedly swept back to the surface by waves or convection cells. The organisms that survive either can swim (the nekton) or else have well-developed methods of flotation (gas, high fat content) or structures that retard their sinking, permitting them to be cycled back to the surface. The entire cycle of such materials as NO_3, PO_4, Ca, Si, and food occurs in open water. Within the oceanic region, four more limited zones are distinguished: the surface, the euphotic zone, the aphotic zone, and the benthic zone.

The ocean surface   The **ocean surface,** as the interface between air and water, is a variable and often temporary habitat in oceans. Only a few insects are known to venture far out to sea, the water-striderlike *Halobates* being the best

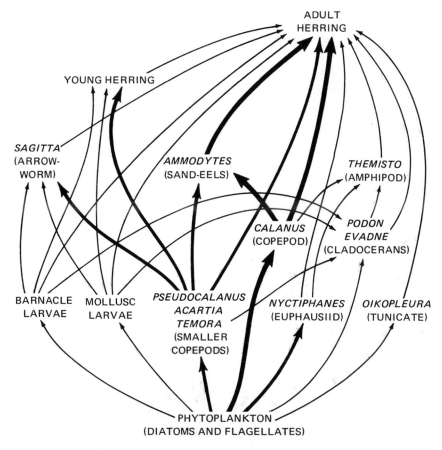

Figure 16.12   Marine food web supporting a population of herring (*Clupea harengus*) in the North Sea. (After Russell-Hunter, 1970.)

known. A few siphonophores and cephalopods, numerous birds, and several mammals occur in this layer. Examples of the siphonophores are the Portuguese man-of-war *Physalia* and the lovely sail-equipped *Velella;* the cephalopods include the chambered *Nautilus* and the paper nautilus *Argonauta.* Rafts of sargasso weed *Sargassum natans* are scattered throughout the Sargasso Sea, and a number of them move northward along the Gulf Stream, carrying with them a closely interdependent community of specialized fish and invertebrates.

The euphotic zone The **euphotic zone,** comparable to the trophogenic stratum of inland waters, extends down to the depth (about 100 m) at which there is just enough light to cause at least some photosynthesis. The vast bulk of all marine life occurs comparatively in this thin layer. Open-water autophytes include planktonic diatoms, dinoflagellates, silicoflagellates, and yellow-green microflagellates. In the northern seas, the predominant diatoms are centric types such as *Chaetoceros, Bacteriastrum,* and *Coscinodiscus,* and dinoflagellates include *Peridinium, Ceratium, Glenodinium,* and *Gonyaulax. Halosphaera* is practically the only green open-water planker, while *Phaeocystis* is a brackish-water chrysophyte alga occurring in gelatinous colonies. These organisms are most productive when in the level with optimum light intensity, and become more sparse and less productive as they are moved up or down from that depth. Dinoflagellates, in general, tolerate more intense light, and are found in greatest numbers in the upper layers. Diatoms seem more concentrated in somewhat deeper parts of the euphotic zone. The microflagellates, lumped with nanno-plankton, seem to occur throughout this zone. The phytoplankton are fed upon by zooplankton, such as copepods, foraminiferans, and radiolarians, and these in turn support populations of larger zooplankters, such as crustaceans *(Euphausia, Crago),* which are eaten by small fish (Figure 16.12). Because of the high diversity of forms, open-water food webs can be very intricate.

The net primary production in the euphotic zone of open oceans has been calculated to amount to about 1000 kcal/m^2/yr, ranging from a high value of 6000 kcal/m^2/yr in certain areas of upwelling down to about 8 kcal/m^2/yr in the Sargasso Sea (Ryther, 1969).

The aphotic zone Deep-water communities lie in the **aphotic zone;** they are not exposed to solar radiant energy and are totally dependent on organic materials that move down from the euphotic zone. The fish of these dark regions are mostly small, rarely exceeding 15 cm in length. They are generally black. Many have luminescent structures. Some have oversized mouths and tiny bodies, features fitted to survival in a cold region of sparse food. There are regions in which certain plankters become more numerous than usual. A layer of sergestid shrimp occurs between 230 and 270 m depth, a layer of euphausiids at about 430 m, and a third of small myctophid fish between 520 and 580 (or 700) m (McConnaughey, 1970). It would seem that, for these organisms to remain suspended at such specific depths in uniformly dark and thermally stable water, the density either of the zooplankter or of the food must be such that settlings accumulate at these depths. It is also possible that these are shearing layers between moving strata of water of different densities. The biomass in these layers is greater than in adjacent ones, although the standing crop values are still low relative to those of surface waters.

The benthic zone The **benthic zone** is the bottom. In the deep-sea region it is uniformly cold, dark, and presumably calm. Perhaps surprisingly, the extreme water pressure and cold do not exclude living things, and although they are sparse in numbers and low in biomass, organisms from most marine animal phyla are represented. Plant life in general is virtually absent, and bacterial action is slight. Among the animals represented, holothurians and crinoids are the most common, but invertebrates include sponges, polychaetes, isopods, gastropods, starfish. Vertebrates include tunicates and a number of small fish such as *Chimaera* and *Harriotta*. The food web of this benthic community is fairly simple, filter and detritus feeders rather than algae forming the basic trophic level. Energy is obtained from the rain of detritus and occasional carcasses from above.

The Neritic Region (Waters Over Continental Shelf)

Neritic waters are those over the continental shelves. Depth varies from just zero at low tide level to as much as 200 m at the outer edge of the shelf. Much of the area, however, is less than 80 m deep, and is well within the euphotic zone. Although much of the neritic region has the intense blue color of open oceans, its color grades to aqua, and finally to brownish, in shallower water.

The neritic region is far more productive than the open ocean. It is over the continental shelves that the great fish and mammal migrations occur. These neritic waters differ from oceans in being less constant in chemical and physical features; the farther inshore, the more variable these become. The outflows from major rivers and estuaries dilute the water and add a variety of new factors. Waves and currents sweep up nutrients from the bottom and bring up food which is in the process of settling on the bottom. The major neritic zones are the surface, the euphotic zone, and the benthic zone.

The neritic surface The surface in neritic waters is similar to that of open oceans. However, red tides occasionally occur extending from the vicinity of some estuary along the coastline. Inshore mammals such as seals and sea lions are found, especially near breeding rocks; and seabirds become increasingly common. One such area is the Pribilof Islands, Alaska, a sea lion breeding ground; another is Bonaventure Island, Quebec, a gannet and murre rookery.

The euphotic zone The zone of water overlying the continental shelf is generally shallow enough to be within the euphotic zone. Phytoplankton exists throughout much of the water column, and heavier species swept up from the shallower bottom may mingle with oceanic forms. Plankton of inshore waters often has a higher density and greater diversity than that in open oceanic waters. Pennate diatoms tend to predominate over centric ones in the phytoplankton, and armored dinoflagellates predominate over unarmored forms. They tend to have less ornate flotation structures, and are often tychoplanktonic from the bottom.

The zooplankton of inshore waters consists largely of species that prefer relatively warmer water with somewhat reduced salinity. Vast swarms of larvae of benthic invertebrates and spawn of fish are characteristic. These temporary plankters have been called **meroplankton.** Among inshore water animals, plankton, copepods are especially predominant.

Mean productivity rates for coastal waters have been reported (Ryther, 1969) at about 2000 kcal/m²/yr, about twice that given for the open ocean. In waters off river mouths, values are considerably higher, rising to about 5760 kcal/m²/yr off the Mississippi (Thomas and Simmons, 1960) and to about 10,940 kcal/m²/yr offshore from a Georgia salt-marsh estuary (Thomas, 1966).

The benthic zone Benthic conditions in the neritic region are less rigorous than in the ocean deeps. Some light is present and plant life is not totally absent. Food settling from above is much more available here than in the deep seas, for plankton and nekton are more abundant than in the open ocean and they do not have so far to sink nor so long to decompose. Also, the existence of a bottom adds a suitable habitat for many sessile organisms which are excluded from the upper regions of oceanic waters, and meroplanktonic larvae from these benthic species add considerably to the zooplankton of neritic waters. Life in the neritic benthos is rich, and includes a wide variety of protozoans, sponges, coelenterates, sea worms, mollusks, echinoderms, and chordates. The horseshoe crab *(Limulus)* and lobster *(Homarus)* are common, as are benthic fish such as flounder and rays. In increasingly shallow water, where light supports greater algal growth, beds of attached seaweeds such as *Laminaria* or *Macrocystis* form dense forestlike vegetation. Animal life in the inshore sediments is sometimes differentiated into **epifauna** and **infauna.** The epifauna includes the animals on sediment or among rocks and other structures. The infauna comprises animals that live within the sediments, often in holes they have dug. The fauna changes periodically, especially with sequence of larva production, but generally the benthic communities are quite stable. Productivity, too, is fairly uniform. The standing crop of epifauna foraminifera community at 200 to 350 m depth is 10 to 15 g/m², while in deep-sea benthos it is nearer 1 g/m².

The Littoral Region (Intertidal Zone)

In marine studies, the term **littoral** has been used in several ways, and none agrees with the concept as employed in inland waters. From a broad point of view, it is the tiny rim of the oceans, just a very few meters in height and extent, which is alternately immersed and exposed by tides. Along the shoreline in the intertidal zone, conditions vary from sand beaches to rocky coasts.

The marine beach The life of a marine beach depends on the nature of the substrate. Shingle and cobble beaches seem virtually sterile, perhaps due to grinding of stones over one another caused by waves. Sand beaches are somewhat better inhabited. The small sand crab *Emerita* is common in the sand near the water line in southeastern and western beaches. Also, a number of slender animals from nearly all orders have been reported (Pennak, 1951) in the interstitial spaces among the sand particles. A number of forms migrate across the beach, such as the ghost crab *Ocypoda* (Hedgpeth, 1957). Shorebirds forage along the beach, and in certain regions sea turtles *(Cheloniidae)* lay eggs in the sand. Considerable amounts of seaweed drift ashore in the wash and become stranded by falling tide, sometimes piling up to a depth of a meter or more. These dying algae, often decaying in place, permit heterotrophic decomposition as well as putrefaction. The swash and backwash of the waves can draw organic matter back into the sea or work it down into the sand.

   Rocky shorelines   Rocky coasts, typical of sinking shorelines, tend to develop a pattern of life that is similar throughout much of the world. This is characterized by what are called the "universal tide features" (Stephenson *et al.,* 1949). Between the highest level of storm waves and the lowest level to which water sinks, living things are sorted into zones. On the rocky shorelines of New England (Table 16.3), this involves the uppermost bare rock area (the **supralittoral** zone), then a blackish stainlike band (**upper littoral** zone) 0.5 m or more wide—the accumulated colonies of a blue-green alga *Calothrix crustacea.* Mixed with this are somewhat distinctive patches of blackish lichens *(Verrucaria)* and patches of the feltlike green alga *Blidingia minima* (Johnson *et al.,* 1961). Beneath this zone, a bare-rock belt blends into a barnacle zone, about 1 m wide, occupied mostly by *Balanus balanoides.* Mixed with the barnacles are occasional

TABLE 16.3   Biomass of major intertidal zones at two stations on a protected rock jetty, Galilee, Rhode Island[a]

| Algal zone | Predominant organisms | Biomass, g/m^2 | | | |
		Wet	Dry	Wet	Dry
Black	Calothrix Balanus	100	15.5[b]	102.5	15.9[b]
Feltlike green	Blidingia Calothrix	—	—	152	23.6[b]
Brown	Fucus Ascophyllum	6816	1060	6803	1058[b]
Green	Enteromorpha	192	40	152	23.6[b]
Red	Chondrus Rhodymenia	5284	821[b]	6789.1	1055[b]

[a] Class data, October 1971 and 1972.
[b] Calculated from factor of wet to dry weight of 6.43.

stunted brown algae, usually *Fucus*. A line, usually quite sharp and straight, cuts along the upper margin of this brown algae zone and marks the upper edge of the **midlittoral** zone. This brown or rockweed zone is often a mixture of *F. vesiculosus* and *Ascophyllum nodosum,* frequently with patches of *F. edentatus (F. distichus)*. The broadest and densest vegetation of the intertidal zone, (with the highest biomass) occurs in this region. The midlittoral is often subtended by a fairly narrow belt of a green, bladelike alga *Enteromorpha linza*. It then gives way to a compact turf of the blackish-red Irish moss *Chondrus crispus,* which extends from mean sea level down to where the large, stalked blades of *Laminaria agardhii* and *L. digitata* begin the **sublittoral** or **infralittoral** zone at about mean low water. Beneath this, the zones are obscured by patchy stands of a variety of seaweeds. Similar zones, with different but ecologically equivalent species, have been described on the west coast (Stephenson, 1972). Data on animal life in the rocky littoral is somewhat difficult to find, but mollusk species are known (Dexter, 1945) to occur in a similarly stratified pattern, with the *Littorina* (periwinkle) species in the upper areas and such forms as *Crepidula* (slipper limpet) and *Anomia* (jingle shells) restricted to near water-level (Table 16.4).

Coral Reefs

The term **coral reef**, or **organic reef**, is applied to the rocklike reefs built up of living things, principally corals. The reefs are biologically dominated geological structures of the warmer seas. There is some question as to how they develop, but they consist of accumulations of calcareous deposits of corals and coralline algae with the intervening space cemented with sand which consists largely of shells of foraminifera. Present reefs are living associations growing on this accumulation of the past. They are generally restricted to waters between 23.5°C and 25°C, but occasionally occur in waters which reach as low as 18°C, as in the Bahama Islands. Coral reefs are of different types—fringing, barrier, platform, and atoll; but the living associations are similar on all kinds. Local conditions, especially exposure, affect the type of organisms present more than does the type of reef (Wells, 1957). The reef will often stand like a protective wall around an island or lagoon, creating a body of calm water in the midst of a pounding sea.

The density of life on such organic reefs makes them of special interest to the ecologist. In the vast Pacific Ocean, which is extremely sparse in organisms and low in nutrients, such reefs have an abundance of animal and plant life that makes them seem like oases in a desert. The communities of a reef occupy several distinct zones, which may be considered in terms of their relation to the prevailing winds. From the windward side of an atoll to the leeward, these major zones are the outer reef wall, the algal ridge, the reef flat, the lagoon slope, the lagoon, and the knolls and patch corals.

TABLE 16.4 Vertical distribution of mollusks in the intertidal region of a rocky marine coast, Cape Ann, Massachusetts. Numbers express density of individuals per square meter[a]

Height above spring low-tide level (cm)	Littorina saxatilis	Littorina littorea	Mytilus edulis	Mya arenaria[b]	Littorina obtusata	Thais lapillus	Acamaea testudinalis	Anomia aculeata	Crepidula fornicata
+199	4	4	—	—	—	—	—	—	—
+94	23	112	—	71	—	—	—	—	—
+84	248	2225	23	81	248	35	—	—	—
+68	58	1339	116	0	151	8	—	—	—
+59	31	387	132	8	341	15	—	—	—
+43	—	704	31	0	163	70	—	—	—
+25	—	1300	—	0	77	15	8	—	—
+15	—	83	—	—	—	—	15	15	8
0	—	387	—	15	—	—	—	—	45

[a] Data from Dexter, 1945.
[b] Seeds only.

The **outer wall** or seaward slope of an atoll on the windward side consists of a mixture of the massive boulder-shaped coralline red alga *Porolithon* and such sturdy corals as *Millepora*. The growth expands seaward and upward, but stops at the surface of the water, so the reef grows gradually wider but no higher.

The **algal ridge** is the uppermost edge of the outer reef wall. It is the region where pounding surf breaks continuously, and consists of large, often jagged rocklike coralline algae, mostly *Porolithon*. The ridge is generally broken by surge channels. Much of the water that breaks onto the reef drains back through these channels. Corals are usually absent on the algal ridge, but colonize extensively just back of the ridge on the reef flat in a coral-algal zone.

The broad top of a reef is referred to as a **reef flat**, although it is not really flat, but rather somewhat higher near the windward edge. Here, an encrusting red alga, *Lithothamnion,* may solidify the structure into a rocklike pavement. Regardless of whether the ridge consists of *Lithothamnion* or a reef rock, a thin turf covers this part of the reef. It is usually no more than 5 to 15 mm thick, consisting primarily of the wiry red alga *Jania* with an admixture of other sturdy species. At low water this algal turf appears thin and dry, but this appearance is misleading, as the turf is in fact quite productive; treatment with acetone will cause the turf to yield an intense rich green solution of concentrated chlorophyll.

As the reef gradually tilts downward on the lagoon side, it forms another habitat, the **lagoon slope**. This is a well-protected region with abundant sunlight, warmth, constantly flowing water covering the reef (except at high tide), and nutrients from the rich algal turf. The lagoon slope is densely populated by corals of many kinds, and reef animal life here reaches its maximum in variety and abundance.

The **lagoon** itself is protected on all sides by the wall-like reef, open only by way of a few narrow passes on its leeward side to the more turbulent waters of the open sea. Lagoons vary in size from less than one km across to lengths of more than 100 km. The depth varies, with several Pacific atolls having lagoons about 30 m deep. Water flows into the lagoon over the reef and flows out through the passes.

Phytoplankton in lagoons is extremely sparse, and consists mostly of typical marine species, such as the diatom *Chaetoceros* and the dinoflagellate *Ceratium*. Nektonic animal life is sparse, an occasional fish being observed, but at times swarms of jellyfish or ctenophores occur in the lagoon, and schools of dolphins and flying fish are seen near the passes. The lagoon floor consists of soft coral sand, barren or covered with plant life such as encrusted green alga *(Halimeda)* or sea grass *(Thalassia)*. The sponge *Phyllospongia* may be common. Psammophilic crustaceans, mollusks, and echinoderms occur in the sand.

The densest concentration of lagoon life, comparable to that on the lagoon slope of the reef, occurs around **knolls** and **patch corals**. Knolls are tall slender hills arising from deep water toward the surface. Patch corals are lower, isolated clumps of coral. Knolls are covered with a very great diversity of corals, and thousands of species of fish, coelenterates, and echinoderms have been reported on them. The fish of this habitat are often superbly colorful. The food web in the "coral garden" is complex because of the many species involved.

The gross primary production of the coral reef is very high, calculated at about 20,000 kcal/m²/yr or about 10 g/m²/day. This value exceeds that for biomass by a factor of about 12.5. Such high productivity values are especially remarkable since most of the production in coral stands is by the dense layers of the mutualistic dinoflagellate *Symbiodinium,* which lives in the ectoderm of coral tissue and also, to some extent, by the cylinder of filamentous green algae (mostly *Ostreobium quecketii)* living just inside the coral skeleton. In contrast to these high reef values, those obtained for lagoon plankton in Palau (Motoda, 1969) averaged only about 0.09 g/m²/day, or about twice the value obtained in the open ocean.

STRATIFICATION IN AQUATIC COMMUNITIES

In the first part of this chapter we considered some selected communities of lakes and streams with emphasis on their plant and animal components, and we studied them in terms of their organization as generally predictable assemblages of organisms. We then turned to trophic aspects of the communities and recognized within each its organization with respect to food webs and energy relationships. A third concept embodying organization within the structure of communities is that of **community stratification.**

Community stratification pertains to the horizontal or vertical arrangement of populations, their activities (responses to environmental processes), or their effects on the environment. We have previously considered many cases of both vertical and horizontal stratification in aquatic communities. Thus we need now only to recall some of them in illustration of this very outstanding ecological principle. Thermal stratification, so characteristic of most lakes of the temperate zone in summer, serves to bring about vertical zonation of the lake community. Associated with this phenomenon we have seen the stratification of a lake with respect to heat content, dissolved substances, and, indeed, the animal and plant inhabitants. Community stratification resulting from thermal stratification breaks down during vernal and autumnal overturn, but is re-established the following summer, attesting to the pliant nature of organic communities. Associated with water transparency and light transmission is the layering of a body of water with respect to production and decomposition of organic matter; we have termed these layers the tropholytic and trophogenic zones. In these layers we see stratification of organismal activity and the effects of the organisms on the environment.

Within a given community, certain species populations may occur in strata, usually as responses to tolerance to the effects of limiting factors. In the plankton community of Wisconsin lakes during summer stratification each layer contains a rather distinct population of cladocerans, which is modified, at any given time of day or night, by light intensity. This arrangement lends to each stratum a certain amount of temporary internal integrity. Some populations of the nekton community exhibit sharp vertical stratification as illustrated by two species of fishes in Norris Reservoir, Tennessee, shown in Figure 16.13. The sauger *(Stizostedion canadense)* is inherently a "cool-water" species and less tolerant of warm temperatures. The large-mouth bass *(Micropterus salmoides)* thrives best in waters of warmer climates. During the winter months of uniformly low vertical temperatures, both species exist together. During summer stratification, however, the species effectively segregate according to the most favorable temperature conditions. Since, as we already know, thermal stratification is correlated with depth, the fishes also exhibit, during summer stratification, a marked layering of population density. Note, however, that in early September the sauger population was "forced" into warmer waters by an oxygen minimum which developed with a density current passing through the impoundment. Parenthetically, we might point out that here is an example of the operation of the principle of limiting factors; the sauger population appears to be "preferring" to move into the less optimal temperature regime in order to avoid the low oxygen. These data have practical application in that fishermen may adjust their line length and depth of bait according to vertical temperature and the desired species of fish. Newspapers in the vicinity of Norris Reservoir have published these data in the interest of better fishing.

Horizontal stratification of substrates and their communities is generally characteristic of natural bodies of water. We have considered the general ten-

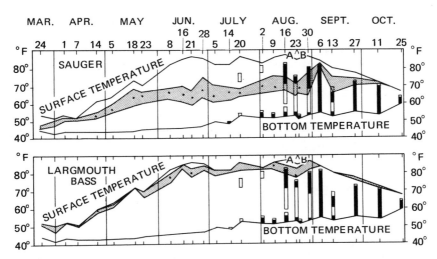

Figure 16.13 Vertical stratification of two fish species in relation to temperature, in Norris Reservoir, Tennessee, 1943. The shaded area shows the temperature range in which the middle 50 percent of individuals in the sample were caught. Vertical bars indicate the concentration of dissolved oxygen: open areas indicate a range of 1.6 to 3.0 ppm; solid areas indicate 0.0 to 1.5 ppm. Note that in early September the sauger responded to low oxygen values (accompanying a density current) by moving into the warmer upper waters. (After Dendy, 1945.)

dency of higher plants to occur in rather distinct zones about the shallow waters of lakes and stream. We have learned, furthermore, of the relative distinctiveness of the plankton of the vegetated zone and that of the adjacent limnetic zone. Similarly, we have noted how benthic communities grade horizontally from the littoral to the nonvegetated bottom of the deep region. Broadly considered, however, the benthos is also influenced by factors related to depth as a vertical dimension. Thus, benthic communities may often be influenced by two dimensions, vertical and horizontal. This is not always the case, however. We have seen, for example, that in the shallow waters of lakes, community composition is often related to the sedimentary nature of the substrate, which may be independent of depth.

Longitudinal succession of communities in streams is now appreciated and accepted as principle. In this impressive linear arrangement of plant and animal assemblages, we observe horizontal zonation clearly structured by environmental factors attributable to stream flow, and by evolutionary adaptations of the organisms. The principles of zonation also operate in estuarine communities. Horizontal zonation of plankton, nekton, and benthos from the freshwater upper reaches to the more saline mouth of the estuary is usually clearly demonstrable.

PERIODISM IN AQUATIC COMMUNITIES

Periodism in communities refers to the recurrence of events, activities, and changes in the component species populations which, in turn, are ultimately felt in the nature of the community. A given recurrence may be sporadic and irregular, or it may be cyclic and regular. Cyclic phenomena are commonplace in nature and, indeed, are obvious and important in aquatic communities. The most conspicuous cycles are those associated with the earth's daily rotation, the 28-day movement of the moon around the earth, and the yearly orbiting of the earth about the sun. These periodisms are best recognized in terms of activities and events related to diel (daily), lunar (tidal), and annual phenomena. Endogenous and long-term cycles may also play a part.

DAILY CYCLES

Although relatively short-termed, **diel** (24-hr) cycles in aquatic communities can often exert considerable influence on community metabolism and structure. The response of plants to light manifested in photosynthesis and primary production is one of the most important daily cycles in nature. In addition to the production of energy-containing substance, this process also is related to nocturnal-diurnal cycles in the chemical composition of natural waters. Recall that at night organismal respiration generally exceeds photosynthesis, while during the day the reverse occurs, resulting in shifts in a number of chemical relationships. Here is a cycle with widespread effects throughout the entire ecosystem.

Daily cycles of plankton activity are prominent in lakes and in the sea. Many species of zooplankton migrate toward the surface during the night and sink bottomward during the day. Rate and extent of movement vary among the species (Table 16.5), resulting in considerable modification of vertical structure of the plankton community. This cyclic change in structure is further complicated by the fact that numerous plankton-feeding animals migrate with the zooplankton. Although light intensity (absolute, or variations in) is doubtless involved in these migrations, the complete explanation is not yet known.

LUNAR CYCLES

Cycles of activity which are associated with 28-day phases of the moon are less well known in inland waters. In estuaries and the seas, however, many events are correlated with tides, and thus indirectly with moon-earth gravitational effects. An oft-cited example is that of the grunion (*Leuresthes tenuis*), a fish of the Pacific coast. These fish appear at night in the surf along the California coast precisely at the time of the highest high tides of the lunar cycle. During the second through the fourth night of this series of ''spring'' tides the fish go onto the shore with the farthest-reaching wave. Here, in the sand, the eggs are deposited and fertilized. Being at this high point, the eggs are not disturbed by waves again until ready for hatching, about two weeks after spawning. At this time, the waves of another spring tide series wash the rapidly hatching eggs and young fish back into the sea. In this adaptation the entire spawning activity and development of eggs of the grunion are tuned to tidal cycles.

TABLE 16.5 Vertical movements of zooplankton in Lake Lucerne[a]

Species	Extent of vertical movement (m)	Time of 1 m descent (min)	Time of 1 m ascent (min)	
			noon to dusk	dusk and later
Daphnia longispina, adults	10 to 40	5	25	6.3
D. longispina, young	5 to 60	3.1	17.5	4.3
Bosmina coregoni, young	7 to 55	4.5	26	7.5
Diaptomus gracilis, adults	13 to 30	12	60	15
D. gracilis, young	20 to 37	20	60	12
Cyclops strenuus, young	12 to 46	7	60	5.5

[a] Data from Allee *et al.,* 1949.

ANNUAL CYCLES

Annual cycles of plant and animal activity, usually associated with a given season, are familiar to all, and thus we will not dwell at length upon them here. The seasonal emergence of midges, craneflies, and mayflies, and the "blackfly season" attest to reproduction and development regulated by annual cycles. In coastal areas, "runs" of salmon or shad into freshwater streams occur with regularity each year. Winter quiescence and summer activity in lakes and streams are reflections of seasonal changes in metabolic activity of the organisms, these changes being related to annual temperature cycles of the environment.

Probably no phenomenon in fresh waters illustrates seasonal changes in community structure and metabolism more vividly than does the annual cycle of overturn and stagnation in temperate lakes. We have already considered the immensity and complexity of the dynamics in this process. This example serves to illustrate how seasonal variations add complications to the community *per se,* as well as to the concept of community ecology.

ENDOGENOUS CYCLES

One of the many complicating factors of community periodism rests on the fact that, as we might have concluded from the preceding discussion, not all cycles are regulated by environmental factors. It has been shown that a great number of population and community activities are regulated by cyclic behavior patterns possessed innately by the organisms. These patterns are termed **endogenous,** and, although synchronized with environmental events, they may function independently of external factors. In other words, endogenous periodisms continue to be manifested by the organisms even if the environmental factors are changed or made uniform. Color changes in the fiddler crab (*Uca*) are synchronized to diel or tidal cycles of a given area. When the cycles are changed, for example by transporting the crabs from the east coast of North America to the west coast, or the clues such as light or temperature are removed, the color-change rhythm persists. Studies of these activities suggest the presence of a "biological clock" within the organism. This mechanism has been called an "endogenous self-sustaining oscillation," and abbreviated ES-SO.

LONG-TERM CYCLES

One further kind of periodism should be included in our consideration. It involves the long-term cycles which constitute a major principle in geomorphology, and which we might term **geologic periodism.** During the earth's history, mountains have been thrown up and eroded in cyclic fashion, glaciers have advanced and receded periodically and continents have formed and moved about.

Such processes as these greatly influence the nature of lake basins, stream channels, and their waters and the communities developed in them. An example of the operation of geologic cycles is seen in a section of earth cut by the Caloosahatchee River in southern Florida. Within a vertical distance of a few meters, several alternating layers of freshwater material and marine fossils (mostly mollusks) record times of fluctuating sea level during the Pleistocene.

ECOLOGICAL SUCCESSION

Our preceding considerations have pointed vividly to the fact that communities change. We have seen change in respect to metabolism, stratification, and periodism. From a broader point of view a community has a beginning, and then undergoes a series of ecological changes until it reaches a relatively stable state. The orderly sequence of progressive change of communities over a given space is termed **ecological succession** (Chapter 12).

The history of a lake offers a classic example of ecological succession and involves many concepts and principles which we have already learned. A typical "young" lake of the temperate zone is oligotrophic; it has little organic content and a poorly developed littoral zone. In time, erosion of the basin and inflow of nutrient substances contribute to eutrophication. Bottom sediments become enriched in organic debris, the littoral develops a zonation of higher plants, and the biota of the limnetic increases. Organic debris and inwashed sediments accumulate in the shore zone, resulting in the lakeward encroachment of the littoral. In time, the lake becomes more shallow; plants which formerly rimmed the shore extend farther and farther lakeward until they meet in the center. The earlier littoral zone is now a marsh, and the central part of the original lake is a pond. Continued filling eventually displaces the water, and, through successional stages involving different plant and animal communities, a forest develops in the area formerly occupied by the lake.

As we learned in Chapter 3, the sediments of lakes usually contain a history of the events and conditions in the ecological succession leading up to the present. Such a history has been developed for Linsley Pond, Connecticut, and is shown in Figure 16.14 (see also Figure 3.15). From a pioneer stage as an oligotrophic lake, eutrophication has proceeded, accompanied by deposition of gyttja to fill the bottom, and by ecological succession of community components, as shown in the figure. The early oligotrophic stage (16.14a) was characterized by relatively high transparency and a thin layer of bottom sediments of silty composition in which the dipteran *Calopsectra* (= *Tanytarsus)* was predominant; phytoplankton and zooplankton were increasing, and the littoral vegetation was sparse. The stage of maximum production (16.14b) was accompanied by the deposition of gyttja, decreased transparency, and a great abundance of plankton and littoral vegetation; plankton blooms were common. Through continued eutrophication, the present pond stage was formed (16.14c). Transpar-

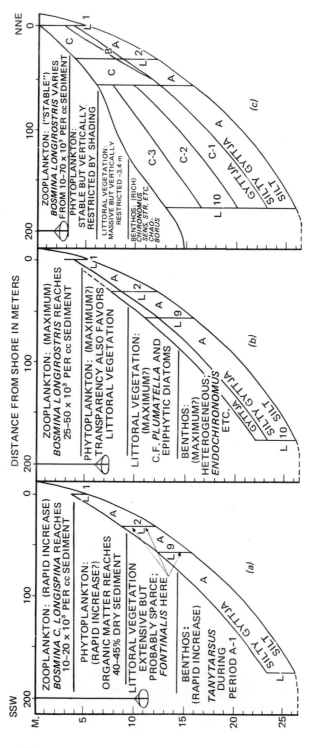

Figure 16.14 Linsley Pond, Connecticut: a schematic representation of bottom profiles and a summary of ecological history. The original oligotrophic pond (*a*) was located in an area dominated by spruce-fir forest. Later, at the time of maximum production of organic matter (*b*), the pond became densely vegetated and eutrophic. At present (*c*) the pond is heavily sedimented but ecologically stable. Estimated Secchi disk transparencies are shown at the upper left of each figure. Vertical dimension is shown at ten times the horizontal scale. (After Deevey, 1942.)

ency is generally low, the bottom consists of a thick deposition of gyttja, littoral vegetation is dense but narrowly restricted vertically, and the dipterans *Tendipes* (= *Chironomus*) and *Chaoborus* predominate in the benthos. Community structure and metabolism are relatively stable at present. Although considered primarily with reference to sedimentation, Figure 3.14 (Chapter 3) gives another example of ecological succession.

From broad-scaled sequences of changes, such as described above, ecological succession can be seen in a great variety of levels and in differing degrees of magnitude. A simple laboratory hay infusion exhibits a marked succession of populations with age. Various population assemblages within a body of water may undergo seasonal succession in a somewhat cyclic fashion. Figures 12.11 and 16.4 illustrate succession in aquatic microfauna.

We indicated above that change is characteristic of communities and also, although somewhat paradoxically, that ecological succession leads to a relatively stable climax. Climax communities tend to be stable and in equilibrium with the environment as long as no intrinsic changes within the populations, or external disturbances, occur. But internal changes, such as fires, diseases, and the effects of man, and external disturbances, such as climate modification, erosion, and the intrusion of new species, do take place.

Community change is ultimately inevitable because the species populations comprising the communities are not static in themselves. Over long periods of time and many generations of organisms, evolution brings about changes which are bound to influence the composition of the community. Both the rate and direction that evolution takes may be strongly influenced by the physical, chemical, and biological features of the environment acting, on the one hand, to limit species development, or, on the other, to provide new conditions in which organisms with novel modifications may flourish. It seems clear that to understand the momentary relationships and, further, the evolutionary aspects of aquatic communities, a thorough appreciation of the ecology of natural waters is mandatory. In the foregoing pages of this book we have attempted to set forth the principles of that ecology.

SUGGESTED FOR FURTHER READING

Hart, C.W. Jr., and S. L. H. Fuller, editors. 1974. *Pollution ecology of freshwater invertebrates.* Academic Press, New York. Ten chapters, outlining the biology, systematics, and pollution ecology of the major groups of freshwater invertebrates. Unusually clear and thorough.

Hynes, H.B.N. 1960. *The biology of polluted waters.* Liverpool University Press, Liverpool. A comprehensive overview, intended for the concerned nonspecialist, of the biological effects of human influences on natural waters, particularly rivers. The first chapters give a general picture of biological relationships in unpolluted waters; later chapters detail the physical, chemical, and biotic effects that follow the introduction of chemical effluents (including poisons), organic matter, salts, heat, and so forth. The author examines the undesired side effects of systems of freshwater management, and concludes with an overall assessment of the problem as it appeared at the time of writing. The many examples, clearly set forth, help to elucidate the various topics and make the book of continuing value to the student.

Klots, E.B. 1966. *The new field book of freshwater life.* G.P. Putnam's Sons, New York. A recent addition to Putnam's excellent series of natural history field guides. The 700 illustrations are well integrated with a text that identifies the major species of freshwater plants and animals and also sets forth clearly some of the most important relationships among many of the various species. Especially useful are the dichotomous keys to the major insect genera.

Russell, F. 1961. *Watchers at the pond.* Alfred A. Knopf, New York. A narrative account of the animal life and natural history in and around a temperate-zone freshwater pond during the course of one year. Although the author is a non-scientist, his descriptions and ecological insight have made this volume a classic work in the popular literature of ecology.

Bibliography

The journal name abbreviations given here closely follow the *Style Manual for Biological Journals,* 2nd ed. (AIBS, 1964). The complete names of journals can readily be found from the listings in *Biosis* (Biological Abstracts) or *Access* (The Chemical Abstracts).

A.A.A.S. 1939. *Problems in lake biology.* Pub. No. 10, American Association for the Advancement of Science, Washington, D. C. 142 pp.

Adams, G. F., and J. Wycoff. 1971. *Landforms.* Golden Press, New York. 160 pp.

Ahlstrom, E. H. 1943. A revision of the rotatorian genus *Keratella* with descriptions of three new species and five new varieties. *Bull. Amer. Mus. Nat. Hist.* 80:411-457.

Alexander, W. B., B. A. Southgate, and R. Bassindale. 1932. The salinity of the water retained in the muddy foreshore of an estuary. *J. Mar. Biol. Ass. U.K.,* N. S., 18:297-298.

_____. 1935. Survey of the River Tees. Part II. The estuary—chemical and biological. *Water Pollution Res. Tech. Pap.,* No. 5, Dep. Sci. Ind. Research, London.

Allee, W.C. 1951. *Cooperation among animals.* Henry Schuman, Inc. New York. 233 pp.

Allee, W. C., and E. Bowen. 1932. Studies in animal aggregations: mass protection against colloidal silver among goldfishes. *J. Exp. Zool.* 61:185-207.

Allee, W. C., E. Bowen, J. Welty, and R. Oesting. 1934. The effect of homotypic conditioning of water on the growth of fishes, and chemical studies of the factors involved. *J. Exp. Zool.* 68:183-213.

Allee, W.C., A. E. Emerson, O. Park, T. Park, and K. P. Schmidt. 1949. *Principles of animal ecology.* W. B. Saunders Co., Philadelphia, Pa. 837 pp.

Allee, W. C., P. Frank, and M. Berman. 1946. Homotypic and heterotypic conditioning in relation to survival and growth of certain fishes. *Physiol. Zool.* 19:243-258.

Allee, W. C., and K. P. Schmidt. 1951. *Ecological animal geography,* 2nd ed. John Wiley & Sons, Inc., New York. 715 pp.

Amos, W. H. 1967. *The life of the pond.* McGraw-Hill, New York.

_____. 1970. *The infinite river: a biologist's vision of the world of water.* Random House, New York.

Anderson, G. C. 1958. Some limnological features of a shallow saline meromictic lake. *Limnol. Oceanogr.* 3:259-269.

Andrewartha, H. G., and L. C. Birch. 1954. *The distribution and abundance of animals.* Univ. Chicago Press, Chicago. 782 pp.

Antevs, E. 1922. The recession of the last ice sheet in New England. *Amer. Geog. Soc. Res. Ser.* 11:1-10.

A.P.H.A. 1971. *Standard methods for the examination of water and sewage,* 13th ed. American Public Health Association, New York, 874 pp. (Revised periodically.)

Arber, A. 1920, 1963. *Water plants: a study of aquatic angiosperms.* University Press, Cambridge. 436 pp. (Reprinted 1963 by J. Cramer, Weinheim.)

Arnold, A. F. 1901, 1968. *The sea-beach at ebb-tide.* The Century Co., New York. 490 pp. (Republished by Dover Publications, Inc., New York, 1968.)

Atwood, W. W. 1940. *The physiographic provinces of North America.* Ginn and Co., Boston.

Bailey, R. M., H. E. Winn, and C. L. Smith. 1954. Fishes from the Escambia River, Alabama and Florida, with ecologic and taxonomic notes. *Proc. Acad. Nat. Sci. Philadelphia* 106:109-164.

Bainbridge, G., G. Evans, and O. Rackham, eds. 1966. *Light as an ecological factor.* Blackwell Scientific Publications, Oxford.

Bamforth, S. S. 1958. Ecological studies on the planktonic protozoa of a small artificial pond. *Limnol. Oceanogr.* 3:398-412.

Barlow, J. P. 1955. Physical and biological processes determining the distribution of zooplankton in a tidal estuary. *Biol. Bull.* 109:211-225.

Barnes, H. 1957. Nutrient elements. Pages 297-344 *in* J. W. Hedgpeth, ed. *Treatise on marine ecology and paleocelogy.* Vol. I. *Ecology.* Mem. 67, The Geological Society of America, New York.

Barnes, R. D. 1968. *Invertebrate zoology,* 2nd ed. W. B. Saunders Co., Philadelphia. 743 pp.

Bassindale, R. 1938. The intertidal fauna of the Mersey Estuary. *J. Mar. Biol. Ass. U.K.* 23:83-98.

_____. 1942. The distribution of amphipods in the Severn Estuary and Bristol Channel. *J. Anim. Ecol.* 2:131-144.

_____. 1943. Studies on the biology of the Bristol Channel. XI. The physical environment and the intertidal fauna of the southern shores of the Bristol Channel and the Severn Estuary. *J. Ecol.* 31:1-29.

_____. 1943. A comparison of the varying salinity conditions of the Tees and Severn Estuaries. *J. Anim. Ecol.* 12:1-10.

Batchelor, G.K. 1967. *Introduction to fluid dynamics.* University of Cambridge Press, Cambridge, England.

Baylor, E. F., and W. H. Sutcliffe, Jr. 1963. Dissolved organic matter in seawater as a source of particulate food. *Limnol. Oceanogr.* 8:369-371.

Beadle, L. C. 1943. Osmotic regulation and the faunas of inland waters. *Biol. Rev.* 18:172-183.

Bennett, I. L., Chairman. 1967. *The world food problem, a report of the President's Science Advisory Committee Panel on World Food Supply.* The White House, Washington, D. C. 3.

Berner, L. M. 1950. The mayflies of Florida. *Univ. Florida Stud., Biol. Sci. Ser.,* No. 4.

_____. 1951. Limnology of the lower Missouri River. *Ecology* 32:1-12.

_____. 1954. The occurrence of a mayfly nymph in brackish water. *Ecology* 35:98.

Bick, G. H., L. E. Hornuff, and E. N. Lambremont. 1953. An ecological reconnaissance of a naturally acid stream in southern Louisiana. *J. Tennessee Acad. Sci.* 28:221-231.

Birge, E. A., and C. Juday. 1914. A limnological study of the Finger Lakes of New York. *Bull. U.S. Bur. Fisheries* 35:525-609.

_____. 1921. Further limnological study of the Finger Lakes of New York. *Bull. U.S. Bur. Fisheries* 37:211-252.

_____. 1922. The inland lakes of Wisconsin. The plankton: 1. Its quantity and chemical composition. *Bull. Wisconsin Geol. Nat. Hist. Surv.* 64:1-222.

_____. 1932. Solar radiation and inland lakes, fourth report. Observations of 1931. *Trans. Wisconsin Acad. Sci. Arts & Lett.* 27:523-562.

_____. 1934. Particulate and dissolved organic matter in inland lakes. *Ecol. Monogr.* 4:440-474.

Bissell, H. J. 1963. Lake Bonneville: geology of southern Utah Valley, Utah. *Geol. Surv. Prof. Pap.* 257-B:101-129.

Black, A. P., and E. Brown. 1951. Chemical character of Florida's water. *Florida Water Surv. Res. Pap.*, No. 6. Florida State Board Conserv. 119 pp.

Black, E. C., F. E. J. Fry, and V. S. Black, 1954. The influence of carbon dioxide on the utilization of oxygen by some fresh-water fish. *Can. J. Zool.* 32:408-420.

Blair, W.F., N.P. Blair, P. Brodkorb, F.R. Cagle, and G.A. Moore. 1957. *Vertebrates of the United States.* McGraw-Hill, New York.

Bloom, A. L. 1969. *The surface of the Earth.* Prentice-Hall, Inc., Englewood Cliffs, N. J.

Blum, H. F. 1955. *Time's arrow and evolution.* Princeton Univ. Press, Princeton, N.J.

Blum, J.L. 1956. The ecology of river algae. *Bot. Rev.* 22:291-341.
_____. 1957. An ecological study of the algae of the Saline River, Michigan. *Hydrobiologia* 9:361-408.

Borecky, G. W. 1956. Population density of the limnetic *Cladocera* of Pymatuning Reservoir. *Ecology* 37:719-727.

Bormann, F. H., and G. E. Likens. 1967. Nutrient cycling. *Science* 155:424-428.

Bourn, W. S., and C. Cottam. 1950. Some biological effects of ditching tidewater marshes. *U.S. Fish Wildl. Serv. Res. Rep.* 19:1-30.

Bourrelley, P. 1966-1970. *Les algues d'eau douce.* Editions N. Boubee & Cie, Paris. 1-3:508, 433, 512 pp.

Boyce, S. G. 1954. The salt spray community. *Ecol. Monogr.* 24:29-67.

Boycott, A. E. 1936. The habits of fresh water Mollusca in Britain. *J. Anim. Ecol.* 5:116-186.

Brander, B. 1966. *The River Nile.* National Geographic Society, Washington, D.C. 208 pp.

Braun, E., and D. Cavagnaro. 1971. *Living water.* American West Publishing Company, Palo Alto, California.

Brescia, F., J Arents, H. Meislich, and A. Turk. 1966. *Fundamentals of chemistry, a modern introduction.* Academic Press, New York. 816 pp.

Brezonik, P. L., and G. F. Lee. 1968. Denitrification as a nitrogen sink in Lake Mendota. *Wisconsin Environ. Sci. Tech.* 2:120-125.

Brock, T.M. 1966. *Principles of microbial ecology.* Prentice-Hall, Inc. Englewood Cliffs, N. J. 306 pp.

Brockman, C. F., R. Merrilees, and H. S. Zim. 1968. *Trees of North America.* Golden Press, New York. 280 pp. (Paperback).

Broekhuysen, G. J. 1935. The extremes in the percentage of dissolved oxygen to which the fauna of a *Zostera* field in the tidal zone at Nieuwdiep can be exposed. *Arch. N'eerl. Zool.* 1:339-346.

Brown, C. J. D. 1933. A limnological study of certain freshwater Polyzoa with special reference to their statoblasts. *Trans. Amer. Microscop. Soc.* 52:271-316.

Brown, F. A., Jr. M. Fingerman, M. I. Sandeen, and H. M. Webb. 1953. Persistent diurnal and tidal rhythms of color change in the fiddler crab, *Uca pugnax. J. Exp. Zool.* 123:29-60.

Brown, L. 1971. *Energy and the environment.* Charles E. Merrill division of Bell & Howell Company, Columbus, Ohio.

Brunn, A. F. 1957. Deep sea and abyssal depths. Pages 641-672 *in* J.W. Hedgepeth, ed. *Treatise on marine ecology and paleoecology.* Vol. I. *Ecology.* Mem. 67, The Geological Society of America, New York.

Brunson, R. B. 1950. An introduction to the taxonomy of the Gastrotricha with a study of eighteen species from Michigan. *Trans. Amer. Microscop. Soc.* 69:325-352.

Bryson, R. A., and C. R. Stearns. 1959. A mechanism for the mixing of the waters of Lake Huron and South Bay, Manitoulin Island. *Limnol. Oceanogr.* 4:246-251.

Bue, C. D. 1963. Principal lakes of the United States. *Geol. Surv. Circ.* 476. 22 pp.

Bullock, T. H. 1955. Compensation for temperature in the metabolism and activity of poikilotherms. *Biol. Rev.* 30:311-342.

Butcher, R. W. 1933. Studies on the ecology of rivers. 1. On the distribution of macrophytic vegetation in the rivers of Britain. *J. Ecol.* 21:58-91.

Buzzati-Traverso, ed. 1958. *Perspectives in marine biology*. Univ. California Press, Berkeley.

Cairnes, J., Jr. 1968. We're in hot water. *Scientist and Citizen* 10(8):187.

Carpenter, P. L. 1967. *Microbiology*. W. B. Saunders Co., Philadelphia. 476 pp.

Carriker, M. R. 1950. *A preliminary list of the literature on the ecology of the estuaries, with emphasis on the middle Atlantic coast of the United States* (mimeographed). Rutger Univ., New Brunswick, N.J.

_____. 1951. Ecological observations on the distribution of oyster larvae in New Jersey estuaries. *Ecol. Monogr.* 21:19-38.

Carson, R. L. 1950. *The sea around us*. The New American Library of World Literature, Inc., New York. 176 pp. (Paperback)

_____. 1961. *The sea around us*. Rev. ed. Oxford University Press, New York.

Caspers, H. 1948. Oekologische Untersuchungen uber die Wattentierwelt im Elbe-Estuar. *Verh. Deutsch. Zool. Kiel* 1948:350-359.

_____. 1957. Black Sea and Sea of Azov. Pages 801-889 *in* J. W. Hedgpeth, ed. *Treatise on marine ecology and paleoecology*. Vol. I. *Ecology*. Mem. 67, The Geological Society of America, New York.

Chandler, D. C. 1964. The St. Lawrence Great Lakes. *Int. Ass. Theor. Appl. Limnol.* 15:59-75.

Chapman, C. F. 1968. *Piloting, seamanship and small boat handling*. Motor Boating, New York. 664 pp.

Chapman, V. J. 1940. Studies in salt-marsh ecology. Sect. VI, VII. Comparison with marshes on the east coast of North America. *J. Ecol.* 28:118-152.

Chu, S. P. 1943. The influence of the mineral composition of the medium on the growth of planktonic algae. *J. Ecol.* 31:109-148.

_____. 1946. The utilization of organic phosphorus by phytoplankton. *J. Mar. Biol. Ass. U.K.* 26:285-295.

Churchill, M. A. 1958. Effects of impoundments on oxygen resources. Pages 107-129 *in* U.S.P.H.S., *Oxygen relationships in streams*. U. S. Public Health Service, Washington, D. C.

Clark, J. R. 1969. Thermal pollution and aquatic life. *Sci. Amer.* 220(3):18-27.

Clarke, F. W. 1924. The composition of river and lake waters of the United States. *U.S. Geol. Surv. Prof. Pap.*, No. 135.

Clarke, G. L. 1939. The utilization of solar energy by aquatic organisms. Pages 27-38 *in* A.A.A.S., *Problems in lake biology*. Publ. No. 10, American Association for the Advancement of Science, Washington, D.C.

_____. 1954. *Elements of ecology*. John Wiley & Sons, Inc., New York.

Clarke, G. L., and E. J. Denton. 1962. Light and animal life. Pages 456-468 *in* M. H. Hill, ed. *The sea*. Vol. I. Wiley—Interscience, New York.

Clarke, G. L., and H. R. James. 1939. Laboratory analysis of the selective absorption of light by sea water. *J. Opt. Soc. Amer.* 29:43-55.

Clements, F. E., and V. E. Shelford. 1939. *Bio-ecology*. John Wiley & Sons, Inc., London. 425 pp.

Coe. W. R. 1932. Season of attachment and rate of growth of sedentary marine organisms at the pier of the Scripps Institution of Oceanography, La Jolla, California. *Bull. Scripps Inst. Oceanogr., Tech. Ser.* 3:37-86.

Coffin, C. C., F. R. Hayes, L. H. Jodrey, and S. G. Whiteway. 1949. Exchange of materials in a lake as studied by the addition of radioactive phosphorus. *Can. J. Res.*, D;27:207-222.

Coker, R. E. 1954. *Streams, lakes, ponds*. Univ. North Carolina Press, Chapel Hill. 327 pp.

Collier, A., and J. W. Hedgpeth. 1950. An introduction to the hydrography of tidal waters of Texas. *Pub. Inst. Mar. Sci. Univ. Texas* 1:123-194.

Collier, B. D., G. W. Cox, A. W. Johnson, and P. C. Miller. 1973. *Dynamic ecology*. Prentice-Hall, Inc., Englewood Cliffs, N.J. 563 pp.

Connell, J. H. 1961. The influence of interspecific competition and other factors on the distribution of the barnacle, *Chthalamus stellatus*. *Ecology* 42:133-146.

Cooke, C. W. 1939. Scenery of Florida. *Bull. Florida Geol. Surv.*, No. 17.

Cooper, L.H.N., and A. Milne. 1938. The ecology of the Tamar Estuary. II. Underwater Illumination. *J. Mar. Biol. Ass. U.K.* 22:509-527.

Cowgill, U.M., G.E. Hutchinson, A. A. Ravek, C. E. Goulden, R. Patrick, and M. Tuskada. 1966. *The history of Laguna de Petenxil*. Mem. Connecticut Acad. Arts & Sci. No. 17.126 pp.

Crombie, A. C. 1947. Interspecific competition. *J. Anim. Ecol.* 16:44-73.

Cummings, K. W. 1967. *Caloric equivalents for studies in ecological energetics* (mimeographed). Pymatuning Laboratory, Univ. Pittsburgh.

Curl. H., Jr. 1957. A source of phosphorus for the western basin of Lake Erie. *Limnol. Oceanogr.* 2:315-320.

Cutbill, J. L., ed. 1971. *Data processing in biology and geology.* Academic Press, London. 346 pp.

Davis, C. A. 1910. Some evidence of recent subsidence of the New England coast. *Science,* N.S., 32:63.

Davis, C. C. 1954. A preliminary study of the plankton of the Cleveland harbor area, Ohio. II. The distribution and quantity of the phytoplankton. *Ecol. Monogr.* 24:321-347.

_____. 1955. *The marine and freshwater plankton.* Michigan State Univ. Press, East Lansing.

Davis, J. H. 1940. The ecology and geologic role of mangroves in Florida. *Pap. Tortugas Lab.* 32:302-412.

Davis, M. B. 1967. Pollen accumulation rates at Rogers Lake during late and postglacial time. *Rev. Paleobot. Palynol.* 2:214-230.

Davis, M. B., and E. S. Deevey. 1964. Pollen accumulation rates: estimates from late-glacial sediments of Rogers Lake, Connecticut. *Science* 145:1293-1295.

Dawson, C. E. 1955. A contribution to the hydrography of Apalachicola Bay, Florida. *Pub. Inst. Mar. Sci. Univ. Texas* 4:15-35.

Dawson, E. Y. 1966. *Marine botany.* Holt, Rinehart & Winston, Inc., New York. 371 pp.

Day, J. H. 1951. The ecology of South African estuaries. Part 1. A review of estuarine conditions in general. *Trans. Roy. Soc. S. Africa* 33:53-91.

Deevey, E. S., Jr. 1940. Limnological studies in Connecticut. V. A contribution to regional limnology. *Amer. J. Sci.* 238:717-741.

_____. 1942. Studies on Connecticut lake sediments. III. The biostratonomy of Linsley Pond. *Amer. J. Sci.* 240:313-324.

_____. 1950. The probability of death. *Sci. Amer.* 182:58-60.

_____. 1969. Cladoceran populations of Rogers Lake, Connecticut, during late- and postglacial times. *Int. Ass. Theor. Appl. Limnol.* 17:56-63.

Deevey, E. S., Jr., M. S. Gross, G. E. Hutchinson, and H.L. Kraybill. 1954. The natural C^{14} contents of materials from hard-water lakes. *Proc. Nat. Acad. Sci. Washington* 40:285-288.

Defant, A. 1958. *Ebb and flow: the tides of earth, air, and water.* University of Michigan Press, Ann Arbor, Michigan.

Dendy, J. S. 1945. Fish distribution, Norris Reservoir, Tennessee, 1943. II.

Depth distribution of fish in relation to environmental factors, Norris Reservoir. *J. Tennessee Acad. Sci.* 20:114-135.

Dewey, J. F. 1972. Plate tectonics. *Sci. Amer.* 226(5):56-68.

Dexter, R. W. 1945. Zonation of the intertidal marine mollusks at Cape Ann, Massachusetts. *Nautilus* 58:138-142.

_____. 1947. The marine communities of a tidal inlet at Cape Ann, Massachusetts: A study in bio-ecology. *Ecol. Monogr.* 17:261-294.

Dice, L. R. 1952. *Natural communities.* Univ. Michigan Press, Ann Arbor, Mich. 547 pp.

Dineen, C. F. 1953. An ecological study of Minnesota pond. *Amer. Midl. Nat.* 50:349-376.

Dobrin, M. B., B. Perkins, Jr., and B. L. Snavely. 1949. Subsurface construction of Bikini Atoll as indicated by a seismic-refraction survey. *Bull. Geol. Soc. Amer.* 60:807-828.

Domogalla, B. P., C. Juday, and W. H. Peterson. 1925. The forms of nitrogen found in certain lake waters. *J. Biol. Chem.* 63:269-285.

Dorsey, N. E. 1940. *Properties of ordinary water-substance in all its phases: water-vapor, water, and all the ices.* Amer. Chem. Soc. Monogr. Ser., No. 81. Reinhold Publishing Corp., New York.

Doty, M. S. 1957. Rocky intertidal surfaces. Pages 535-585 *in* J. W. Hedgpeth, ed. *Treatise on marine ecology and paleoecology.* Vol. I. *Ecology.* Mem. 67, The Geological Society of America, New York.

Doty, M. S., and J. Newhouse. 1954. The distribution of marine algae in estuarine waters. *Amer. J. Bot.* 41:508-515.

Doudoroff, P., and C. E. Warren. 1957. Biological indices of water pollution, with special reference to fish populations. Pages 144-163 *in* U.S.P.H.S., *Biological problems in water pollution.* U.S. Public Health Service, Washington, D. C.

Drost-Hansen, W. 1969. Allowable thermal pollution limits—a physico-chemical approach. *Chesapeake Sci.* 10:281-288.

Dugdale, R. C., V. A. Dugdale, J. Neess, and J. Goering. 1959. Nitrogen fixation in lakes. *Science* 130:859-860.

Dugdale, V. A., and R. C. Dugdale. 1965. Nitrogen metabolism in lakes. III Tracer studies of the assimilation of inorganic nitrogen sources. *Limnol. Oceanogr.* 10:53-57.

Eddy, S. 1925. Fresh water algal succession. *Trans. Amer. Microscop. Soc.* 44:138-147.

_____. 1927. The plankton of Lake Michigan. *Illinois Nat. Hist. Surv. Bull.* 17:203-232.

_____. 1934. A study of fresh-water plankton communities. *Illinois Biol. Monogr.* 12:1-93.

Edgren, R. A., M. K. Edgren, and L. H. Tiffany. 1953. Some North American turtles and their epizoophytic algae. *Ecology* 34:733-740.

Edmondson, W. T. 1944. Ecological studies of sessile Rotatoria. Part I. Factors affecting distrbution. *Ecol. Monogr.* 14:32-66.

_____. 1956. The relation of photosynthesis by phytoplankton to light in lakes. *Ecology* 37:161-174.

_____. 1957. Trophic relations of the zooplankton. *Trans. Amer. Microscop. Soc.* 76:225-245.

Edmondson, W. T., ed. 1959. *Fresh water biology,* 2nd ed. John Wiley & Sons, Inc., New York. 1248 pp.

Eggleton, F. E. 1931. A comparative study of the benthic fauna of four northern Michigan lakes. *Pap. Michigan Acad. Sci.* 20:609-644.

_____. 1956. Limnology of a meromictic, interglacial, plunge-basin lake. *Trans. Amer. Microscop. Soc.* 75:334-378.

Einsele, W. 1938. Uber chemische und kolloid-chemische Vorgange in Eisen-phosphatsystemen unter limnochemischen und limnogeologischen Gesichtspunkten. *Arch. Hydrobiol.* 33:361-387.

Ekman, S. 1953. *Zoogeography of the sea.* Sedgwick and Jackson, London. 542 pp.

Ellis, M. M. 1936. Erosion silt as a factor in aquatic environments. *Ecology* 17:29-42.

Ellis, M. M., B. A. Westfall, and M. D. Ellis. 1946. Determination of water quality. *U. S. Fish Wildl. Serv. Res. Rap.,* No. 9. 22 pp.

Elton, C. 1946. Competition and structure of ecological communities. *J. Anim. Ecol.* 15:54-68.

Emerson, R., and L. Green. 1938. Effect of hydrogen-ion concentration on *Chlorella* photosynthesis. *Pl. Physiol.* 13:157-158.

Emery, K. O., and R. E. Stevenson. 1957. Estuaries and lagoons. Pages 673-749 *in* J. W. Hedgpeth, ed. *Treatise on marine ecology and paleoecology.* Vol. I. *Ecology.* Mem. 67, The Geological Society of America, New York.

Eyster, C. 1958. Bioassay of water from a concretion-forming marl lake. *Limnol. Oceanogr.* 3:455-458.

Fairbridge, R. W., ed. 1966. *The encyclopedia of oceanography*. Van Nostrand Reinhold Co., New York.

Fanning, O. 1971. *Opportunities in oceanographic careers*. Universal Publishing and Distributing Corp., New York. 144 pp.

FAO. 1968. *Yearbook of fishery statistics*. Food and Agriculture Organization of the United Nations, Rome.

Fassett, N. C. 1957. *A manual of aquatic plants*. Revised ed. with revised appendix. Univ. Wisconsin Press, Madison. 405 pp.

Fenneman, N. M. 1938. *Physiography of eastern United States*. McGraw-Hill Book Co., Inc., New York.

Ferguson, G. E., C. W. Lingham, S. K. Love, and R. O. Vernon. 1947. Springs of Florida. *Bull. Florida Geol. Surv.* 31:1-197.

Fleming, R. H. 1940. The composition of plankton and units for reporting populations and production. *Proc. 6th Pacific Sci. Congr.* (Calif., 1939). 3:535-540. (Univ. California Press, Berkeley).

Flint, R. F. 1947. *Glacial geology and the Pleistocene epoch*. John Wiley & Sons, Inc., New York. 589 pp.

_____. 1957. *Glacial and Pleistocene geology*. John Wiley & Sons, Inc., New York. 589 pp.

Flint, R. F., and W. A. Gale. 1958. Stratigraphy and radiocarbon dates at Searles Lake, California. *Amer. J. Sci.* 256:689-714.

Forbes, S. A. 1887. The lake as a microcosm. *Bull. Peoria (Ill.) Sci. Ass.* (Reprinted 1925 in *Illinois Nat. Hist. Surv. Bull.* 15:537-550.)

Forel, F. A. 1869. Introduction a l'etude de la faune profonde du lac Léman. *Bull. Soc. Vaud., Lausanne.* 10:217-233.

Forel, F. A. 1892-1904. *Le Léman, monographie limnologique*. F. Rouge, Lausanne. Tome 1. Geographie, hydrographie, geologie, climatologie, hydrologie. 1892. 543 pp. Tome 2. Mecanique, chimie, thermique, optique, acoustique. 1895. 651 pp. Tome 3, Livre 1. Biologie. 409 pp. Livre 2. Histoire. 305 pp.

Forrest, H. 1959. Taxonomic studies on the hydras of North America. VII. Description of *Chlorohydra hadleyi,* new species, with a key to the North American species of hydras. *Amer. Midl. Nat.* 62:440-448.

Frank, P. W. 1952. A laboratory study of intraspecies and interspecies competition in *Daphnia puliceria* and *Simocephalus vetulus*. *Physiol. Zool.* 25:178-204.

_____, 1957. Coactions in laboratory populations of two species of *Daphnia*. *Ecology* 38:510-519.

Freeman, O. W., ed. 1951. *Geography of the Pacific*. John Wiley & Sons, New York. 573 pp.

Frey, D. G. 1950. Carolina bays in relation to the North Carolina coastal plain. *J. Elisha Mitchell Sci. Soc.* 66:44-52.

_____, 1955. Längsee: A history of meromixis. *Mem. Inst. Ital. Idrobiol.*, Suppl. 8:141-161.

_____, 1955. Distributional ecology of the Cisco *(Coregonus artedii)* in Indiana. *Invest. Indiana Lakes & Streams* 4:177-208.

_____, 1960. The ecological significance of caldoceran remains in lake sediments. *Ecology* 41:684-699.

_____, ed. 1963. *Limnology in North America*. Univ. Wisconsin Press, Madison. 734 pp.

_____, 1964. Remains of animals in Quaternary lake and bog sediments and their interpretations. *Arch. Hydrobiol. Beih. Ergebn. Limnol.* 2:1-114.

Fuller, M. L. 1912. New Madrid earthquake. *U.S. Geol. Surv. Bull.*, No. 494.

Gardner, E. J. 1965. *History of biology*. Burgess Publishing Co., Minneapolis. 376 pp.

Gates, F. C. 1926. Plant successions about Douglas Lake, Cheboygan County, Michigan. *Bot. Gaz.* 82:170-182.

Gaufin, A. R., and C. M. Tarzwell. 1955. Environmental changes in a polluted stream during winter. *Amer. Midl. Nat.* 54:78-88.

Gause, G. P. 1934. *The struggle for existence*. Williams & Wilkins Co., Baltimore. 163 pp.

Gerking, S. D. 1953. Evidence for the concepts of home range and territory in stream fishes. *Ecology* 34:347-365.

Gifford, C. E., and E. P. Odum. 1961. Chlorophyll *a* content of intertidal zones on a rocky shoreline. *Limnol. Oceanogr.* 6:83-85.

Gilbert, G. K. 1890. Lake Bonneville. *U. S. Geol. Surv. Monogr.* 1.

Goldberg. E. D. 1963. The ocean as a chemical system. Pages 3-25 *in* M. N. Hill, ed. *The sea*. Vol. II. Wiley-Interscience, New York.

Golley, F. B. 1961. Energy values of ecological materials. *Ecology* 43:581-584.

Golterman, H. L., ed. 1969. *Methods for chemical analysis of fresh waters*. Blackwell Scientific Publications, Oxford, England. 166 pp.

Gorham, E. 1957. Chemical composition of Nova Scotian waters. *Limnol. Oceanogr.* 2:12-21.

_____. 1957. The ionic composition of some lowland lake waters from Cheshire, England. *Limnol. Oceanogr.* 2:22-27.

_____. 1958. Observations on the formation and breakdown of the oxidized microzone at the mud surface in lakes. *Limnol. Oceanogr.* 3:291-298.

Goulden, C. E. 1964. The history of the cladoceran fauna of Esthwaite Water (England) and its limnological significance. *Arch. Hydrobiol.* 60(1):5-52.

_____. 1969. Interpretive studies of caldoceran microfossils in lake sediments. *Int. Ass. Theor. Appl. Limnol.* 17:43-55.

Granhall, U., and A. Lundgren. 1971. Nitrogen fixation in Lake Erken. *Limnol. Oceanogr.* 16:711-719.

Greenberg, B. 1947. Some relations between territory, social hierarchy and leadership in the green sunfish *(Lepomis cyanellus)*. *Physiol. Zool.* 20:269-299.

Gregory, K. J., and D. E. Walling. 1973. *Drainage basin form and process: a geomorphological approach.* John Wiley & Sons, New York.

Grinnell, J. 1917. The niche-relationships of the California thrasher. *Auk* 34:427-433.

Gunter, G. 1938. Notes on invasion of fresh water by fishes of the Gulf of Mexico, with special reference to the Mississippi-Atchafalaya system. *Copeia* 2:69-72.

_____. 1942. A list of the fishes of the mainland of North and Middle America recorded from both fresh water and sea water. *Amer. Midl. Natl.* 28:305-326.

_____. 1945. Studies on marine fishes of Texas. *Pub. Inst. Mar. Sci. Univ. Texas* 1:1-190.

_____. 1947. Paleoecological import of certain relationships of marine animals to salinity. *J. Paleontol.* 21:77-79.

_____. 1957. Paleoecological import of certain relationships of marine animals to salinity. *J. Paleontol.* 21:77-79.

Haeckel, E. H. P. A. 1870. *Natürliche Schöpfungsgeschichte.* G. Reimer, Berlin. 659 pp.

Hansen, K. 1959. The terms gyttja and dy. *Hydrobiologia* 13:309-315.

Hardin, G. 1960. The competitive exclusion principle. *Science* 13:1292-1297.

Hardy, A. 1965. *The open sea: its natural history.* Part I. The world of plankton.

335 pp. Part II. The fish and fisheries. 322 pp. Houghton Mifflin Co., Boston,

Harrington, H. K., and F. J. Myers. 1926. The rotifer fauna of Wisconsin. III. A revision of the genera *Lecane* and *Monostyla*. *Trans. Wisconsin Acad. Sci.* 22:315-423.

Harris, E. 1957. Radiophosphorus metabolism in zooplankton and microorganisms. *Can. J. Zool.* 35:769-782.

Harshberger, J. W. 1909. The vegetation of the salt marshes and of the salt and fresh water ponds of northern coastal New Jersey. *Prod. Acad. Nat. Sci. Philadelphia* 61:373-400.

Hart, C. W. Jr., and S.L.H. Fuller, editors. 1974. *Pollution ecology of freshwater invertebrates*. Academic Press, New York.

Hartley, P.H.T., and G.M. Spooner. 1938. The ecology of the Tamar Estuary. I. Introduction. *J. Mar. Biol. Ass. U.K.* 27:501-508.

Harvey, H. W. 1950. On the production of living matter in the sea off Plymouth. *J. Mar. Biol. Ass. U.K.* 29:97-137.

_____. 1955.*The chemistry and fertility of sea-water*. Cambridge Univ. Press, Cambridge. 224 pp.

Hasler, A. D. 1938. Fish biology and limnology of Crater Lake. *J. Wildl. Manage.* 2:94-103.

_____. 1947. Eutrophication of lakes by domestic drainage. *Ecology* 28:383-395.

Hayes, F. R. 1955. The effect of bacteria on the exchange of radiophosphorus at the mud-water interface. *Int. Ass. Theor. Appl. Limnol.* 12:111-116.

_____. 1957. On the variation in bottom fauna and fish yield in relation to trophic level and lake dimensions. *J. Fish. Res. Board Can.* 14:1-32.

Hayes, F. R., J. A. McCarter, M. L. Cameron, and D. A. Livingstone. 1952. On the kinetics of phosphorus exchange in lakes. *J. Ecol.* 40:202-216.

Hayes, F. R., and J. E. Phillips. 1958. Lake water and sediment. IV. Radiophosphorus equilibrium with mud, plants and bacteria under oxidized and reduced conditions. *Limnol. Oceanogr.* 3:459-475.

Heath, R. A. 1973. Flushing of coastal embayments by changes in atmospheric conditions. *Limnol. Oceanogr.* 18:849-862.

Hedgpeth, J. W. 1947. The Laguna Madre of Texas. *Trans. 12th N. Amer. Wildl. Conf.* pp. 364-380.

_____. 1951. The classification of estuarine and brackish waters and the hydrographic climate. *Geol. Soc. Amer., Rep. Comm. on Treatise on Mar. Ecol. & Paleoecol.* 1951:49-56.

————. 1953. An introduction to the zoogeography of the northwestern Gulf of Mexico with reference to the invertebrate fauna. *Pub. Inst. Mar. Sci. Univ. Texas* 3:107-224.

————. 1957. Classification of marine environments. Pages 17-28 *in* J. W. Hedgpeth, ed. *Treatise on marine ecology and paleoecology.* Vol. I. *Ecology.* Mem. 67, The Geological Society of America, New York.

————. 1957. Concepts of marine ecology. Pages 29-52 *in* J. W. Hedgpeth, ed. *Treatise on marine ecology and paleoecology.* Vol. I. *Ecology.* Mem. 67, The Geological Society of America, New York.

————. 1957. Sandy beaches. Pages 587-608 *in* J. W. Hedgpeth, ed. *Treatise on marine ecology and paleoecology.* Vol. I *Ecology.* Mem. 67, The Geological Society of America, New York.

————. 1957. *Treatise on marine ecology and paleoecology.* Vol. I. *Ecology.* Mem. 67, The Geological Society of America, New York. 1296 pp.

Hegner, R. W., and J. G. Engelmann. 1968. *Invertebrate zoology,* 2nd ed. Macmillan Co., New York.

Hela, I., C. A. Carpenter, and J. K. McNulty. 1957. Hydrography of a positive, shallow, tidal bar-built estuary (Report on the hydrography of the polluted area of Biscayne Bay). *Bull. Mar. Sci. Gulf & Carib.* 7:47-99.

Henderson, L. J. 1913. *The fitness of the environment.* Macmillan Co., New York.

Hendricks, S. B. 1955. Necessary, convenient, commonplace. Pages 9-14 *in* U.S.D.A., *Yearbook of agriculture.* U. S. Department of Agriculture, Washington, D. C.

Henson, E. B. 1959. Evidence of internal wave activity in Cayuga Lake, New York. *Limnol. Oceanogr.* 4:441-447.

Hesse, E., W. C. Allee, and K. P. Schmidt, 1937, 1951. *Ecological animal geography.* John Wiley & Sons, Inc. New York. 597 pp (2nd ed. 1951. 715 pp.)

Heusser, C. J. 1960. Late-Pleistocene environments of north Pacific North America. *Amer. Geogr. Soc., Spec. Pub.* No. 35.

Heyerdahl, T. 1950. *Kon-Tiki.* Rand McNally, New York. 240 pp. (Paperback. Permabook, Affiliated Publishers, Inc., New York.)

————. 1971. *The Ra expeditions.* George Allen & Unwin, Ltd., London. 334 pp.

Hicks, S. D. 1959. The physical oceanography of Narrangansett Bay. *Limnol. Oceanogr.* 4:316-327.

Hill, M. N., ed. 1962. *The sea.* Vol. I. *Physical oceanography.* Wiley-Interscience, New York. 864 pp.

————, ed. 1963. *The sea.* Vol. II. *Composition of sea-water.* Wiley-Interscience, New York. 554 pp.

Hjulstrom, F. 1935. Studies of the morphological activity of rivers as illustrated by the River Fjris. *Bull. Univ. Upsala Geol. Inst.,* No. 25:221-527.

Holmes, R. W. 1970. The Secchi disc in turbid coastal waters. *Limnol. Oceanogr.* 15:688-694.

Hooper, F. F. 1954. Limnological features of Weber Lake, Cheboygan County, Michigan. *Pap. Michigan Acad. Sci.* 39:229-240.

Hopkins, D. M. 1949. Thaw lakes and thaw sinks in the Imruk Lake area, Seward Peninsula, Alaska. *J. Geol.* 57:119-131.

Hopkins, S. H. 1957. B. Parasitism. Pages 413-428 *in* J. W. Hedgpeth, ed. *Treatise on marine ecology and paleoecology.* Vol. I. *Ecology.* Mem. 67, The Geological Society of America, New York.

Hornuff, L. E. 1957. A survey of four Oklahoma streams with reference to production. *Oklahoma Fish. Res. Lab. Rep.,* No. 63.

Hough, J. L. 1958. *Geology of the Great Lakes.* Univ. Illinois Press, Urbana.

————. 1962. Geologic framework. Pages 3-27 *in* H. J. Pincus, ed. *Great Lakes basin.* Publ. No. 71, American Association for the Advancement of Science, Washington, D. C.

Hrbácvek, J. 1959. Circulation of water as a main factor influencing the development of helmet in *Daphnia cucullata* Sars. *Hydrobiologia* 13:170-185.

Hulbert, E. M. 1956. Distribution of phosphorus in Great Pond, Massachusetts. *J. Mar. Res.* 15:181-192.

Hunt, C. A., and R. M. Garrels. 1972. *Water: the web of life.* W. W. Norton & Co., Inc., New York. 208 pp (Paperback).

Hutchinson, G. E. 1938. On the relation between the oxygen deficit and the productivity and typology of lakes. *Int. Rev. Hydrobiol.* 36:336-355.

————. 1944. Limnological studies in Connecticut. VII. A critical examination of the supposed relationships between phytoplankton periodicity and chemical changes in lake waters. *Ecology* 25:3-26.

————. 1957. *A treatise on limnology.* Vol. I. *Geography, physics, and chemistry.* John Wiley & Sons, Inc., New York. 1015 pp.

_____. 1967. *A treatise on limnology*. Vol. II. *Introduction to lake biology and the limnoplankton*. John Wiley & Sons, Inc., New York. 1115 pp.

Hutchinson, G. E., E. Bonatti, U. M. Cowgill, C. E. Goulden, E. A. Leventhal, M. E. Mallett, F. Margaritora, R. Patrick, A. Ravek, W. A. Robak, E. Stella, J. B. Wart-Perkins, and T. R. Wellman. 1970. Ianula: an account of the history and development of the Lago di Monterosi, Latium, Italy. *Trans. Amer. Phil. Soc.* 64(4):1-178.

Hutchinson, G. E., E. S. Deevey, Jr., and A. Wollack. 1939. The oxidation-reduction potential of lake waters and its ecological significance. *Proc. Nat. Acad. Sci.*, Washington 25:87-90.

Hyman, L. H. 1929. Taxonomic studies on the hydras of North America. I. General remarks and descriptions of *Hydra americana*. *Trans. Amer. Microscop. Soc.* 48:242-255.

_____. 1930. Taxonomic studies on the hydras of North America. II. The characters of *Pelmatohydra oligactis* (Pallas). *Trans. Amer. Microscop. Soc.* 49:322-329.

_____. 1940-1967. *The invertebrates*. Vol. I-VI. McGraw-Hill Book Co., Inc., New York. Vol. I, 726 pp.; II, 550 pp.; III, 572 pp.; IV, 763 pp.; V, 783 pp.; VI, 791 pp.

Hynes, H. B. N. 1960. *The biology of polluted waters*. Liverpool University Press, Liverpool.

_____. 1970. *The ecology of running waters*. University of Toronto Press, Toronto, Canada.

Idyll, C. P., ed. 1972. *Exploring the ocean world,* 2nd ed. Thomas Y. Crowell Company, New York.

Irwin, W. H. 1945. Methods of precipitating colloidal soil particles from impounded waters of central Oklahoma. *Bull. Oklahoma A. & M. Coll.*, No. 42:1-16.

Iselin, C. O. 1952. *Marine fouling and its prevention*. U. S. Naval Institute, Annapolis, Md. 388 pp.

Jackson, D. F., and W. A. Dence. 1958. Primary productivity in a dichothermic lake. *Amer. Midl. Nat.* 59:511-517.

Johnson, D. W. 1919. *Shore processes and shoreline development*. John Wiley & Sons, Inc., New York. 584 pp.

_____. 1925. *The New England-Acadian shoreline*. John Wiley & Sons, Inc., New York.

Johnson, T. W., and F. K. Sparrow. 1961. *Fungi in oceans and estuaries*. J. Cramer, Weinheim. 668 pp.

Juday, C. 1922. Quantitative studies of the bottom fauna in the deeper waters of Lake Mendota. *Trans. Wisconsin Acad. Sci.* 20:461-493

_____. 1940. The annual energy budget of an inland lake. *Ecology* 21:438-450.

_____. 1942. The summer standing crop of plants and animals in four Wisconsin lakes. *Trans. Wisconsin Acad. Sci.* 29:1-82.

Juday, C., and E. A. Birge. 1932. Dissolved oxygen and oxygen consumed in the lake waters of northeastern Wisconsin. *Trans. Wisconsin Acad. Sci.* 27:415-486.

Juday, C., E. A. Birge, and V. W. Meloche. 1938. Mineral content of the lake waters of northeastern Wisconsin. *Trans. Wisconsin Acad. Sci.* 31:223-276.

Kanwisher, J. W. 1963. On the exchange of gases between the atmosphere and the sea. *Deep-Sea Res.* 10:195-207.

Keeton, W. T. 1967, 1972. *Biological science.* W. W. Norton & Co., Inc., New York. 888 pp. (2nd ed. 1972).

Keirn, M. A., and P. L. Brezonik. 1971. Nitrogen fixation by bacteria in Lake Mize, Florida, and in some lacustrine sediments. *Limnol. Oceanogr.* 16:720-731.

Kendeigh, S. C. 1974. *Ecology, with special reference to animals and man.* Prentice-Hall, Inc., Englewood Cliffs, N.J. 474 pp.

Kennedy, C. H. 1915. Notes on the life history and ecology of dragonflies (Odonata) of Washington and Oregon. *Proc. U.S. Nat. Mus.* 49:259-345.

Ketchum, B. H. 1947. The biochemical relations between marine organisms and their environment. *Ecol. Monogr.* 17:309-315.

_____. 1950. The exchanges of fresh and salt waters in tidal estuaries. *Proc. Colloq. on Flushing of Estuaries,* U. S. Off. Naval Res., Washington, D.C.

_____. 1951. Flushing of tidal estuaries. *Sew. Industr. Wastes* 23(2):198-209.

_____. 1953. Circulation in estuaries. *Proc. 3rd Conf. Coastal Eng.,* Contrib. No. 642 from Woods Hole Oceanogr. Inst., pp. 65-76.

_____. 1954. Relation between circulation and planktonic populations in estuaries. *Ecology* 35:191-200.

Kimball, H. H. 1928. Amount of solar radiation that reaches the surface of the earth on the land and on the sea, and methods by which it is measured. *Monthly Weath. Rev.* 56:393-399.

Klots, E. B. 1966. *The new field book of freshwater life.* G. P. Putnam's Sons, New York.

Kofoid, C. A. 1903. The plankton of the Illinois River, 1894-1899, with introductory notes upon the hydrography of the Illinois River and its basin. Part I. Quantitative investigations and general results. *Bull. Illinois State Lab. Nat. Hist.*, No. 6:95-629.

_____. 1908. The plankton of the Illinois River, 1894-1899, with introductory notes upon the hydrography of the Illinois River and its basin. Part II. Constituent organisms and their seasonal distribution. *Bull. Illinois State Lab. Nat. Hist.*, No. 8:1-355.

Kolkowitz. R., and M. Marsson. 1909. Okologie der fierischen Saprobein. *Int. Rev. Ges. Hydrobiol. Hydrogr.* 2:126-152.

Kormondy, E. J. 1969. *Concepts of ecology.* Prentice-Hall, Inc., Englewood Cliffs, N. J. 209 pp. (Paperback.)

Kossinna, E. 1921. *Die Tiefen des Weltsmeeres.* Reine 9. Berlin Univ., Inst. f. Meereskunde, Veroff., N. F., A. Geogr.-Naturwiss. 70 pp.

Kozhov. M. 1963. *Lake Baikal and its life.* Dr. W. Junk, The Hague. 344 pp.

Krebs, C. J. 1972. *Ecology: the experimental analysis of distribution and abundance.* Harper and Row, Publishers, New York.

Krenkel, R. H., and F. L. Parker, eds. 1969. *Biological aspects of thermal pollution.* Proc. Nat. Symp. Therm. Poll., Vanderbilt Univ. Press, Nashville. 407 pp.

Kriss, A. E., I. E. Mishustina, N. Mitskevich, and E. V. Zemstova. 1964. *Microbial population of oceans and seas.* (English transl. by K. Syers; ed. by G. E. Ofgg, 1967.) Edward Arnold (Publishers) Ltd., London.

Krogh, A. 1939. *Osmotic regulation in aquatic animals.* Cambridge Univ. Press, Cambridge.

Krogh, A., and E. Lange. 1932. Quantitative Untersuchungen uber Plankton, Kolloide und gelöste organische und anorganische Substanzen in dem Furesse. *Int. Rev. Hydrobiol.* 26:20-53.

Kuenen, P. H. 1950. *Marine geology.* John Wiley & Sons, Inc., New York.

_____. 1955. *Realms of water.* John Wiley & Sons, Inc., New York.

Ladd, H. S. 1951. Brackish-water and marine assemblages of the Texas coast, with special reference to mollusks. *Pub. Inst. Mar. Sci. Univ. Texas* 2:125-164.

Langbein, W. B. 1961. The salinity and hydrology of closed lakes. *U. S. Geol. Surv., Prof. Pap.* No. 410. 20 pp.

Langlois, T. H. 1954. *The western end of Lake Erie and its ecology.* J. W. Edwards Pub., Inc., Ann Arbor, Michigan. 479 pp.

Langmuir, I. 1938. Surface motion of water induced by wind. *Science* 87:119-123.

Lanjouw, J., Comm. Chairm. 1966. *International code of botanical nomenclature.* International Bureau for Plant Taxonomy and Nomenclature, Utrecht. 399 pp.

Lauff, G. H., ed. 1967. *Estuaries.* A.A.A.S. Publ. No. 83. American Association for the Advancement of Science, Washington, D. C. 757 pp.

Layne, J. N. 1958. Observation on freshwater dolphins in the upper Amazon. *J. Mammal.* 39:1-22.

Leet, L. D., and S. Judson. 1958. *Physical geology,* 2nd ed. Prentice-Hall, Inc., Englewood Cliffs, N. J.

LeGrand, H. E. 1953. Streamlining of the Carolina bays. *J. Geol.* 61:263-274.

Leopold, L. B. 1974. *Water: a primer.* W. H. Freeman, San Francisco.

Leopold, L. B., and S. Davis, with the eds. of *Life.* 1966. *Water.* Time Incorporated, New York.

Leopold, L. B., and T. Maddock. 1953. *The hydraulic geometry of stream channels and some physiographic implications.* U.S. Geol. Surv. Prof. Pap., No. 252.

Leopold, L. B., and J. P. Miller. 1956. *Ephemeral streams—hydraulic factors and their relation to the drainage net.* U. S. Geol. Surv. Prof. Pap. No. 282-A.

Lewis, J. R., and H. T. Powell. 1960. Aspects of the intertidal ecology of rocky shores in Argyll, Scotland. I. General description of the area. *Trans. Roy. Soc. Edinburgh* 64:45-74.

Liebig, J. 1840. *Chemistry in its application to agriculture and physiology,* 4th ed. Taylor & Walton, London.

Likens, G. E., and F. H. Bormann. 1974. Linkages between terrestrial and aquatic ecosystems. *BioScience* 24(8):447-456.

Likens, G. E., ed. 1972. *Nutrients and eutrophication: the limiting-nutrient controversy.* Allen Press, Lawrence, Kansas.

Lind, O. T. 1974. *Handbook of common methods in limnology.* C.V. Mosby Co., St. Louis. 154 pp.

Lindeman, R. L. 1942. The trophic-dynamic aspect of ecology. *Ecology* 23:399-418.

Lobeck, A. K. 1956. *Things maps don't tell us.* The Macmillan Co., New York. 159 pp.

Longwell, C. R., A. Knopf, and R. F. Flint. 1939. *A textbook of geology. Part I–Physical geology,* 2nd ed. John Wiley & Sons, Inc., New York.

Macan, T. T. 1974. *Freshwater ecology,* 2nd ed. John Wiley & Sons, New York.

Macan, T. T., and E. B. Worthington. 1951. *Life in lakes and rivers.* New Naturalist Series. Wm. Collins Sons & Co., Ltd., London. 272 pp.

McAtee, W. L. 1941. *Wildlife of the Atlantic coastal marshes.* U. S. Fish Wildl. Serv. Cir., No. 11.

McCombie, A. M. 1953. Factors influencing the growth of phytoplankton. *J. Fish. Res. Board Can.* 10:253-282.

McConnaughey, B. H. 1970. *Introduction to marine biology.* The C. V. Mosby Co., St. Louis. 449 pp.

McGee, W. J. 1893. A fossil earthquake. *Bull. Geol. Soc. Amer.* 4:411-415.

MacGinitie, G. E. 1935. Ecological aspects of a California marine estuary. *Amer. Midl. Nat.* 16:629-795.

————. 1939. Littoral marine communities. *Amer. Midl. Nat.* 21:28-55.

————. 1939. Some effects of fresh water on the fauna of a marine harbor. *Amer. Midl. Nat.* 21:681-686.

MacGinitie, G. E., and N. MacGinitie. 1949. *Natural history of marine animals.* McGraw-Hill Book Co., Inc., New York.

McLusky, D. S. 1971. *Ecology of estuaries.* Heinemann Educational Books, London.

Mann, K. H. 1958. Annual fluctuations in sulphate and bicarbonate hardness in ponds. *Limnol. Oceanogr.* 3:418-422.

Margalef, R. 1958. Temporal succession and spatial heterogeneity in phyto-plankton. Pages 323-349 *in* A. A. Buzzati-Traverso, ed. *Perspectives in marine biology.* University of California Press, Berkeley.

————. 1963. (Margalef's model of succession, Fig. 6.1) Page 79 *in* E. P. Odum. *Ecology.* Holt, Rinehart & Winston, Inc., New York.

————. 1963. Succession of populations. *Adv. Frontiers of Plant Sci.* 2:137-188. (Inst. Adv. Sci. & Culture, New Delhi, India).

————. 1967. The food web in the pelagic environment. *Helgoland. Wiss. Meeresunters.* 15:548-559.

————. 1968. *Perspectives in ecological theory.* Univ. Chicago Press, Chicago. 111 pp.

Marmer, H. A. 1926. *The tide.* Appleton, Inc., New York.

Martin, D. F. 1968. *Marine chemistry.* Vol. 1. *Analytical methods.* Marcel Dekker, Inc., New York. 280 pp.

Mathieson, A. C., and R. L. Burns. 1971. Ecological studies of economic red algae. *J. Exp. Mar. Biol. Ecol.* 7:197-206.

Meglitsch, P. A. 1972. *Invertebrate zoology,* 2nd ed. Oxford Univ. Press, New York. 834 pp.

Meinzer, O. E., ed. 1942. *Physics of the Earth.* IX. *Hydrology.* McGraw-Hill Book Co., Inc., New York.

Melton, F. A., and W. Schriever. 1933. The Carolina bays—are they meteorite scars? *J. Geol.* 41:52-66.

Metcalf, Z. P. 1930. Salinity and size. *Science* 72:526-527.

Meyer, B. S., and D. B. Anderson. 1952. *Plant physiology,* 2nd ed. D. Van Nostrand Co., Inc., New York. 784 pp.

Millard, N. A. H., and A. D. Harrison. 1954. The ecology of South African estuaries. Part V: Richard's Bay. *Trans. Roy. Soc. S. Africa* 34(1):157-174.

Miller, D. E. 1936. A limnological study of *Pelmatohydra* with special reference to their quantitative and seasonal distribution. *Trans. Amer. Microscop. Soc.* 55:126-193.

Miller, W. R., and F. E. Egler. 1950. Vegetation of the Wequetequock-Pawcatuck tidal marshes, Connecticut. *Ecol. Monogr.* 20:141-172.

Mills, E. A. 1913. *The beaver world.* Houghton Mifflin, Boston. 146 pp.

Milne, A. 1938. The ecology of the Tamar Estuary. III. Salinity and temperature conditions in the lower estuary. *J. Mar. Biol. Ass. U.K., N.S.,* 22:529-542.

————. 1940. The ecology of the Tamar Estuary. IV. The distribution of the fauna and flora on buoys. *J. Mar. Biol. Ass. U.K.,* N. S. 24:69-87.

Milsum, J. H. 1966. *Biological control systems analysis.* McGraw-Hill Book Co., Inc., New York. 466 pp.

Miner, R. W. 1950. *Field book of seashore life.* G. P. Putnam's Sons, New York. 888 pp.

Ministry of Tourism, Nassau, Bahamas. 1972. *The yachtmans guide.* Tropic Isle Publishers, Inc., Coral Gables, Fla. 316 pp.

Möbius, K. 1877, 1883. *Die Auster und die Austernwirthschaft.* Wiegandt, Hempel & Parry, Berlin. 126 pp. (English transl. by H. J. Rice, 1883. The oyster and oyster culture. Rep. U. S. Fish. Comm. 1880:683-751.)

Mock, C. R. 1966. Natural and altered estuarine habitats of penaeid shrimp. *Proc. Gulf Carib. Fish. Inst.* 19:86-98.

Moore, H. B. 1958. *Marine ecology*. John Wiley & Sons, Inc., New York. 493 pp.

Moore, W. C. 1950. Limnological studies of Louisiana lakes. I. Lake Providence. *Ecology* 31:86-99.

Mortimer, C. H. 1941. The exchange of dissolved substances between mud and water in lakes. *J. Ecol.* 29:280-329.

_____. 1942. The exchange of dissolved substances between mud and water in lakes. *J. Ecol.* 30:147-201.

_____. 1952. Water movements in lakes during summer stratification; evidence from the distribution of temperature in Windermere. *Phil. Trans. Roy. Soc. London,* Ser. B, 236:355-404.

_____. 1953. The resonant response of stratified lakes to wind. *Schweiz. Z. Hydrol.* 15:94-151.

_____. 1954. Models of the flow pattern in lakes. *Weather* 9:177-184.

_____. 1954. An explorer of lakes. Pages 165-211 *in* G. C. Sellery. *E. A. Birge*. Univ. Wisconsin Press, Madison.

Motoda, S. 1939. Submarine illumination, silt content and quantity of food plankton of reef corals in Iwayama Bay, Palao. *Palao Trop. Biol. Sta. Stud.* 1:637-649.

Moyle, J. B. 1949. Some indices of lake productivity. *Trans. Amer. Fish. Soc.* 76(1946):322-334.

Muenscher, W. C. 1944, 1973. *Aquatic plants of the United States.* Compton Publ. Assoc., Ithaca, New York. 374 pp. (Reprinted by Cornell Univ. Press, 1973.)

Murphy, R. C. 1962. The oceanic life of the Antarctic. *Sci. Amer.* 1962:187-210.

Murray, J., and J. Hjort. 1912. *The depths of the oceans.* Macmillan & Co., Ltd., London. 821 pp.

Myers, G. S. 1949. Usage of anadromous, catadromous and allied terms for migratory fishes. *Copeia* 1949:89-97.

Nash, C. B. 1947. Environmental characteristics of a river estuary. *J. Mar. Res.* 6:147-174.

National Academy of Scientices, 1969. *Eutrophication: causes, consequences, correctives.* National Acad. Sciences, Washington, D.C.

National Academy of Sciences. 1969. *Eutrophication: causes, consequences, correctives.* National Academy of Sciences, Washington, D. C. 661 pp.

Naylor, E. 1965. Effects of heated effluents upon marine and estuarine organisms. *Adv. Mar. Biol.* 3:63-103.

Needham, J. 1930. On the penetration of marine organisms into fresh water. *Biol. Zentr.* 50:504-509.

Needham, J. G., and M. J. Westfall. 1955. *A manual of dragonflies of North America.* Univ. California Press, Berkeley.

Neil, J. H. 1957. Investigations and problems in Ontario. Pages 184-187 *in* U.S.P.H.S., *Biological problems in water pollution.* U.S. Public Health Service, Washington, D. C.

Nelson, T. 1947. Some contributions from the land in determining conditions of life in the sea. *Ecol. Monogr.* 17:337-346.

Newcombe, C. E., W. A. Horne, and B. B. Shepherd. 1939. Studies on the physics and chemistry of estuarine waters in Chesapeake Bay. *J. Mar. Res.* 2:87-116.

Nicol, E. A. T. 1935. The ecology of a salt-marsh. *J. Mar. Biol. Ass. U.K.* 20:203-261.

Niering, W. A. 1966. *The life of the marsh.* McGraw-Hill, New York.

Noel, M. S. 1954. Animal ecology of a New Mexico springbrook. *Hydrobiologia* 6:120-135.

Norris, R. M. 1953. Buried oyster reefs in some Texas bays. *J. Paleontol.* 27:569-576.

Nybakken, J. W. 1971. *Readings in marine ecology.* Harper & Row, Publishers, New York. 544 pp.

Odum, E. P. 1963. *Ecology.* Holt, Rinehart & Winston, Inc., New York. 152 pp. (Paperback).

—————. 1971. *Fundamentals of ecology,* 3rd ed., W. B. Saunders Co., Philadelphia. 547 pp.

Odum, E. P., S. G. Marshall, and T. G. Marples. 1965. The caloric content of migrating birds. *Ecology* 46:901-904.

Odum, H. T. 1952. The Carolina bays and a Pleistocene weather map. *Amer. J. Sci.* 250:263-270.

—————. 1953. Dissolved phosphorus in Florida waters. *Rep. Florida Geol. Surv.* 9:1-40.

_____. 1956. Primary production in flowing waters. *Limnol. Oceanogr.* 1:102-167.

_____. 1957. Trophic structure and productivity of Silver Springs, Florida. *Ecol. Monogr.* 27:55-112.

_____. 1960. Analysis of diurnal oxygen curves for the assay of reaction rates and metabolism in polluted marine bays. Pages 547-555 *in* E. A. Pearson, ed. *International conference on waste disposal in the marine environment.* Pergamon Press, New York.

_____. 1967. Energetics of world food production. Pages 55-94 *in* I. L. Bennett (chairman). *The world food problems.* Vol. 3. Report of the President's Science Advisory Committee Panel on World Food Supply. The White House, Washington, D. C.

_____. 1971. *Environment, power, and society.* Wiley-Interscience, New York. 331 pp. (Paperback).

Odum, H. T., and D. K. Caldwell. 1955. Fish respiration in the natural oxygen gradient of an anaerobic spring in Florida. *Copeia* 1955:104-106.

Odum, H. T., B. J. Copeland, and E. A. McMahan. 1969. *Coastal ecological systems of the United States.* Rep. to Federal Water Pollution Control Administration (mimeographed).

Odum, H. T., and E. P. Odum. 1955. Trophic structure and productivity of a windward coral reef community on Eniwetok Atoll. *Ecol. Monogr.* 25:291-320.

Odum, W. E. 1970. Insidious alteration of the estuarine environment. *Trans. Amer. Fish. Soc.* 99:836-847.

Ohle, W. 1934. Chemische und physikalische Untersuchungen norddeutscher Seen. *Arch. Hydrobiol.* 26:386-464.

Ortmann, A. E. 1902. The geographical distribution of freshwater decapods and its bearing upon ancient geography. *Proc. Amer. Phil. Soc.* 41:267-400.

Osterhout, W. J. W. 1933. Permeability in large plant cells and in models. *Ergebn. Physiol.* 35:967-1021.

Ostwald, W. 1902. Zur theorie des planktons. *Biol. Cent.* 22:596-605.

Palmer, M. D., and J. B. Izatt. 1972. Lake movements with partial ice cover. *Limnol. Oceanogr.* 17:403-409.

Parker, B., and G. Barsom. 1970. Biological and chemical significance of surface microlayers in aquatic ecosystems. *BioScience* 20(2):87-93.

Patrick R. 1949. A proposed biological measure of stream conditions, based on a survey of the Conestoga Basin, Lancaster County, Pennsylvania. *Proc. Acad. Natl. Sci. Philadelphia* 101:277-341.

Patten, B. C., ed. 1971. *Systems analysis and simulation in ecology.* Vol. 1. Academic Press, New York.

————, ed. 1972. *Systems analysis and simulation in ecology.* Vol. 2. Academic Press, New York.

Pauling, L. C. 1960. *Nature of the chemical bond and the structure of molecules and crystals: an introduction to modern structural chemistry.* Cornell Univ. Press, Ithaca.

Pearse, A. S. 1950. *The emigrations of animals from the sea.* Sherwood Press, Dryden, N. Y. 210 pp.

Pearse, A. S., and G. Gunter. 1957. Salinity. Pages 129-158 *in* J. W. Hedgpeth, ed. *Treatise on marine ecology and paleoecology.* Vol. I. *Ecology.* Mem. 67, The Geological Society of America, New York.

Pearson, E. A., ed. 1960. *International conference on waste disposal in the marine environment.* Pergamon Press, New York.

Penfound, W. T. 1956. Primary production of vascular plants. *Limnol. Oceanogr.* 1:92-101.

Penfound, W. T., and E. S. Hathaway. 1938. Plant communities in the marshlands of southeastern Louisiana. *Ecol. Monogr.* 8:1-56.

Pennak, R. W. 1944. Diurnal movements of zooplankton organisms. *Ecology* 25:387-403.

————. 1949. Annual limnological cycles in some Colorado lakes. *Ecol. Monogr.* 19:233-267.

————. 1951. Comparative ecology of the interstitial fauna of freshwater and marine beaches. *Ann. Biologiques* 27(6):217-248.

————. 1953. *Fresh-water invertebrates of the United States.* Ronald Press Co., New York. 769 pp.

————. 1955. Comparative limnology of eight Colorado mountain lakes. *Univ. Colorado Ser. Biol.* 2:1-75.

————. 1955. Persistent changes in the dominant species composition of limnetic entomostracan populations in a Colorado mountain lake. *Trans. Amer. Microscop. Soc.* 2:116-118.

————. 1957. Species composition of limnetic zooplankton communities. *Limnol. Oceanogr.* 2:222-232.

Percival, E. 1929. A report on the fauna of the estuaries of the River Tamar and the River Lynker. *J. Mar. Biol. Ass. U.K.* 16:81-108.

Perraton, C. 1953. Salt marshes of the Hampshire-Sussex border. *J. Ecol.* 41:240-247.

Phleger, F. B., and W. R. Walton. 1950. Ecology of marsh and bay foraminifera, Barnstable, Mass. *Amer. J. Sci.* 248:274-295.

Picken, L. R. 1937. The structure of some protozoan communities. *J. Ecol.* 25:368-384.

Pierce, E. L. 1947. An annual cycle of the plankton and chemistry of four aquatic habitats in northern Florida. *Univ. Florida Stud., Biol. Sci. Ser.,* Vol. IV, No. 3.
_____. 1958. The Chaetognatha of the inshore waters of North Carolina. *Limnol. Oceanogr.* 3:166-170.

Pincus, H. J., ed. 1962. *Great Lakes basin.* A.A.A.S., Publ. No. 71. American Association for the Advancement of Science. Washington, D. C.

Pittendrigh, C. S. 1958. Perspectives in the study of biological clocks. Pages 239-268 *in* A. A. Buzzati-Traverso, ed. *Perspectives in marine biology.* Univ. California Press, Berkeley.

Pomeroy, L. R. 1959. Algal productivity in Georgia salt marshes. *Limnol. Oceanogr.* 4:386-397.

Potzger, J. E. 1956. Pollen profiles as indicators in the history of lake filling and bog formation. *Ecology* 37:476-483.

Powell, A. W. B. 1939. Notes on the importance of recent animal ecology as a basis of paleoecology. *Proc. 6th Pacific Sci. Cong.,* pp. 607-617.

Pratt, D. M. 1966. Competition between *Skeletonema costatum* and *Olisthodiscus luteus* in Narragansett Bay and in culture. *Limnol. Oceanogr.* 11(4):447-455.

Prescott, G. W. 1951. *Algae of the western Great Lakes.* Cranbrook Institute of Science, Bloomfield Hills, Michigan. 966 pp.
_____. 1954. *How to know the fresh water algae.* Wm. C. Brown Co., Dubuque, Iowa.
_____. 1962. *Algae of the western Great Lakes area,* 2nd ed. Wm. C. Brown Co., Dubuque, Iowa. 977 pp.
_____. 1968. *The algae: a review.* Houghton Mifflin Co., Boston, Massachusetts. 436 pp.
_____. 1969. *How to know the aquatic plants.* Wm. C. Brown Co., Dubuque, Iowa. 171 pp.

Pritchard, D. W. 1951. The physical hydrography of estuaries and some applications to biological problems. *Trans. N. Amer. Wildl. Conf.* 16:368-376.

_____. 1952. Salinity distribution and circulation in the Chesapeake Bay estuarine system. *J. Mar. Res.* 11:106-123.

_____. 1953. Distribution of oyster larvae in relation to hydrographic conditions. *Proc. Gulf & Carib. Fish. Inst.*, 5th Session.

_____. 1956. The structure of a coastal plain estuary. *J. Mar. Res.* 15:32-42.

Proctor, V. W. 1962. Viability of *Chara* oospores taken from migratory water birds. *Ecology* 43:528-529.

_____. 1966. Dispersal of desmids by waterbirds. *Phycologia* 5(4): 227-232.

Prouty, W. F. 1952. Carolina bays and their origin. *Bull. Geol. Soc. Amer.* 63:167-224.

Provasoli, L. 1958. Nutrition and ecology of protozoa and algae. *Ann. Rev. Microbiol.* 12:279-308.

Ragotzkie, R. A., and R. A. Bryson. 1953. Correlations of currents with the distribution of adult Daphnia in Lake Mendota. *J. Mar. Res.* 12:157-172.

Rainwater, F. H., and L. L. Thatcher. 1960. *Methods for collection and analysis of water samples.* U.S. Geological Survey, Washington, D.C. water-supply paper 1454. Washington, D.C. 301 pp.

Randall, C. W., and J. O. Ledbetter. 1966. Bacterial air pollution from activated sludge units. *Amer. Ind. Hyg. Ass. J.* 27:506-519.

Raney, E. C., and W. H. Massman. 1953. The fishes of the Pamunkey River, Virginia. *J. Washington Acad. Sci.* 43:424-432.

Rawson, D. S. 1930. The bottom fauna of Lake Simcoe and its role in the ecology of the lake. *Univ. Toronto Stud., Pub. Ontario Fish. Res. Lab.,* No. 40.

_____. 1939. Some physical and chemical factors in the metabolism of lakes. Pages 9-26 *in* Pub. No. 10, A.A.A.S., *Problems in lake biology.* American Association for the Advancement of Science, Washington, D. C.

_____. 1944. The calculation of oxygen saturation values and their correction for altitude. *Limnol. Soc. Amer. Spec. Pub.,* No. 15.

_____. 1951. The total mineral content of lake waters. *Ecology* 34:669-672.

_____. 1956. Algal indicators of lake types. *Limnol. Oceanogr.* 1:18-25.

Redeke, H. 1933. Ueber den jetztigen Stand unserer Kentnisse der Flora und des Brackwassers. *Int. Ass. Theor. Appl. Limnol.* 6:46-61.

Reid, D. M. 1930. Salinity interchange between sea water in sand and overflowing fresh water at low tide, I. *J. Mar. Biol. Ass. U. K.* 16:609-614.

_____. 1932. Salinity interchange between salt water in sand and overflowing fresh water at low tide, II. *J. Mar. Biol. Ass. U. K.* 18:299-306.

Reid, G. K. 1950. Food of the black crappie, *Pomoxis nigro-maculatus* (LeSueur), in Orange Lake, Florida. *Trans. Amer. Fish. Soc.* 79:145-154.

_____. 1952. Some considerations and problems in the ecology of floating islands. *Quart. J. Florida Acad. Sci.* 15:63-66.

_____. 1954. An ecological study of the Gulf of Mexico fishes, in the vicinity of Cedar Key, Florida. *Bull. Mar. Sci. Gulf & Carib.* 4:1-94.

_____. 1955. A summer study of the biology and ecology of East Bay, Texas, *Texas J. Sci.* 7:316-343.

_____. 1957. Biologic and hydrographic adjustment in a disturbed Gulf coast estuary. *Limnol. Oceanogr.* 2:198-212.

_____. 1967. *Ecology of intertidal zones.* Rand McNally & Co., Chicago, 85 pp.

Reid, G. K., and D. Squibb. 1971. Limnological cycles in a phosphatic limestone lake. *Quart. Jour. Florida Acad. Sci.* 3(1):17-47.

Reid, G. K., and H. S. Zim. 1967. *Pond life.* Golden Press, New York, 160 pp.

Reimers, N. 1958. Conditions of existence, growth, and longevity of brook trout in a small, high-altitude lake of the eastern Sierra Nevada. *California Fish Game* 44:319-333.

Reimers, N., J. A. Maciolek, and E. P. Pister. 1955. Limnological study of the lakes in Convict Creek basin, Mono County, California. *U.S. Fish Wildl. Serv. Fish. Bull.* No. 103.

Richards, F. A., and N. Corwin. 1956. Some oceanographic applications of recent determinations in the solubility of oxygen in sea water. *Limnol. Oceanogr.* 1:263-267.

Richman, S. 1958. The transformation of energy by *Daphnia pulex. Ecol. Monogr.* 18:273-291.

Ricker, W. E. 1934. *An ecological classification of certain Ontario streams.* Pub. Ontario Fish. Res. Board, Biol. Ser., No. 37. 114 pp.

_____. 1937. Physical and chemical characteristics of Cultus Lake, British Columbia. *J. Biol. Board Can.* 3:363-402.

_____. 1952. The benthos of Cultus Lake. *J. Fish. Res. Board Can.* 9:204-212.

Ricketts, E. F., and J. Calvin. 1968. Between Pacific tides, 4th ed. (Revised by J. W. Hegpeth.) Stanford Univ. Press, Stanford, California. 614 pp.

Rigler, F. H. 1956. A tracer study of the phosphorus cycle in lake water. *Ecology* 37:550-562.

Riley, G. A. 1937. The significance of the Mississippi River drainage for biological conditions in the northern Gulf of Mexico. *J. Mar. Res.* 1:60-74.

_____. 1946. Factors controlling phytoplankton populations on Georges Bank. *J. Mar. Res.* 6:54-71.

_____. 1956. Oceanography of Long Island Sound, 1952-54. IX. Production and utilization of organic matter. *Bull. Bingham Oceanogr. Coll.* 15:324-344.

_____. 1957. Phytoplankton of the north central Sargasso Sea, 1950-52. *Limnol. Oceanogr.* 2:252-270

_____. 1963. Organic aggregates in sea water and the dynamics of their formation and utilization. *Limnol. Oceanogr.* 8:372-381.

_____. 1963. Theory of food-chain relations in the oceans. Pages 438-463 *in* M. N. Hill, ed. *The sea.* Vol. II. Wiley-Interscience, New York.

_____. 1965. A mathematical model of regional variations in plankton. *Limnol. Oceanogr.* (Suppl. 10):R202-215.

Riley, G. A., H. Stommel, and D. F. Bumpus. 1949. Quantitative ecology of the plankton of the western Atlantic. *Bull. Bingham Oceanogr. Coll.* 12:1-169.

Riley, J. P., and G. Skirrow, eds. 1965. *Chemical oceanography.* Academic Press, Inc., New York. Vol. 1, 712.; Vol. 2, 508 pp.

Robeck, G. C., C. Henderson, and R. G. Palange. 1954. *Water quality studies on the Columbia River.* U. S. Public Health Service, Washington, D. C.

Rodeheffer, T. 1941. The movements of marked fish in Douglas Lake, Michigan. *Pap. Michigan Acad. Sci.* 26:265-280.

Rodhe, W. 1948. Environmental requirements of fresh water plankton algae. *Symb. Bot. Upsaliensis,* No. 10.

Rogick, M. D. 1935. Studies of freshwater bryozoa. II. The bryozoa of Lake Erie. *Trans. Amer. Microscop. Soc.* 54:245-263.

Round, F. W. 1965. *The biology of the algae.* Edward Arnold (Publishers) Ltd., London. 269 pp.

Rubey, W. W. 1951. Geologic history of sea water. An attempt to state the problem. *Bull. Geol. Soc. Amer.* 62:1111-1148.

Russell, F. 1961. *Watchers at the pond.* Alfred A. Knopf, New York. 264 pp.

Russell, I. C. 1895. *Lakes of North America.* Ginn & Co., Boston, Massachusetts.

Russell-Hunter, W. D. 1970. *Aquatic productivity: an introduction to some basic aspects of biological oceanography and limnology.* The Macmillan Co., New York. 306 pp. (Paperback).

Ruttner, F. 1930. Das Plankton des Lunzer Untersees. *Int. Rev. Hydrobiol.* 23:1-287.

_____. 1963. *Fundamentals of limnology,* 3rd ed. (transl. by D. G. Frey and F. E. J. Fry). Univ. Toronto Press, Toronto. 295 pp.

Ryther, J. H. 1956. The measurement of primary production, *Limnol. Oceanogr.* 1:72-84.

_____. 1969. Photosynthesis and fish production in the sea. *Science* 166:73-76.

Saur, J. F. T., and E. R. Anderson. 1956. The heat budget of a body of water of varying volume. *Limnol. Oceanogr.* 1:247-251.

Scagel, R. F., R. J. Bandoni, G. E. Rouse, W. B. Schofield, J. R. Stein, and T. M. C. Taylor. 1966. *An evolutionary survey of the plant kingdom.* Wadworth Publishing Company, Inc., Belmont, Calif. 658 pp.

Scheffer, V. B. 1969. The year of the whale. Charles Scribner's Sons, New York. 244 pp.

Schneller, M. V. 1955. Oxygen depletion in Salt Creek, Indiana. *Invest. Indiana Lakes & Streams* 4:163-175.

Schubel, J. R. 1968. Turbidity maximum of the northern Chesapeake Bay. *Science* 161:1013-1015.

Schwoerbel, J. 1970. *Methods of hydrobiology.* Pergamon Press, Oxford. 200 pp.

Sculthorpe, C. D. 1967. *The biology of aquatic vascular plants.* St. Martin's Press, New York. 610 pp.

Segerstrale, S. G. 1957. Baltic Sea. Pages 751-800 in J. W. Hedgpeth, ed. *Treatise on marine ecology and paleoecology.* Vol. I *Ecology.* Mem. 67, The Geological Society of America, New York.

Sellery, G. C. 1956. *E. A. Birge.* Univ. Wisconsin Press, Madison.

Shand, S. J. 1946. Dust devils? Parallelism between the South African salt pans and the Carolina bays. *Sci. Monthly* 62-95.

Shaprio, J. 1957. Chemical and biological studies on the yellow organic acids of lake water. *Limnol. Oceanogr.* 2:161-179.

Shelford, V. E. 1913. *Animal communities in temperate America.* Univ. Chicago Press, Chicago. 368 pp.

Shelford, V. E., A. O. Weese, L. A. Rice, D. I. Rasmussen, and A. Maclean. 1935. Some marine biotic communities of the Pacific coast of North America: Part I, general survey of the communities. *Ecol. Monogr.* 5:249-293.

Shepard, F. P. 1937. Revised classification of marine shorelines. *J. Geol.* 45:602-624.

_____. 1973. *Submarine geology,* 3rd ed. Harper & Row, New York. 557 pp.

Shepard, F. P., F. B. Phleger, and T. H. van Andel, eds. 1960. *Recent sediments, northwestern Gulf of Mexico.* Amer. Ass. Petrol Geol., Tulsa, Oklahoma.

Silliman, R. P., and J. S. Gutsell. 1958. Experimental exploitation of fish populations. *U.S. Fish Wildl. Serv. Fish. Bull.* 133:215-252.

Simpson, G. G. 1944. *Tempo and mode in evolution.* Columbia Univ. Press, New York.

Singer, S. F. 1968. Waste heat management. *Science* 159:1184.

Skulberg, O. M. 1970. Importance of algal cultures for assessment of eutrophication of Oslofjord. *Helgoländ. Wiss. Meeresunters.* 20:111-125.

Slack, K. V. 1955. A study of the factors affecting stream productivity by the comparative method. *Invest. Indiana Lakes & Streams* 4:3-47.

Sloan, W. C. 1956. The distribution of aquatic insects in two Florida springs. *Ecology* 37:81-98.

Slobodkin, L. B. 1954. Population dynamics in *Daphnia obtusa* Kurz. *Ecol. Monogr.* 24:69-88.

_____. 1961. *Growth and regulation of animal populations.* Holt, Rinehart, and Winston, New York. 184 pp.

Smalley, A. E. 1959. The growth cycle of *Spartina* and its relation to the insect populations in the marsh. *Proc. Salt Marsh Conf., Sapelo Island, Georgia,* 1958. Marine Institute, Univ. Georgia, Athens, pp. 99-100.

Smayda, T. J. 1970. The suspension and sinking of phytoplankton in the sea. *Oceanogr. Mar. Biol. Ann. Rev.* 8:353-414.

Smith, G. M. 1944. *Marine algae of the Monterey Peninsula.* Stanford Univ. Press, Stanford. 622 pp.

_____. 1950. *The fresh-water algae of the United States,* 2nd ed. McGraw-Hill Book Co., New York. 719 pp.

_____. 1951. *Manual of phycology.* Chronica Botanica Co., Waltham, Massachusetts.

_____. 1955. *Cryptogamic botany.* Vol. I. *Algae and fungi,* 2nd ed. McGraw-Hill Book Co., Inc., New York. 546 pp.

Smith, R. I. 1953. The distribution of the polychaete, *Neanthes lighti,* in the Salinas River estuary, California. *Biol. Bull.,* No. 105:335-347.

Smith, R. L. 1966. *Ecology and field biology.* Harper & Row, New York. 686 pp.

_____. 1972. *The ecology of man: an ecosystem approach.* Harper & Row, New York. 436 pp.

Southwick, C. H. 1956. The logistic theory of population growth—past and present attitudes. *Turtox News* 37:131-133.

Spencer, R. S. 1956. Studies in Australian estuarine hydrology. II. The Swan River. *Australian J. Mar. & Freshw. Res.* 7:193-253.

Spooner, G. M. 1947. The distribution of *Gammarus* species in estuaries. *J. Mar. Biol. Ass. U.K.* 27:1-52.

Stankovic, S. 1960. *The balkan Lake Ohrid and its living world.* Dr. W. Junk, The Hague. 357 pp.

Steen, H. 1958. Determinations of the solubility of oxygen in pure water. *Limnol. Oceanogr.* 3:423-426.

Stembridge, J. H., R. B. Withers, and B. J. Garnier. 1958. *The world, a general regional geography—Australian and New Zealand edition.* Oxford Univ. Press, London. 584 pp.

Stephenson, T. A., and A. Stephenson. 1949. The universal features of zonation between tide-marks on rocky coasts. *J. Ecol.* 37:289-305.

_____. 1972. *Life between tidemarks on rocky shores.* W. H. Freeman & Co., San Francisco. 425 pp.

Stickney, A. P. 1959. Ecology of the Sheepscot River Estuary. *U.S. Fish Wildl. Serv. Spec. Sci. Rep.,* No. 309:1-21.

Stockner, J. G., and W. W. Benson. 1967. The succession of diatom assemblages in the recent sediments of Lake Washington. *Limnol. Oceanogr.* 12:513-532.

Stokes, G. G. 1851. On the effect of the internal friction of fluids on the motion of pendulums. *Trans. Cambridge Phil. Soc.* 9(II):8-14.

Stommel, H., and H. Farmer. 1952. On the nature of estuarine circulation. Part III. Pages 52-63 of Ch. 7 *in* W. H. O. I., Tech. Rep. (mimeographed). Woods Hole Oceanographic Institution, Woods Hole, Massachusetts.

Storer, T. I. 1951. *General Zoology,* 2nd ed. McGraw-Hill Book Co., New York. 832 pp.

Strahler, A. N. 1966. *Introduction to physical geography.* John Wiley & Sons, Inc., New York. 455 pp.

Strahler, A. N., and A. H. Strahler. 1973. *Environmental geoscience.* John Wiley & Sons, New York. 455 pp.

Streeter, H. W., and E. B. Phelps. 1958. A study of the pollution and natural purification of the Ohio River. III. Factors concerned in the phenomena of oxidation and reaeration. *U. S. Publ. Health Serv. Publ. Health Bull.,* No. 146.

Strickland, J. D. H., and T. R. Parsons. 1968. *A practical handbook of seawater analysis.* Fisheries Research Board of Canada, Bull. 167. Fisheries Research Board of Canada, Ottawa. 311. pp.

Stumm, W., and J. J. Morgan. 1970. *Aquatic chemistry—an introduction emphasizing chemical equilibria in natural waters.* Wiley-Interscience, New York. 583 pp.

Sublette, J. R. 1955. The physico-chemical and biological features of Lake Texoma (Denison Reservoir), Oklahoma and Texas. Preliminary study. *Texas J. Sci.* 7:164-182.

_____. 1957. The ecology of the macroscopic bottom fauna in Lake Texoma (Denison Reservoir), Oklahoma and Texas. *Amer. Midl. Nat.* 57:371-402.

Sverdrup, H. U., M. W. Johnson, and R. H. Fleming. 1942. *The oceans, their physics, chemistry, and general biology.* Prentice-Hall, Inc., New York. 1087 pp.

Sylvester, R. O. 1958. Water quality studies in the Columbia River basin. *U. S. Fish Wildl. Serv. Spec. Sci. Rep.,* No. 239, 134 pp.

Tansley, A. G. 1935. The use and abuse of vegetational concepts and terms. *Ecology* 16:284-307.

_____. 1939. *The British islands and their vegetation.* Cambridge Univ. Press, Cambridge.

Taylor, W. R. 1937. *Marine algae of the northeastern coast of North America.* Univ. Michigan Press, Ann Arbor. 427 pp.

_____. 1950. *Plants of Bikini and other northern Marshall Islands.* Univ. Michigan Press, Ann Arbor. 227 pp.

_____. 1957. *Marine algae of the northeastern coast of North America,* 2nd ed. Univ. Michigan Press, Ann Arbor. 509 pp.

_____. 1960. *Marine algae of the eastern tropical and subtropical coasts of the Americas.* Univ. Michigan Press, Ann Arbor. 870 pp.

Teal, J. M. 1957. Community metabolism in a temperate cold spring. *Ecol. Monogr.* 27:283-302.

Teal, J., and M. Teal. 1969. *Life and death of the salt marsh*. Ballantine Books, Inc., New York. 274 pp.

Thomas, J. P. 1966. *The influence of the Altamaha River on the primary production beyond the mouth of the river*. M. S. thesis. University of Georgia, Athens.

Thomas, W. H., and E. G. Simmons. 1960. Phytoplankton production in the Mississippi delta. Pages 103-116 in E. P. Shepard, F. B. Phleger, and T. H. van Andel, eds. *Recent sediments, northwestern Gulf of Mexico*. Amer. Ass. Petrol. Geol., Tulsa, Oklahoma.

Thompson, D. H., and F. D. Hunt. 1930. The fishes of Champaign County: study of the distribution and abundance of fishes in small streams. *Illinois Nat. Hist. Surv. Bull.* 19:1-101.

Thorson, G. 1950. Reproductive and larval ecology of marine bottom invertebrates. *Biol. Rev.* 25:1-45.

_____. 1957. Bottom communities (sublittoral or shallow shelf). Pages 461-534 in J. W. Hedgpeth, ed. *Treatise on marine ecology and paleoecology*. Vol. I. *Ecology*. Mem 67, The Geological Society of America, New York.

Thwaites, F. T. 1956. *Outline of glacial geology*. Publ. by author, Madison, Wisconsin.

Tiffany, L. H. 1951. The ecology of freshwater algae. Pages 293-311 in G. M. Smith, *Manual of phycology*. Chronica Botanica Co., Waltham, Massachusetts.

Tiffany, L. H., and M. E. Britton, 1952. *The algae of Illinois*. Univ. Chicago Press, Chicago. 407 pp.

Tilly, L. J. 1968. The structure and dynamics of Cone Spring. *Ecol. Monogr.* 38:169-197.

Todd, D. K., ed. 1970. *The water encyclopedia*. Port Washington. N. Y.

Tonolli, V. 1949. Ripartizione spaziale e migrazioni verticali dello zooplancton—ricerche e considerazioni. *Mem. Inst. Ital. Idrobiol.* 5:211-228.

Tressler, W. L., L. H. Tiffany, and W. P. Spencer. 1940. Limnological studies of Buckeye Lake, Ohio. *Ohio J. Sci.* 40:261-290.

Trueman, A. E. 1938. *The scenery of England and Wales*. Gollancz, London. 351 pp.

Tucker, A. 1957. The relation of phytoplankton periodicity to the nature of the physico-chemical environment with special reference to phosphorus. *Amer. Midl. Nat.* 57:300-370.

Twenhofel, W. H. 1939. *Principles of sedimentation.* McGraw-Hill, New York.

Twitty, V. C. 1959. Migration and speciation in newts. *Science* 130:1735-1743.

U.S. Coast and Geodetic Survey. (issued annually). *Tide tables, high and low water predictions, east coast of North and South America, including Greenland.* Washington, D. C. (Available through its sales agents. Also issued for the west coast.)

U.S. Department of Agriculture. 1955. *Yearbook of Agriculture.* Washington, D. C.

U.S. Public Health Service. 1957. *Biological problems in water pollution.* Washington, D. C.

————. 1958. *Oxygen relationships in streams.* Washington, D. C.

Usinger, R. L. 1957. Marine insects. Pages 1177-1182 *in* J. W. Hedgpeth, ed. *Treatise on marine ecology and paleoecology.* Vol. I. *Ecology.* Mem. 67, The Geological Society of America, New York.

————. 1967. *The life of rivers and streams.* McGraw-Hill, New York.

Vallentyne, J. R., 1957. Principles of modern limnology. *Amer. Sci.* 45:218-244.

————. 1957. The molecular nature of organic matter in lakes and oceans with lesser reference to sewage and terrestrial soils. *J. Fish. Res. Board Can.* 14:33-82.

————. 1957. Carotenoids in a 20,000-year-old sediment from Lake Searles, California. *Arch. Biochem. Biophys.* 70:29-31.

Ventura, L. A. 1973. *The provisional algal assay procedure, evaluations of its methodology and sensitivity in the field.* M. S. Thesis. University of Rhode Island, Kingston. 151 pp.

Verduin, J. 1954. Phytoplankton and turbidity in western Lake Erie. *Ecology* 35:550-561.

————. 1956. Energy fixation and utilization by natural communities in western Lake Erie. *Ecology* 37:40-50.

————. 1956. Primary production in lakes. *Limnol. Oceanogr.* 1:85-91.

Verduin, J., E. E. Whitmer, and B. C. Cowell. 1959. Maximal photosynthetic rates in nature. *Science* 130:268-269.

Voss, G. L. 1972. *Oceanography*. Golden Press, New York. 160 pp.

Wagner, R. H. 1971. *Environment and man*. W. W. Norton & Co., Inc., New York. 491 pp.

Walford, L. A. 1958. *Living resources of the sea*. Ronald Press Co., New York. 321 pp.

Wallace, A. R. 1972. *A narrative of travels on the Amazon and Rio Negro*. Dover Publications, Inc., New York. (Reprint of 1889, 2nd ed.) 363 pp.

Warren, B. 1966. Physical oceanography pp. 623-632 *in* R. W. Fairbridge, ed., *Encyclopedia of oceanography*. Van Nostrand Reinhold, New York.

Watt, K. E. F. 1968. *Ecology and resource management. A quantitative approach*. McGraw-Hill, New York. 450 pp.

Wattenberg, H. 1933. *Ueber die Titrationsalkalinität und den Kalxiumkarbonatgehalt des meerwasser*. Deutsche Atlantische Exped. Meteor. 1925-1927. Wis. Erg., Bd. 8, 2 Teil, pp. 122-231.

Welch, P. S. 1948. *Limnological methods*. Blakiston Co., New York. 381 pp.

_____. 1952. *Limnology,* 2nd ed. McGraw-Hill, New York. 538 pp.

Wells, J. W. 1957. Coral reefs. Pages 609-631 *in* J. W. Hedgpeth, ed. *Treatise on marine ecology and paleoecology*. Vol. I. *Ecology*. Mem. 67, The Geological Society of America, New York.

Wesenberg-Lund, C. 1905. A comparative study of the lakes of Scotland and Denmark. *Proc. Roy. Soc. Edinburg* 25:401-448.

Whipple, G. C. 1927. *The microscopy of drinking water,* 4th ed. John Wiley & Sons, Inc., New York. 586 pp.

Whitford, L. A. 1956. Communities of algae in springs and spring streams of Florida. *Ecology* 37:433-442.

_____. 1960. The current effect and growth of fresh-water algae. *Trans. Amer. Microscop. Soc.* 79(3):302-309.

Whitford, L. A., and G. J. Schumacher. 1973. *A manual of the fresh-water algae in North Carolina,* rev. ed. Sparks Press, Raleigh, N. C. 324 pp.

Whittaker, J. R., and J. R. Vallentyne. 1957. On the occurrence of free sugars in lake sediment extracts. *Limnol. Oceanogr.* 2:98-110.

Whittaker, R. H. 1969. New concepts of kingdoms of organisms. *Science* 163:150-160.

_____. 1970. *Communities and ecosystems*. The Macmillan Co., London. 161 pp.

Wilhm, J. L. 1967. *J. Water Poll. Contr. Fed.* 39:1673-1683.

Wilson, J. N. 1957. Effects of turbidity and silt on aquatic life. Pages 235-239 *in* U.S.P.H.S., *Biological problems in water pollution.* U. S. Public Health Service, Washington, D. C.

Wisler, C. O., and E. F. Brater. 1949. *Hydrology.* John Wiley & Sons, New York.

Wistendahl, W. A. 1958. The flood plain of the Raritan River, New Jersey. *Ecol. Monogr.* 28:129-153.

Wolfe, P. E. 1953. Periglacial frost-thaw basins in New Jersey. *J. Geol.* 61:133-141.

Wolle, F. A. 1887. *Freshwater algae of the United States.* Moravian Book Store, Bethlehem, Pa. 364 pp.

Wolman, M. G., and L. B. Leopold. 1957. River flood plains: some observations on their formation. *U.S. Geol. Surv. Prof. Pap.,* No. 282-C.

Woltereck, R. 1932. Races, associations and stratification of pelagic daphnids in some lakes of Wisconsin and other regions of the United States and Canada. *Trans. Wisconsin Acad. Sci.* 27:487-522.

Wood, K. G. 1953. The bottom fauna of Louisa and Redrock Lakes, Algonquin Park, Ontario. *Trans. Amer. Fish. Soc.* 82(1952):203-212.

Wood, P. W. 1957. Coactions in laboratory populations of two species of Daphnia. *Ecology* 38:510-519.

Wood, R. D. 1959. A naturally occurring visible thermocline. *Ecology* 40:153-154.

_____. 1970. *Hydrobotanical methods.* Published by the author, Kingston, R. I. 176 pp. (2nd ed. 1975, Univ. Park Press, Baltimore, Md.)

Wood, R. D., and P. E. Hargraves. 1969. Comparative benthic plant ecology by SCUBA-monitored quadrats. *Hydrobiologia* 33:561-586.

Wood, R. D., and J. Lutes. 1967. Guide to the phytoplankton of Narragansett Bay, Rhode Island. Published by R. D. Wood, Kingston, R. I. (Available from Bookstore, Memorial Union, Univ. of Rhode Island). 65 pp. (Suppl. by R. D. Wood, 1973. 10 pp.)

Wood, R. D., and M. Villalard-Bohnsack. 1974. Marine algae of Rhode Island. *Rhodora* 76:399-421.

Woods, W. J. 1960. *An ecological study of Stony Brook, New Jersey.* (Ph. D. dissertation.) Rutgers University, New Brunswick, N. J.

Woods Hole Oceanographic Institute. 1951. *Report on a survey of the hydrog-*

raphy of Great South Bay made during the summer of 1950 for the town of Islip, N. Y., Falmouth, Mass. (Mimeographed.)

Wright, H. E., Jr., and D. G. Frey, eds. 1965. *The quaternary of the United States.* Princeton Univ. Press, Princeton, N. J. 922 pp.

Wright, J. C. 1954. The hydrology of Atwood Lake, a flood-control reservoir. *Ecology* 35:305-316.

————. 1965. The population dynamics and production of *Daphnia* in Canyon Ferry Reservoir, Montana. *Limnol. Oceanogr.* 10:583-590.

Yonge, C. M. 1949. *The sea shore.* Wm. Collins Sons & Co., Ltd., London. 311 pp.

Yoshimura, S. 1931. Contributions to the knowledge of the stratification of iron manganese in lake waters of Japan. *Jap. J. Geol. Geogr.* 9:61-69.

Yoshimura, S., and K. Masuko. 1935. Kato- and dichothermy during the autumnal circulation period in small ponds of Miyagi Prefecture, Japan. *Proc. Imp. Acad. Tokoyo* 11:146-148.

Young, F. N., and J. R. Zimmerman. 1956. Variations in the temperature in small aquatic situations. *Ecology* 37:609-611.

Zernitz, E. R. 1932. Drainage patterns and their significance. *J. Geol.* 40:498-521.

Zim, H. S., and L. Ingle. 1955. *Seashores.* Simon & Shuster, New York. 160 pp.

Zinn, D. J., R. D. Wood, and J. Berkowitz. 1957. *Fouling project, final report.* Office of Naval Research, Biology Branch, Contract Nonr-396(06). 36 pp. (Mimeographed.)

ZoBell, C. E. 1946. *Marine microbiology.* Chronica Botanica Co., Waltham, Massachusetts.

ZoBell, C. E., and C. B. Feltham. 1942. The bacterial flora of a marine mud flat as an ecological factor. *Ecology* 23:69-78.

Zottoli, R. 1973. *Introduction to marine environments.* The C. V. Mosby Co., St. Louis. 125 pp.

Index